Automorphic Forms and Lie Superalgebras

Algebra and Applications

Volume 5

Algebra and Applications aims to publish well-written and carefully refereed monographs with up-to-date expositions of research in all fields of algebra, including its classical impact on commutative and noncommutative algebraic and differential geometry, K-theory and algebraic topology, and further applications in related domains, such as number theory, homotopy and (co)homology theory through to discrete mathematics and mathematical physics.

Particular emphasis will be put on state-of-the-art topics such as rings of differential operators, Lie algebras and super-algebras, group rings and algebras, Kac-Moody theory, arithmetic algebraic geometry, Hopf algebras and quantum groups, as well as their applications within mathematics and beyond. Books dedicated to computational aspects of these topics will also be welcome.

Automorphic Forms
and Lie Superalgebras

by

Urmie Ray
Université de Reims, Reims, France

A C.I.P. Catalogue record for this book is available from the Library of Congress.

ISBN-13 978-90-481-7254-2
ISBN-10 1-4020-5010-0 (e-book)
ISBN-13 978-1-4020-5010-7 (e-book)

Published by Springer,
P.O. Box 17, 3300 AA Dordrecht, The Netherlands.

www.springer.com

Printed on acid-free paper

To my Mother

Contents

Preface

I would like to thank Profs. M. Castellet, G. Cornelissen, and E. Friedlander, for their encouragements at the start of this project. I would like to take this occasion to express my gratitude to Prof. R. Borcherds for many stimulating discussions and for his advice and support throughout the years. I am also grateful to him and to Profs. V. Kac, J. Lepowsky and C. Schweigert for reading part of the manuscript and for their valuable comments. The major part of this book was written while I was on leave at the Centre de Recerca Matemàtica, in Barcelona, Spain. I express my thanks to the staff of the institute for their hospitality and to the Ministerio de Educación of the Spanish government for providing financial support during this period.

Most of all, I am indebted to Dr. C.-M. Patris whose unwavering support and trust made it possible for me to finish this book.

Chapter 1

Introduction

In this book, we give an exposition of the theory of Borcherds-Kac-Moody Lie algebras and of the ongoing classification and explicit construction project of a subclass of these infinite dimensional Lie algebras. We try to keep the material as elementary as possible. More precisely, our aim is to present some of the theory developed by Borcherds to graduate students and mathematicians from other fields. Some familiarity with complex finite dimensional semisimple Lie algebras, group representation theory, topology, complex analysis, Fourier series and transforms, smooth manifolds, modular forms and the geometry of the upper half plane can only be helpful. However, either in the appendices or within specific chapters, we give the definitions and results from basic mathematics needed to understand the material presented in the book. We only omit proofs of properties well covered in standard undergraduate and graduate textbooks.

There are several excellent reference books on the above subjects and we will not attempt to list them here. However for the purpose of understanding the classification and construction of Borcherds-Kac-Moody (super)algebras the following are particularly useful. Serre's approach to the theory of finite dimensional semisimple Lie algebras in [Ser1] is conducive to the construction of Borcherds-Kac-Moody Lie algebras as it emphasizes the presentation of finite dimensional semi-simple Lie algebras via generators and relations. For a first approach to automorphic forms and the geometry of the upper half plane, the book by Shimura may be a place to start at [Shi].

We also do not replicate the proofs of properties of Kac-Moody Lie algebras that are treated in depth in the now classical reference book by Kac [Kac14].

Borcherds-Kac-Moody Lie algebras are generalizations of symmetrizable (see Remarks 2.1.10) Kac-Moody Lie algebras, themselves generalizations of finite dimensional semi-simple Lie algebras. That this further level of generality is needed was first shown in the proof of the moonshine theorem given by Borcherds [Borc7], where the theory of Borcherds-Kac-Moody Lie algebras plays a central role. So let us briefly explain what this theorem is about.

1.1   The Moonshine Theorem

The remarkable Moonshine Theorem, conjectured by Conway and Norton and one part of which was proved by Frenkel, Lepowsky and Meurman, and another by Borcherds, connects two areas apparently far apart: on the one hand, the Monster simple group and on the other modular forms. Any connection found between an object which has as yet played a limited abstract role and a more fundamental concept is always very fascinating. Also it is not surprising that ideas on which the proof of such a result are based would give rise to many new questions, thus opening up different research directions and finding applications in a wide variety of areas.

1.1.1   A Brief History

The famous Feit-Thompson Odd order Theorem [FeitT] implying that the only finite simple groups of odd order are the cyclic groups of prime order, sparked off much interest in classifying the finite simple groups in the sixties and seventies. The classification of the building blocks of finite symmetries was completed in the eighties [Gor]: there are 17 infinite families of finite simple groups and 26 which do not belong to any families and are thus called sporadic [Bur]. The Monster M [Gri] is the largest sporadic simple group. It has about 10^{54} elements or more precisely

$$2^{46}.3^{20}.5^9.7^6.11^2.13^3.17.19.23.29.31.41.47.59.71$$

elements. Evidence for the existence of the Monster were first found by B. Fischer and R.L. Griess in 1973. Based on Conway and Norton's conjecture that the dimension of one of its representations is 196883 Fischer, Livingstone and Thorne computed the full character table of M [FisLT]. We remind the reader that a group representation is a group homomorphism from the group to the isomorphism group of a vector space. Then, McKay noticed that the dimensions of its two smallest representations, 1 and 196883, are closely related to the first two coefficients of the Fourier expansions of the normalized modular invariant

$$J = j - 744.$$

Since $j(\tau + 1) = j(\tau)$ for τ in the upper half plane $\mathcal{H} = \{x + iy | y > 0\}$, we can write j as a function of $q = e^{2\pi i\tau}$. Moreover j is holomorphic on $\mathcal{H}$. So, as a function of q, J has a Laurent series in the punctured disc of radius 1 centered at 0:

$$J(q) = j(q) - 744 = q^{-1} + 196884q + ... = \sum_{n=-1}^{\infty} c(n)q^n$$

and $c(-1) = 1$, $c(1) = 196883 + 1$. This is the Fourier series of J. Further details can be found in [Ser2].

At the time it was thought unlikely that there would be any connection between modular forms and the Monster group. Shortly after, McKay and

Thompson found that other coefficients of the q-expansion of J are also simple linear combinations of the dimensions of irreducible representations of the Monster group M [Th]. Now, the dimension of an M-module is the trace of the action of the trivial element 1 of M on it, and any M-module is completely determined by its character, i.e. the trace function. So based on this observation, McKay and Thompson's insight was to forecast that for all $n \in \mathbf{Z}$ there is an M-module $V(n)$ (later called a head representation of M) of dimension $c(n)$ with character H_n such that for all $g \in M$,

$$T(g) = \sum_{n=-1}^{\infty} H_n(g)q^n$$

would be an interesting function. When $c(n)$ was a known linear combination, the M-modules $V(n)$ were to be taken to be the sums of the irreducible representations whose dimensions appeared in the linear combination. Like in the case of $V(1)$, the head representations are obviously not just sums of the trivial module.

Thus their work suggested these series, now known as the McKay-Thompson series, were worth investigating. For $g = 1$, as already noted, we get the function J. Furthermore, they conjectured that M has a natural infinite dimensional representation

$$V = \oplus_{n \in \mathbf{Z}} V(n)$$

with graded dimension $J(q)$. It came as a surprise that the most natural module of a finite group was infinite dimensional.

1.1.2   The Theorem

In 1979, Conway and Norton calculated the first 11 coefficients of the McKay-Thompson series for all $g \in M$. Based on these calculations and other observations, they made the following remarkable conjecture [ConN]:

Moonshine Theorem. *There exists a graded M-module V such that for all $g \in M$, the series $T(g)$ is a normalized Hauptmodul for a genus 0 discrete subgroup of $SL(2, \mathbf{R})$ commensurable with $SL(2, \mathbf{Z})$, where $T : M \to \mathbf{C}$ is the graded character of the module V.*

Thus the Moonshine Theorem reveals a surprising connection between Hauptmoduls and the Monster group (or more precisely its conjugacy classes since characters are class functions) via its head representations.

We remind the reader of the definitions and properties necessary to understand the statement of the theorem. For details, see [Shi]. Two subgroups G and H of $SL(2, \mathbf{R})$ are commensurable if both $| H : G \cap H |$ and $| G : G \cap H |$ are finite. For any discrete subgroup G of $SL(2, \mathbf{R})$ commensurable with $SL(2, \mathbf{Z})$, $\mathcal{H}^*/G$ is a compact Riemann surface, where $\mathcal{H}^* = \mathcal{H} \cup \{\text{cusp points of } G\}$. A cusp of G is a point in $\mathbf{R} \cup \{i\infty\}$ fixed by a parabolic subgroup of G, i.e. a subgroup with a unique fixed point in $\mathbf{R} \cup \{i\infty\}$. The group G is said to be of genus 0 if the corresponding Riemann surface is of genus 0. When this is the case, there exist bijections F from $\mathcal{H}^*/G$ to the Riemann sphere $\mathbf{C} \cup \infty$, and any meromorphic function on $\mathcal{H}$ fixed by G is a rational function of F. If furthermore $F(i\infty) = \infty$, then F is called a Hauptmodul, and in this case there is a smallest positive real number r such that $\begin{pmatrix} 1 & r \\ 0 & 1 \end{pmatrix} \in G$ and so $F(\tau + r) = F(\tau)$. The Hauptmodul F can thus be written as a function of $q = e^{2\pi i \tau / r}$ and has a Laurent expansion (see Corollary C.2.7) in a punctured disc centered at 0: $F(\tau) = \sum_{n \in \mathbf{Z}} a_n q^n$. As G is of genus 0, it can be shown that F has a simple pole at ∞. Hence $a_n = 0$ for all $n \leq -2$. The Hauptmodul F is said to be normalized if $a_0 = 0$ and $a_{-1} = 1$.

Conway and Norton not only conjectured that the graded traces $T(g)$ (for all $g \in M$) were Hauptmoduls, but based on the calculation of the first few coefficients, they gave a list of these Hauptmoduls [ConN]. So in order to prove the conjecture, one needs to first construct a $\mathbf{Z}$-graded M-module (whose existence was conjectured by McKay and Thompson), and then show that the traces associated with it are equal to the given Hauptmoduls.

In 1983, Frenkel, Lepowsky and Meurman [FrenLM1,2] proved the first part: They constructed a new natural M-module V with graded dimension J, which they called the moonshine module. It has a very rich algebraic structure. They thus proved the Thompson-McKay Conjecture. In particular, it is the sum of two vector spaces acted on by a certain involution of the Monster group – one summand being a subspace of the lattice vertex algebra V_L (see section 4.3) and the other of a twisted version of V_L, where L is the Leech lattice. In particular, the centralizer C of this involution preserves the sum. As a result, their work also yields the traces $T(g)$ for all elements $g \in C$ and thus proves the Moonshine Theorem for all conjugacy classes of elements in the subgroup C. However, most elements of the Monster group do not belong to any of these conjugacy classes. In 1992, Borcherds [Borc7] gave a very beautiful proof of the second part – namely finding T_g for all $g \in M$ and not just for $g \in C$ – which, together with Frenkel, Lepowsky and Meurman's work, explains how the modular forms and the Monster group are related and leads to new interesting questions and directions of research worth investigating.

The principal ingredients in Borcherds' proof are vertex algebras (see Definition 4.1.13) and Borcherds-Kac-Moody Lie algebras (see Definition 2.1.7). In [ConN], Conway and Norton had put forward the idea that there may be a Lie superalgebra which in some way "explains" the Moonshine Conjecture (see also Chapter 27 in [ConS3]). The construction of the Monster Lie algebra – the most well known and simplest example of a Borcherds-Kac-Moody Lie algebra which is not a Kac-Moody Lie algebra – (and thus the proof of the conjecture) is based on Frenkel, Lepowsky and Meurman's construction of the Moonshine module, in particular the fact that it has the structure of a vertex algebra acted on by the Virasoro algebra with central charge 24 (see Chapter 4 for definitions). For a more detailed history, see [Lep2] and [Ray4] and for a detailed study, see[Gan].

The two principal items in the classification of some Borcherds-Kac-Moody Lie (super)algebras, apart from the Lie (super)algebras themselves, are generalizations of two main ingredients in the proof of the moonshine theorem, namely vector valued modular forms and G-graded vertex algebras, where G is a finite abelian group. Let us first give an idea of what these infinite dimensional Lie (super)algebras are like.

1.2 Borcherds-Kac-Moody Lie Superalgebras

Borcherds-Kac-Moody Lie algebras were constructed by Borcherds as a generalization of Kac-Moody Lie algebras. They are (mostly infinite dimensional) Lie algebras generalizing the notion of a finite dimensional semisimple Lie algebra. A fundamental aspect of Lie's work on continuous transformations groups was his realization of how infinitesimal transformations could be used to study the former. Thus Lie algebras were originally constructed to study Lie groups. However it is worth noticing that the generalization to infinite dimension has come first through Lie algebras and associated infinite dimensional Lie or algebraic groups were constructed later [Gar2, KacP3,4,5,6,8, Kac13, KacWan1], [Ti1,2,3,4] and for Borcherds-Kac-Moody Lie algebras, this has not yet been completed.

Kac-Moody Lie algebras were constructed concurrently but independently by Kac [Kac1,2] and Moody [Mo2]. Moody generalized a characteristic aspect of finite dimensional semisimple Lie algebra, namely their construction via Chevalley-Serre generators and relations from Cartan matrices to a more general class of matrices. Kac, in his study of $\mathbf{Z}$-graded Lie algebras having polynomial growth (see Definition 2.1.24) found that apart from the finite dimensional Lie algebras and the simple Lie algebras of polynomial vector fields, there is only another type of contragredient Lie algebras (see Definition 2.1.21) with this property, namely Lie algebras which he called affine [Kac2]. The latter were also independently constructed by Moody in [Mo3], where they are called "Euclidean Lie algebras". They are in some sense the first level of generalization to infinite dimension and are a sub-class of Kac-Moody Lie algebras (also defined in [Kac1]). The affine Lie algebras have since then been successfully applied in several areas of mathematics and physics.

We give a brief history of the steps that led Borcherds to generalize further and define Borcherds-Kac-Moody Lie algebras. In his Ph.D. thesis, published in 1985, Borcherds studied properties of the Leech lattice (for elementary definitions concerning integral lattices, see section 3.1) [Borc1]. This is the unique unimodular even lattice of rank 24 without any vectors of norm 2 discovered by J. Leech in 1965 and extensively studied by Conway and Sloane [Con1,2, ConPS, ConS1,2,3]. Borcherds found that the multiplicities of all the norm -2 vectors of the even unimodular Lorentzian lattice $II_{25,1}$ with signature $(25, 1)$, which is the direct sum of the Leech lattice and of the unimodular even Lorentzian lattice $II_{1,1}$ of rank 2, are always less than 324 and in most cases equal to 324.

At about the same time in 1985, Frenkel constructed hyperbolic Kac-Moody Lie algebras of rank at most 26 as Lie subalgebras of the quotient of a space on which vertex operators act, this quotient having the structure of a Lie algebra

[Fren]. Hyperbolic Lie algebras are Kac-Moody Lie algebras (see Definition 2.1.7) with an even Lorentzian root lattice, i.e. a lattice of signature $(n, 1)$. Their study, as will be seen in greater detail in sections 5.3 and 5.4, is of interest for various geometric and number theoretic reasons. Frenkel's insight was to apply the Goddard-Thorn No-Ghost Theorem [GodT] of 1972 from the theory of dual resonance in theoretical physics to this context in order to find upper bounds for the multiplicities of the roots of a hyperbolic Kac-Moody Lie algebra of rank 26 – the critical dimension in the No-Ghost Theorem (see section 5.5). These upper bounds can be expressed in terms of a function of the norm of the roots. More precisely, for a root α, $\text{mult}(\alpha) \leq p_{24}(1 - (\alpha, \alpha)/2)$, where $p_{24}(n)$ is the number of partitions of n into parts of 24 colours. For a root α of norm -2, we get precisely 324 $(= p_{24}(2))$. Note that a root in the context of infinite dimensional Kac-Moody Lie algebras does not necessarily describe a reflection symmetry across a hyperplane whereas this is always the case for finite dimensional semisimple Lie algebras. However for Kac-Moody Lie algebras this is still true for the simple roots generating the root lattice, but no longer in the context of Borcherds-Kac-Moody Lie algebras. See section 2.3 for details on roots.

Furthermore, as was shown by Conway, the Dynkin diagram of this hyperbolic Kac-Moody Lie algebra of rank 26 is the Leech lattice [Con3]. More precisely, each point of the Dynkin diagram (i.e. each simple root) corresponds to a point in the Leech lattice in such a way that the edges between any two points of the diagram only depend on the distance between the corresponding two points in the Leech lattice. Note that in this case there are infinitely many simple roots though this is not true in general for infinite dimensional Kac-Moody Lie algebras (e.g. affine Lie algebras). Conway and Frenkel's results led Borcherds to become interested in investigating this hyperbolic Kac-Moody Lie algebra on the one hand and vertex operators on the other. He thus arrived at his definition of lattice vertex algebras and the construction of the Fake Monster Lie algebra from them and in particular put the No-Ghost Theorem to its full use [Borc5]: The real simple roots of this Lie algebra generate the hyperbolic Kac-Moody Lie algebra of rank 26 and the multiplicities of the roots of the Fake Monster Lie algebra are precisely the upper bounds given by Frenkel – so we get an equality and not just a bound. Studying this Lie algebra led him to the idea of imaginary simple roots. The Fake Monster Lie algebra F is graded by the function p_{24}. For each integer n, let F_n be the Lie subalgebra generated by the vector subspaces of degree at most n. It can be shown that the Fake Monster Lie algebra is generated by root vectors of degree n orthogonal to F_{n-1} (see section 2.5). The corresponding roots are thus the generating – or rather the simple – roots of the root system of F and some of them have non-positive norm, i.e. are imaginary. It is then only one step to generalize the class of Kac-Moody Lie algebras to that of Lie algebras today known as Borcherds-Kac-Moody Lie algebras or generalized Kac-Moody Lie algebras [Borc4].

Finite dimensional Lie algebras have natural automorphisms induced by the symmetries of their Dynkin diagrams – or equivalently of bijections of the indexing set of simple roots leaving the Cartan matrix invariant–, called diagram automorphisms. This remains the case for Kac-Moody Lie algebras. Another reason to go one step further in the generalization process stems from the fact that the class of Lie subalgebras of Kac-Moody Lie algebras fixed by a finite group of diagram automorphisms is not in general a Kac-Moody Lie algebra but a Borcherds-Kac-Moody Lie algebra (see Exercise 2.2.4).

The class of Borcherds-Kac-Moody Lie algebras was enlarged to include Lie superalgebras in 1994 [Ray1]. This also generalizes the class of Kac-Moody Lie superalgebras constructed by Kac [Kac6]. This larger class of Lie superalgebras is needed because hyperbolic reflection groups are, in general, Weyl groups of Borcherds-Kac-Moody Lie superalgebras with non-trivial odd parts (see Definition 2.1.1). This is the case for one of the most interesting known examples constructed by Scheithauer: the Fake Monster Lie superalgebra [Sch1,2].

Borcherds-Kac-Moody Lie superalgebras are special cases of contragredient Lie superalgebras previously constructed by Kac in [Kac6], and they are the Lie superalgebras to which many of Kac's results on Kac-Moody Lie superalgebras can be extended.

We next give a short introduction to the other two main ingredients in the classification project.

1.3 Vector Valued Modular Forms

Modular forms are a central concept in mathematics developed first in the context of elliptic functions, one of whose characteristics is double periodicity. They are complex valued holomorphic functions (see Definition C.2.1) on the upper half plane which transform "nicely" under the action of the isometry group of $\mathcal{H}$ (see section 3.2), namely the group $SL_2(\mathbf{Z})$. They can be generalized to holomorphic functions on the upper half plane with values in a vector space, which transform under the action of the metaplectic group (see section 3.3) – i.e. the double cover of $SL_2(\mathbf{Z})$. The vector space in question is a representation space of the metaplectic group. Some of the most important examples of ordinary modular forms are theta functions of positive definite unimodular lattices. Similarly some of the crucial examples of vector valued modular forms are theta functions of indefinite lattices. These particular lead to the construction of the theta transform of a vector valued modular function.

This is described in sections 3.3 and 3.4, where we give the basis for the theory developed by Borcherds in [Borc9,11]. Modular forms are not only central to the moonshine questions but more generally to the theory of Borcherds-Kac-Moody Lie algebras with Lorentzian root lattice.

1.4 Borcherds-Kac-Moody Lie Algebras and Modular forms

The aim is to classify the "interesting" Borcherds-Kac-Moody Lie (super) algebras. By "interesting" we mean those whose structure can be explicitly described in some concrete manner. Finite dimensional simple Lie algebras and affine Lie algebras are the two classes for which this is so far possible. Calculations seem to indicate that it will be very hard to do so for Kac-Moody Lie algebras which are neither affine nor finite dimensional.

The root lattices of finite dimensional simple Lie algebras are positive definite and that of affine Lie algebras are semi-positive definite (see Exercise 2.3.12). There is good evidence indicating that a class of Borcherds-Kac-Moody Lie superalgebras which should be as explicitly describable have Lorentzian root lattices. The two Borcherds-Kac-Moody Lie algebras at the origin of the general theory of Borcherds-Kac-Moody Lie superalgebras, namely the Monster and Fake Monster Lie algebras, have Lorentzian root lattices (Examples 2.3.11, 2.6.40): the even unimodular Lorentzian lattices $II_{1,1}$ and $II_{1,25}$ of rank 2 and 26 respectively.

Basically, the idea is to try to classify all the Borcherds-Kac-Moody Lie algebras to which one can associate a vector valued modular form which is holomorphic on the upper half plane in such a way that this modular form will contain – or rather induce – all the essential information on the corresponding Borcherds-Kac-Moody Lie algebra and lead to a complete description of it. This was done by Borcherds in [Borc9,11] though Lepowsky and Moody and later Feingold and Frenkel were the first to suggest a relation between hyperbolic Kac-Moody Lie algebras – namely Kac-Moody Lie algebras whose root lattice is Lorentzian – and Hilbert modular forms and Siegel modular forms or Jacobi forms respectively [LepM, FeinF]. In particular, the case of E_{10} is of special interest [KacMW]. Modular forms were already known to be connected to some Borcherds-Kac-Moody algebras since they play a prominent role in the representation theory of affine Lie algebras [FeinL, KacP1,2, KacWak1,2].

More precisely, the (highest weight) representation theory of Borcherds-Kac-Moody Lie superalgebras leads to an equality called the denominator formula (see section 2.6). This formula is central to the theory of these Lie superalgebras as it contains the essential information on the structure of the Borcherds-Kac-Moody Lie superalgebra it is related to. In the case of finite dimensional simple Lie algebras it is the classical Weyl formula and for affine Lie algebras it is equivalent to the famous Macdonald identities [Mac], [Dys]. The latter equivalence is proved in [Kac5], where the Weyl-Kac character formula for arbitrary integrable highest weight modules over symmetrizable Kac-Moody Lie algebras is computed and the denominator formula derived from it. Thus, a connection between modular functions and infinite dimensional Lie algebras first appears in [Kac5]. Moody in [Mo5] independently pointed out that Macdonald identities are equivalent to denominator identities for affine Lie algebras, though he did not prove the character or denominator formulae.

For some Borcherds-Kac-Moody Lie algebras with Lorentzian root lattice, when this formula is known explicitly it gives an infinite product expansion of a function on a hyperbolic space transforming nicely under the action of its isometry group – i.e. an automorphic form on a Grassmannian – with the property that the exponents of the product factors are coefficients of a modular form. Roughly speaking, this automorphic form is the exponential of the theta transform of the modular form. This is how Borcherds-Kac-Moody Lie algebras are associated to automorphic forms and vector valued modular forms (see section 5.3). The basic literature for this material can be found both in the original papers by Borcherds [Borc9,11] and in the survey articles [Borc10,12], [Kon].

1.5 Γ-graded Vertex Algebras

Given a non-degenerate lattice, one can construct its Fock space, which is an infinite dimensional vector space with the following characteristic: to each of its elements, one can associate an infinite family of linear maps satisfying a set of axioms. In other words, the Fock space can be given the structure of a vertex algebra.

The generating formal functions of these families of linear maps are called vertex operators. They were originally defined in theoretical physics, more precisely in the theory of dual resonance, where vertex operators are known as quantum fields. They were introduced by Fubini and Veneziano in 1972 [Man] in dual resonance theory. They first occurred in the mathematics literature in 1978 in the work of Lepowsky and Wilson [LepW] on the basic representations of affine sl_2. A construction of the basic representations for all affine Lie algebras using untwisted vertex operators was done by Frenkel and Kac [FrenK] in 1980. Lepowsky and Wilson's work was extended to all affine Lie algebras by Kac, Kazhdan, Lepowsky and Wilson [KacKazLW] in 1981. Part of the vertex algebra structure first appear in the work of Frenkel and Kac [FrenK] and Frenkel's work. The complete axioms for vertex algebras were then given by Borcherds in [Borc2], a definition heavily influenced by the ideas in [FrenK] and [Fren], and shortly after for vertex operator algebras – structures that defer slightly from vertex algebras – by Frenkel, Lepowsky and Meurman in [FrenLM2] with a different emphasis (see the introduction of [FrenLM2] for a detailed history). The axiomatic side has been studied by the above but also by several others, among whom H. Li [Li2], [LepL] and Y.-Z. Huang [Hua3], [FrenHL]. The simplest definition of a vertex algebra, equivalent to Borcherds', which is now sometimes used, was given by Dong and Lepowsky in [DonL].

The construction of Γ-graded vertex algebras is a generalization given by Dong and Lepowsky in 1993 [DonL] of Borcherds' construction of lattice vertex algebras [Borc2]. These more general structures are fundamental in the construction of the Fake monster Lie superalgebra, hence in the case when the odd part of the Borcherds-Kac-Moody Lie superalgebra is non-trivial.

1.6 A Construction of a Class of Borcherds-Kac-Moody Lie (super)algebras

Classifying all the Borcherds-Kac-Moody Lie (super)algebras whose structure can be obtained from an associated vector valued modular form is only a first step. We want not only to find a way of describing these Borcherds-Kac-Moody Lie superalgebras but to construct them in a "concrete way", i.e. in as explicit a manner as the affine and finite dimensional semisimple Lie algebras.

The vertex operator representations of affine Lie algebras inspired a construction of the simply laced finite dimensional Lie algebras, i.e. of type A, D, E, all of whose roots have the same norm [FrenK]. The point of this construction is that it yields an explicit Chevalley basis [Ser1] for the Lie algebra (see Definition 2.1.7). The least complicated Borcherds-Kac-Moody Lie algebra that is not a Kac-Moody Lie algebra is the Monster Lie algebra (Example 2.2.11). However it is not the most typical example and has a very complex and rich structure [FrenLM1,2], which will not be discussed in this book. All the other known Borcherds-Kac-Moody Lie superalgebras can also be constructed from lattices but to find an explicit basis for the Lie superalgebras in question, the procedure is more complicated and indirect and their construction is based on other objects: lattice vertex superalgebras for the Lie algebra case and Γ-graded vertex algebras for Lie superalgebras with non-trivial odd parts. This is in particular the case for the Fake Monster Lie superalgebra constructed by N. Scheithauer [Sch1,2]. The latter – together with its twisted versions [FucRS], [Sch3] – is the only known example of a Borcherds-Kac-Moody Lie superalgebra with non-trivial odd part.

More precisely, in all known cases (except the Monster one), a Borcherds-Kac-Moody Lie algebra can be realized as a quotient of a particular subspace of the (bosonic) lattice vertex algebra derived from its root lattice when the lattice is even. There is a natural action of the Virasoro algebra on this vertex algebra and the subspace in question is determined by it. The Lie superalgebra bracket on the quotient is given by the linear maps associated to each element of the vertex algebra and in order to show these are Borcherds-Kac-Moody Lie algebras, the Goddard-Thorn No-Ghost Theorem [GodT] is needed. The application of this result from theoretical physics in the context of Lie algebras is a highly exceptional insight of Frenkel [Fren] and Borcherds [Borc2]. For more details, see section 5.5. When the root lattice is odd, know examples can be constructed from the tensor product of the bosonic and of the fermionic lattice vertex superalgebras. This time, we need to use the action of the Neveu-Schwarz superalgebra on it (the Virasoro algebra is the even part of this Lie superalgebra) and the construction in this case is also possible because of the No-Ghost Theorem.

The next three chapters are independent from each other and present the background material to the classification: Borcherds-Kac-Moody Lie (super) algebras, vector valued modular forms, and Γ-graded vertex algebras. In the fifth chapter, we first explain how to associate a vector valued modular form to some Borcherds-Kac-Moody Lie algebras with Lorentzian root lattice and show there is a low upper bound on their rank. In fact, this gives good hope that the classification and construction project can be completed. In sections 5.2–5.4, in order not to present technical tools of greater complexity, we assume some restrictions on the Lorentzian root lattices. The more general case is left as exercises. In section 5.5, we give the method for constructing these Borcherds-

Kac-Moody Lie algebras from even lattice vertex algebras. We do not consider the case of odd lattices or more generally of Borcherds-Kac-Moody Lie superalgebras with non-trivial odd parts in chapter 5 as the technical complications do not serve the purpose of a book aimed at introducing the reader to the above topics. Some of these questions are left as exercises in chapter 4 and section 5.5. The generality in which the material of chapters 2 and 4 is presented gives the reader the background to proceed in this direction, in particular to understand the construction of the Fake monster Lie superalgebra given by Scheithauer in [Sch1,2].

Chapter 2

Borcherds-Kac-Moody Lie Superalgebras

In this chapter we define Borcherds-Kac-Moody Lie superalgebras and explain what their structure is. We will abbreviate this name to BKM. We will take **C** to be the base field instead of **R** as in [Borc4]. It is necessary to do so for the construction of BKM Lie superalgebras with non-trivial odd parts (see [Sch1,2] for the construction of the Fake monster Lie superalgebra) and also for that of Lie superalgebras closely related to BKM superalgebras, for example affine Lie superalgebras (see Example 2.1.26). BKM algebras were first defined by Borcherds in [Borc4]. The interested reader can also read an account of the theory of BKM algebras in [Kac14, Chapter 11] and in [Jur1,2], where some oversights are corrected. We do not repeat proofs that remain the same as in the Kac-Moody Lie algebra case and are presented in [Kac14].

2.1 Definitions and Elementary Properties

So what is a BKM superalgebra? First of all it is a Lie superalgebra.

Definition 2.1.1. *A Lie superalgebra is a* $\mathbf{Z}_2$*-graded algebra* $G = G_{\bar{0}} \oplus G_{\bar{1}}$ *with a Lie bracket satisfying*

$$[a,b] = -(-1)^{d(a)d(b)}[b,a] \quad and \quad [a[b,c]] = [[a,b]c] + (-1)^{d(a)d(b)}[[a,c]b],$$

where for any homogeneous element $x \in G_{\bar{n}}$*,* $n = 0, 1$*,* $d(x) := \bar{n}$*. The subspace* $G_{\bar{0}}$ *(resp.* $G_{\bar{1}}$*) is called the even (resp. odd) part of* G*.*

Therefore a Lie algebra is a Lie superalgebra with trivial odd part. The most obvious example of a Lie superalgebra is that of linear maps on a $\mathbf{Z}_2$-graded vector space.

13

Example 2.1.2. Let $V = V_{\bar{0}} \oplus V_{\bar{1}}$ be a $\mathbf{Z}_2$-graded vector space. Consider the associative algebra $gl(V)$ of endomorphisms of V. It has a natural $\mathbf{Z}_2$-grading:

$$gl(V)_{\bar{0}} = \{f \in gl(V) : f(V_{\bar{n}}) \subseteq V_{\bar{n}}, \bar{n} \in \mathbf{Z}_2\},$$

$$gl(V)_{\bar{1}} = \{f \in gl(V) : f(V_{\bar{n}}) \subseteq V_{\overline{n+1}}, \bar{n} \in \mathbf{Z}_2\}.$$

The Lie bracket is defined as follows:

$$[x,y] = \begin{cases} xy - yx, & \text{if } x \text{ or } y \in gl(V)_{\bar{0}} \\ xy + yx, & \text{if } x, y \in gl(V)_{\bar{1}}. \end{cases}$$

Note that for any $x \in gl(V)$, if $x = \begin{pmatrix} x_{0,0} & x_{0,1} \\ x_{1,0} & x_{1,1} \end{pmatrix}$ with respect to a homogeneous basis (i.e. a basis $v_1, \cdots, v_n$ such that $v_1, \cdots v_m \in V_{\bar{0}}$ and $v_{m+1}, \cdots v_n \in V_{\bar{1}}$), the supertrace of x is defined to be

$$\text{str}\,(x) := tr(x_{0,0}) - tr(x_{1,1}).$$

In particular any Lie superalgebra G is a Lie sub-superalgebra of $gl(G)$ via the adjoint action ad, where G is considered as a $\mathbf{Z}_2$-graded vector space.

As mentioned in the introduction, the best way to think of a BKM super-algebra is to consider it as a generalization of a finite dimensional semisimple Lie algebra. So let us consider the characteristic properties of the latter and see how these can be weakened in a natural way without losing their essence, so as to apply to a larger class of Lie superalgebras. We start by reminding the reader of some basic facts about finite dimensional Lie algebras.

Definition 2.1.3.

(i) A sub-superalgebra I of a Lie superalgebra G is an ideal if $[I, G] \leq I$.

(ii) A Lie superalgebra is said to be simple if its only ideals are (0) and itself.

(iii) A semisimple Lie superalgebra is a direct sum of finitely many simple Lie superalgebras.

Definition 2.1.4. The Killing form on a finite dimensional Lie superalgebra G is defined as follows: for $x, y \in G$,

$$K(x,y) = \text{str}\,(\text{ad}\,(x)\text{ad}\,(y)).$$

Exercises 2.1.2 and 2.3.12 tell us that a finite dimensional semisimple Lie algebra can be characterized in two different ways.

Lemma 2.1.5. Let G be a finite dimensional Lie algebra. The following properties are equivalent.

(a) The Lie algebra G is semisimple.

(b) The Lie algebra G is generated by elements e_i, f_i (called Chevalley generators), $1 \leq i \leq n$ satisfying the Chevalley-Serre relations:

(i) $[e_i, f_j] = \delta_{ij} h_i$;

(ii) $[h_j, e_i] = a_{ij} e_i$, $[h_j, f_i] = -a_{ij} f_i$ where $a_{ij} = K(h_i, h_j)$ is the symmetric Cartan matrix;

(iii) $(\mathrm{ad}\,(e_i))^{1-\frac{2a_{ij}}{a_{ii}}} e_j = 0 = (\mathrm{ad}\,(f_i))^{1-\frac{2a_{ij}}{a_{ii}}} f_j$.

(c) The Killing form is non-degenerate on the Lie algebra G.

The generalization from finite dimension to infinite dimension was made via property (b) in [Kac2] and [Mo2]. Later, the generalization from Kac-Moody Lie algebras to BKM algebras is again via property (b) in [Borc4]. Thus, the definition of a BKM superalgebra is based on the Chevalley-Serre construction of finite dimensional semi-simple Lie algebras by generators and relations. In section 2.2, we will consider the characterization via property (c).

Let I be either a finite set, in which case we often identify it with $\{1, \cdots, n\}$ or a countably infinite set, in which case we sometimes identify it with $\mathbf{N}$. Let $S \subseteq I$ be a subset of I.

The set I will be indexing the simple roots (or equivalently generators) of the BKM superalgebra and S will be indexing the odd generators.

Let $H_{\mathbf{R}}$ be a real vector space with a non-degenerate symmetric real valued bilinear form $(.,.)$ and elements $h_i, i \in I$ such that

(i) $(h_i, h_j) \le 0$ if $i \ne j$,

(ii) If $(h_i, h_i) > 0$, then $\frac{2(h_i, h_j)}{(h_i, h_i)} \in \mathbf{Z}$ for all $j \in I$.

(iii) If $(h_i, h_i) > 0$ and $i \in S$, then $\frac{(h_i, h_j)}{(h_i, h_i)} \in \mathbf{Z}$ for all $j \in I$.

Let $H = H_{\mathbf{R}} \otimes_{\mathbf{R}} \mathbf{C}$. Let A be the symmetric real valued matrix with entries $a_{ij} = (h_i, h_j)$.

Remark 2.1.6. The elements h_i of the subspace H need not be linearly independent, nor even distinct. If the matrix A is non-degenerate, then elementary linear algebraic arguments imply that the vectors h_i are linearly independent.

The vector space H may be considered as an abelian Lie algebra.

Definition 2.1.7. *The Borcherds-Kac-Moody Lie superalgebra $G = G(A, H, S)$ associated to the matrix A and the abelian Lie algebra H is the Lie superalgebra generated by the abelian Lie algebra H and elements $e_i, f_i, i \in I$ satisfying the following defining relations:*

(1) $[e_i, f_j] = \delta_{ij} h_i$;

(2) $[h, e_i] = (h, h_i) e_i$, $[h, f_i] = -(h, h_i) f_i$;

(3) $\deg e_i = 0 = \deg f_i$ if $i \notin S$, $\deg e_i = 1 = \deg f_i$ if $i \in S$;

(4) $(\mathrm{ad}\,(e_i))^{1-\frac{2a_{ij}}{a_{ii}}} e_j = 0 = (\mathrm{ad}\,(f_i))^{1-\frac{2a_{ij}}{a_{ii}}} f_j$ if $a_{ii} > 0$ and $i \ne j$;

(5) $(\mathrm{ad}\,(e_i))^{1-\frac{a_{ij}}{a_{ii}}} e_j = 0 = (\mathrm{ad}\,(f_i))^{1-\frac{a_{ij}}{a_{ii}}} f_j$ if $i \in S$, $a_{ii} > 0$ and $i \ne j$;

(6) $[e_i, e_j] = 0 = [f_i, f_j]$ if $a_{ij} = 0$.

The matrix A is called the generalized symmetric Cartan matrix of the Lie superalgebra G and H is a generalized Cartan subalgebra of G. If $a_{ii} > 0$ for all $i \in I$, then G is said to be a Kac-Moody Lie superalgebra.

When $S = \emptyset$, $G = G(A, H, S)$ is a Lie algebra and for simplicity's sake we will write $G = G(A, H)$. The next Corollary on the derived Lie sub-superalgebra of a BKM superalgebra follows immediately from Definition 2.1.7.

Corollary 2.1.8. *Let $G = G(A, H, S)$ be a BKM superalgebra, then its derived Lie sub-superalgebra $G' = [G, G]$ is the BKM superalgebra $G(A, \langle h_i : i \in I \rangle, S)$.*

The more one generalizes, the lesser the restrictions imposed on the generalized Cartan matrix. For finite dimensional Lie algebras, the Cartan matrix A is positive definite (see Proposition 2.2.8) and this forces $a_{ii} > 0$ for all $i \in I$ and $|a_{ij}| \leq 3$. For Kac-Moody Lie superalgebras, $a_{ii} > 0$ and $2a_{ij}/a_{ii} \in \mathbf{Z}$ for all $i, j \in I$. For BKM superalgebras, $a_{ii} \leq 0$ is allowed and the corresponding rows and columns can be random.

Remark 2.1.9. Note that in some research papers, in the definition of a BKM superalgebra, the bilinear form on H is not supposed to be non-degenerate. In that case, condition (2) of Definition 2.1.7 implies that the kernel of the bilinear form on H in contained in the centre of the BKM superalgebra G. So we can just quotient out G by its centre to get a BKM superalgebra according to the above definition.

In the original definition of a Kac-Moody Lie algebra [Kac1,2], the elements h_i are assumed to be linearly independent by definition. As we will see later this is not the case for the most interesting BKM algebras that are not Kac-Moody Lie algebras, in particular for the Fake monster Lie algebra (see Example 2.6.41).

Hence, in the case of Kac-Moody Lie algebras, the above definition is more general than the one in [Kac14] though according to both the bilinear form on H must be non-degenerate (see chapters 1,2 in [Kac14]). This subtle difference is well illustrated in the next example.

Example 2.1.10. Let E_1 and F_1 be the Chevalley generators of the simple Lie algebra sl_2. The corresponding loop algebra has underlying space $G = \mathbf{C}[t, t^{-1}] \otimes_{\mathbf{C}} sl_2$ and Lie bracket defined as $[t^n \otimes x, t^m \otimes y] = t^{n+m} \otimes [x, y]$ for $x, y \in sl_2$. It is generated by the element $e_1 = 1 \otimes E_1$, $e_2 = t \otimes F_1$, $f_1 = 1 \otimes F_1$ and $f_2 = t^{-1} \otimes E_1$ satisfying the defining relations given in Definition 2.1.7 with

$$h_1 = [e_1, f_1] = -[e_2, f_2] = -h_2.$$

Thus $G = G(A, H)$, where

$$H = \mathbf{C}h_1$$

and the generalized Cartan matrix is

$$A = \begin{pmatrix} 2 & -2 \\ -2 & 2 \end{pmatrix}.$$

Hence, the bilinear form being non-degenerate on the generalized Cartan subalgebra, the loop algebra G is a BKM algebra. Since $a_{ii} > 0$ for $i = 1, 2$, it is therefore a Kac-Moody Lie algebra according to Definition 2.1.10. However it is not a Kac-Moody Lie algebra according to [Kac14] since the vectors h_1 and h_2 are not linearly independent.

The Affine Lie algebra associated to sl_2 is the central extension of G by a 1-dimensional centre generated by c. Its generalized Cartan subalgebra is isomorphic to $H \oplus \mathbf{C}c$ and condition (2) of Definition 2.1.7 implies that $\mathbf{C}c$ is the kernel of the bilinear form. Therefore, the latter is degenerate and so the affine Lie algebra is not a Kac-Moody Lie algebra. This follows from the uniqueness of the triangular decomposition given in section 2.4.

To get the extended affine Lie algebra $\hat{G}$ given by the simple Lie algebra sl_2, a derivation d is added to the previous central extension. It is defined as $\hat{G} = G(A, \hat{H})$, where $\hat{H}$ is isomorphic to $H \oplus \mathbf{C}c \oplus \mathbf{C}d$. The vector $h_1 \in H$ is as above a generator of the vector space H and $h_2 \in \hat{H}$ is now equal to $h_2 = c - h_1$. The bilinear form on $\hat{H}$ given by the above matrix A and $(c, d) = 1$, $(c, c) = 0 = (d, d)$, $(d, h_1) = 0$.

For details about this example, the reader is referred to [Gar1] where the derivation d was first introduced in the context of untwisted affine Lie algebras, to [GarL] where an appropriate family of d operators are given in the general case, and to [Kac14, Chapter 7].

Remark 2.1.11. In Definition 2.1.7, the generalized Cartan matrix A is assumed to be symmetric. The reader may be more familiar with the definition in which the positive diagonal entries of the matrix are all equal to 2. In this case, the generalized Cartan matrix — let us call it B — satisfies the conditions

(i') $b_{ij} \leq 0$ if $i \neq j$;
(ii') $b_{ij} \in \mathbf{Z}$ if $b_{ii} = 2$;
(iii') if $b_{ij} = 0$, then $b_{ji} = 0$;
(iv) there exists a diagonal matrix D with positive entries such that the matrix $A = DB$ is symmetric.

Condition (iv) says that the matrix B is *symmetrizable*. The definition of the Kac-Moody Lie algebra $G(B, H, S)$ corresponding to the matrix B in [Mo2] is similar to Definition 2.1.7. It is the Lie algebra with generators H, e'_i, f'_i such that

$$h'_i = [e'_i, f'_i],$$
$$[h'_i, e'_j] = b_{ji} \quad [h'_i, f'_j] = -b_{ji}$$
$$(\mathrm{ad}\,(e'_i))^{1-b_{ij}} e'_j = 0 = (\mathrm{ad}\,(f'_i))^{1-b_{ij}} f'_j.$$

However, we may as well take the symmetric version of the generalized Cartan matrix for reasons of simplicity since the generators of $G(B, H, S)$ are multiples of those of $G(A, H, S)$ and thus both Lie superalgebras are isomorphic (see Exercise 2.1.3). Obviously there can be several symmetric matrices equivalent to the matrix A (e.g. any positive multiple of A). For example the "usual" Cartan matrix for the simple finite dimensional Lie algebra C_2 is $\begin{pmatrix} 2 & -2 \\ -1 & 2 \end{pmatrix}$ and one of its symmetric versions is $\begin{pmatrix} 2 & -2 \\ -2 & 4 \end{pmatrix}$.

Condition $(ii)'$ for the matrix B is indeed equivalent to condition (ii) for the matrix A. To see this, set d_i to be the (i, i)-th entry of the diagonal matrix D.

From the above definition, we get $a_{ij} = d_i b_{ij}$ for all $i, j \in I$. In particular if $b_{ii} = 2$ then, $a_{ii} = 2d_i$ and so $\frac{2a_{ij}}{a_{ii}} = b_{ij} \in \mathbf{Z}$.

The definition of the generalized Cartan matrix of an arbitrary Kac-Moody Lie superalgebra as given in [Kac7,14] only satisfies conditions (i')-(iii') and is not necessarily symmetrizable. However, in this case there is no bilinear form on H given by the Cartan matrix. There are as yet no known examples of a non-symmetrizable Kac-Moody Lie algebra though some important results have been proved in the non-symmetrizable case by Kumar [Kum1,2]. The bilinear form on H induces a non-degenerate invariant symmetric bilinear form on G and without the existence of such a bilinear form, it is hard to get results on its structure and representation theory. The two most important classes of Kac-Moody Lie algebras (known so far), namely the finite dimensional semisimple and the affine Kac-Moody Lie superalgebras, both have symmetrizable Cartan matrices (see section 2.2 and for more details [Kac14, Chapters 1,2,6,7,8] for the Lie algebra case and [Kac3,4] for the more general Lie superalgebra case).

Remark 2.1.12. Keeping the notation of Remark 2.1.11, the Cartan matrix of a finite dimensional simple Lie algebra has the following property: for $i \neq j$, either $b_{ij} = -1$ or $a_{ji} = -1$. Hence, in this case the graph with nodes labeled by the indexing set I and with $b_{ij} b_{ji} = \frac{4a_{ij}^2}{a_{ii}a_{jj}}$ edges linking the i-th and j-th node with an arrow pointing towards the i-th node if $|a_{ij}| > 1$ contains all the information given by the Cartan matrix. This graph is the well known Dynkin diagram.

It can be generalized to Kac-Moody Lie superalgebras. If $i \in S$, then the corresponding node is set to be black and if $i \notin S$, it is white. However there may exist indices $i, j \in I$ for which both the integers $\frac{2|a_{ij}|}{a_{ii}} > 1$ and $\frac{2|a_{ij}|}{a_{jj}} > 1$. So the i-th and j-th nodes are linked with one edge labeled by the ordered pair $\frac{2|a_{ij}|}{a_{ii}}$ (resp. $\frac{|a_{ij}|}{a_{ii}}$) if $i \notin S$ (resp. $i \in S$) and $\frac{2|a_{ij}|}{a_{jj}}$ (resp. $\frac{|a_{ij}|}{a_{jj}}$) if $j \notin S$ (resp. $j \in S$). So this generalized Dynkin diagram contains the same information as the corresponding generalized Cartan matrix.

However, in the more general context of BKM superalgebras, there may be indices $i \in I$ such that $a_{ii} \leq 0$ and then the scalar $\frac{2a_{ij}}{a_{ii}}$ can take any real value, not necessarily an integer, and in the case when $a_{ii} = 0$, it is not defined. So we cannot associate any useful Dynkin diagram type graphs with most BKM superalgebras.

The next result explains Conditions (4) and (5) of Definition 2.1.7. Set

$$\tilde{G} = \tilde{G}(A, H, S)$$

to be the Lie superalgebra generated by the Lie subalgebra H and elements e_i, f_i, $i \in I$ satisfying relations $(1) - (3)$ of Definition 2.1.7.

Proposition 2.1.13.

(i) *Suppose that $a_{ii} \neq 0$. If $i \in I \backslash S$, then the Lie subalgebra*

$$S_i = \mathbf{C}f_i \oplus \mathbf{C}h_i \oplus \mathbf{C}e_i$$

of the BKM superalgebra G is isomorphic to sl_2 and if $i \in S$, then the Lie sub-superalgebra $S_i = \mathbf{C}[f_i, f_i] \oplus \mathbf{C}f_i \oplus \mathbf{C}h_i \oplus \mathbf{C}e_i \oplus \mathbf{C}[e_i, e_i]$ is isomorphic to $sl(0,1)$. Moreover, for all $i \in I \backslash S$ (resp. $i \in S$) such that $a_{ii} > 0$, considered as a S_i-module via the adjoint action, the vector space $\tilde{G}$ decomposes into finite dimensional S_i-modules if and only if condition (4) (resp. (5)) holds.

(ii) *Suppose that $a_{ii} = 0$. Then, the Lie (sub)-superalgebra $S_i = \mathbf{C}f_i \oplus \mathbf{C}h_i \oplus \mathbf{C}e_i$ is isomorphic to the three dimensional Heisenberg algebra (resp. superalgebra) if $i \in I \backslash S$ (resp. $i \in S$).*

Hence a Kac-Moody Lie superalgebra is generated by copies of the 3-dimensional simple Lie algebra sl_2, one for each even simple root, and of the 5-dimensional simple Lie superalgebra $sl(0,1)$ (see Proposition 2.1.23 for the definition), one for each odd simple root, and the adjoint actions of each of these sl_2's and $sl(0,1)$'s on G decomposes into finite dimensional representations. A BKM superalgebra is generated by copies of sl_2, $sl(0,1)$, and of the 3-dimensional Heisenberg (super)-algebra, and the action of these sl_2's and $sl(0,1)'s$ decompose into finite dimensional representations if and only if they correspond to positive diagonal entries in the generalized Cartan matrix.

From now on, we will assume that the generalized Cartan matrix A is indecomposable. The general case easily follows but would make the statement of definitions and properties less clear without adding any new ideas or technical difficulty. So let us remind the reader of the definition of an indecomposable matrix and justify our assumption:

Definition 2.1.14.

(i) *A square matrix A is said to be indecomposable if there do not exist square matrices B and C such that $A = \begin{pmatrix} B & 0 \\ 0 & C \end{pmatrix}$.*

(ii) *The indexing set I is said to be connected if it cannot be written as the union of two subsets I_1 and I_2 such that $(h_i, h_j) = 0$ for all $i \in I_1$ and $j \in I_2$.*

Lemma 2.1.15. *Let $H_1, H_2 \leq H$ be subspaces of H such that $H_1 + H_2 = H$. Suppose that there exist subsets $I_1, I_2 \leq I$ such that $I_1 \cup I_2 = I$, $\langle h_i : i \in I_j \rangle \leq H_j$ for $j = 1, 2$, and $(h_i, h_j) = 0$ for all $i \in I_1$, $j \in I_2$. Then, the generalized symmetric Cartan matrix $A = (h_i, h_j)_{i,j \in I}$, $A = \begin{pmatrix} A_1 & 0 \\ 0 & A_2 \end{pmatrix}$, where for $i = 1, 2$, A_i is a generalized symmetric Cartan matrices with indexing set I_i; and*

$$G(A, H, S) = G(A_1, H_1, S_1) + G(A_1, H_2, S_2).$$

If moreover $H_1 \oplus H_2 = H$, then

$$G(A, H, S) = G(A_1, H_1, S_1) \oplus G(A_1, H_2, S_2).$$

We leave the proof of this result for the reader to check.

Remark 2.1.16. In many natural examples, the sum $H_1 + H_2$ is not direct. This is trivially the case for the Heisenberg algebras (resp. superalgebras). These are BKM algebras (resp. superalgebras) with $H = \mathbf{C}c$, $I = \mathbf{Z}$, $S = \emptyset$ (resp. $I = S$) $h_i = c$ for all $i \in I$, and $A = 0$.

The assumption that A is indecomposable eliminates the most degenerate cases: if a column or a line of the matrix A is uniformly 0, then $A = 0$. In this case, $I = \{1\}$ and the derived Lie superalgebra G' is the 3-dimensional Heisenberg (super)algebra.

Kac defined a Kac-Moody Lie algebra as a quotient of the Lie superalgebra $\tilde{G}$ by the unique maximal ideal intersecting the Cartan subalgebra trivially. The Gabber-Kac Theorem shows that this definition is equivalent to the original one of Moody by generators and relations [GabK, Chapter 9 in Kac14]. The same remains true in the more general context of BKM superalgebras [Ray2]. Here we only state this result as we need to develop some representation theory for this (see Exercise 2.6.4).

Theorem 2.1.17. *The ideal of $\tilde{G}$ generated by the elements*

(6) $(\operatorname{ad} e_i)^{\frac{-2a_{ij}}{a_{ii}}+1} e_j$, $(\operatorname{ad} f_i)^{\frac{-2a_{ij}}{a_{ii}}+1} f_j$, *for* $a_{ii} > 0$, $i \in I \setminus S$;

(7) $(\operatorname{ad} e_i)^{\frac{-a_{ij}}{a_{ii}}+1} e_j$, $(\operatorname{ad} f_i)^{\frac{-a_{ij}}{a_{ii}}+1} f_j$, *for* $a_{ii} > 0$, $i \in S$; *and*

(8) $[e_i, e_j]$, $[f_i, f_j]$ *if* $a_{ij} = 0$

is the maximal ideal R intersecting the Cartan subalgebra H trivially.

This Theorem is crucial for the structure of G. An immediate consequence is the following important result, without which for example the fundamental fact that generalized Cartan matrices uniquely characterize BKM superalgebras (see section 2.4) would not hold:

Corollary 2.1.18. *The centre of the BKM superalgebra G is contained in the generalized Cartan subalgebra H.*

Also, Theorem 2.1.17 together with exercise 2.1.5 give the well known triangular decomposition of a BKM superalgebra [Kac14, Theorem 1.2 and Borc6].

Corollary 2.1.19. $G = N_- \oplus H \oplus N_+$ *as vector spaces, where N_+ (resp. N_-) is the Lie sub-superalgebra generated by the elements e_i (resp. f_i), $i \in I$.*

Definition 2.1.20. *Keeping the notation of Corollary 2.1.19, the Lie sub-superalgebra $B_+ = H \oplus N_+$ (resp. $B_- = N_- \oplus H$) is a (generalized) positive (resp. negative) Borel sub-superalgebra.*

Finite dimensional and affine BKM superalgebras

The finite dimensional simple Lie superalgebras were classified by Kac in the 70s in [Kac3]. For details, the reader is referred to this paper. Some are sub-superalgebras of finite dimensional general linear Lie superalgebras (see Example 2.1.2). All finite dimensional semisimple Lie algebras and affine Lie algebras are BKM algebras. However this is no longer true for most finite dimensional and affine Lie superalgebras. Some do not even have associated Cartan matrices. Some can be defined via generators of the same type as in Definition 2.1.7 and Cartan matrices but in general the Chevalley-Serre presentation given in Definition 2.1.7 is not sufficient. When the odd part is non-trivial, there are usually more relations [LeiS]. In general they have more than one (non-equivalent) Cartan matrix (see Definition 2.4.1), i.e. non-equivalent Dynkin diagrams (see section 2.2) and these Cartan matrices do not satisfy conditions (i)-(iii) of generalized Cartan matrices given above. The list of the finite dimensional contragredient simple classical Lie superalgebras and for each, all the possible Dynkin diagrams is given in [Kac3, §2.5]. Those with non-trivial odd part are of type $A(m,n)$, $B(m,n)$, $C(n)$, $D(m,n)$, $D(2,1;\alpha)$, $F(4)$ or $G(3)$. As a Dynkin diagram contains the same information as the corresponding generalized Cartan matrix, from this list, it is easy to find those that are BKM superalgebras [Ray3]. Before stating this result, we first define contragredient Lie algebras. These are constructions introduced by Kac in [Kac6], of which BKM superalgebras form a subclass. They allowed him to construct the most important examples of simple finite-dimensional Lie superalgebras.

Definition 2.1.21.

(i) *A Lie superalgebra G is $\mathbf{Z}$-graded if $G = \oplus_{i \in \mathbf{Z}} G_i$, where $[G_i, G_j] \leq G_{i+j}$.*

(ii) *For any matrix A indexed by the set I, a contragredient Lie superalgebra $G = G(A, S)$ is the minimal $\mathbf{Z}$-graded Lie superalgebra such that the vector space G_{-1} (resp. G_1) is generated by the elements f_i (resp. e_i) satisfying conditions $(1) - (3)$ of definition 2.1.7 and $G_0 = [G_1, G_{-1}]$.*

In the above definition, minimal means that G is the epimorphic image of every $\mathbf{Z}$-graded Lie superalgebra with G_{-1}, G_0 and G_1 of the above type.

Lemma 2.1.22. *A BKM superalgebra whose generalized Cartan subalgebra is generated by the vectors h_i, $i \in I$, is the quotient of a contragredient by a central Lie subalgebra $(\subset G_0)$.*

Proof. This result follows from Theorem 2.1.17 by setting $\deg(e_i) = 1$ and $\deg(f_i) = 1$. $\qquad\square$

Corollary 2.1.23. *A simple finite dimensional Lie superalgebra G is a BKM superalgebra if and only if G is contragredient of type $A(m,0) = sl(m+1,1)$, $A(m,1) = sl(m+1,2)$, $B(0,n) = osp(1,2n)$, $B(m,1) = osp(2m+1,2)$, $C(n) = osp(2,2n-2)$, $D(m,1) = osp(2m,2)$, $D(2,1;\alpha)$ for $\alpha \neq 0, -1$, $F(4)$, and $G(3)$.*

Proof. The above do have a Cartan subalgebra H and elements h_i, $i \in I$ with respect to which the generalized Cartan matrix is respectively:

$$
\begin{pmatrix}
0 & -1 & 0 & \cdots & 0 \\
-1 & 2 & -1 & \cdots & 0 \\
0 & -1 & 2 & \cdots & 0 \\
\vdots & \vdots & \vdots & \ddots & \vdots \\
0 & 0 & 0 & \cdots & 2
\end{pmatrix},
\begin{pmatrix}
0 & -1 & 0 & \cdots & 0 & 0 \\
-1 & 2 & -1 & \cdots & 0 & 0 \\
0 & -1 & 2 & \cdots & 0 & 0 \\
\vdots & \vdots & \vdots & \ddots & \vdots & \vdots \\
0 & 0 & 0 & \cdots & 2 & -1 \\
0 & 0 & 0 & \cdots & -1 & 0
\end{pmatrix},
$$

$$
\begin{pmatrix}
1 & -1 & 0 & \cdots & 0 \\
-1 & 2 & -1 & \cdots & 0 \\
0 & -1 & 2 & \cdots & 0 \\
\vdots & \vdots & \vdots & \ddots & \vdots \\
0 & 0 & 0 & \cdots & 2
\end{pmatrix},
\begin{pmatrix}
0 & -1 & 0 & \cdots & 0 & 0 \\
-1 & 2 & -1 & \cdots & 0 & 0 \\
0 & -1 & 2 & \cdots & 0 & 0 \\
\vdots & \vdots & \vdots & \ddots & \vdots & \vdots \\
0 & 0 & 0 & \cdots & 2 & -1 \\
0 & 0 & 0 & \cdots & -1 & 1
\end{pmatrix},
$$

$$
\begin{pmatrix}
0 & -1 & 0 & \cdots & 0 & 0 \\
-1 & 2 & -1 & \cdots & 0 & 0 \\
0 & -1 & 2 & \cdots & 0 & 0 \\
\vdots & \vdots & \vdots & \ddots & \vdots & \vdots \\
0 & 0 & 0 & \cdots & 2 & -2 \\
0 & 0 & 0 & \cdots & -2 & 4
\end{pmatrix},
\begin{pmatrix}
2 & 0 & -1 & \cdots & 0 & 0 \\
0 & 2 & -1 & \cdots & 0 & 0 \\
-1 & -1 & 2 & \cdots & 0 & 0 \\
\vdots & \vdots & \vdots & \ddots & \vdots & \vdots \\
0 & 0 & 0 & \cdots & 2 & -1 \\
0 & 0 & 0 & \cdots & -1 & 0
\end{pmatrix},
$$

$$
\begin{pmatrix}
2 & -1 & 0 \\
-1 & 0 & -\alpha \\
0 & -\alpha & 2\alpha
\end{pmatrix} \text{ for } \alpha > 1,
\begin{pmatrix}
0 & -1 & 0 & 0 \\
-1 & 2 & -2 & 0 \\
0 & -2 & 4 & -2 \\
0 & 0 & -2 & 4
\end{pmatrix},
\begin{pmatrix}
0 & -1 & 0 \\
-1 & 2 & -3 \\
0 & -3 & 6
\end{pmatrix},
$$

where in all cases $i \in S$ if and only if $a_{ii} = 0$ or 1.

The converse is left as an exercise for the reader.   □

Note that the above list of generalized Cartan matrices together with the uniqueness of the generalized Cartan matrix except for $A(1,0)$ (see Theorem 2.4.8) imply that the only finite dimensional simple Kac-Moody Lie superalgebras with non-trivial odd part are of type $B(m,1)$ whereas all finite dimensional semisimple Lie algebras are Kac-Moody Lie algebras.

Affine Lie algebras are well known and have several definitions. Affine BKM superalgebras are defined following the original construction of affine Lie algebras via their growth rate [Leu, Kac7].

Definition 2.1.24. *A* $\mathbf{Z}$-*graded Lie superalgebra* $G = \oplus_{i \in \mathbf{Z}} G_i$ *is of finite or polynomial growth if there is a polynomial* p *such that for each* $i \in \mathbf{Z}$, $\dim G_i \leq p(|i|)$.

Definition 2.1.25. *An affine Lie superalgebra is a contragredient BKM superalgebra of infinite dimension but finite growth.*

The affine Lie algebras have been classified by Kac in [Kac3]; in particular, it follows from his paper that affine Lie algebras are BKM algebras and in particular Kac-Moody algebras (i.e. $a_{ii} > 0$ for all $i \in I$). This does not remain true in the Lie superalgebra setup [Leu]. The affine Lie superalgebras have been classified by van de Leur in [Leu], using the methods developed in [Kac1].

Proposition 2.1.26. *An affine BKM superalgebra is either a Kac-Moody Lie superalgebra or has degenerate generalized symmetric Cartan matrix* $A = (0)$.

We reproduce the list of affine Kac-Moody Lie superalgebras given in [Kac7]. They are of type $B^{(1)}(0,n)$, $B^{(1)}(0,1)$ $A^{(2)}(0,2n-1)$, $n \geq 3$, $A^{(2)}(0,3)$, $C^{(2)}(n+1)$, $n \geq 2$, $C^{(2)}(2)$, $A^{(4)}(0,2n)$, $n \geq 2$, and $A^{(4)}(0,2)$. The first six respectively have Cartan matrices

$$
\begin{pmatrix}
4 & -2 & \cdots & 0 & 0 \\
-2 & 2 & \cdots & 0 & 0 \\
\vdots & \vdots & \ddots & \vdots & \vdots \\
0 & 0 & \cdots & 2 & -1 \\
0 & 0 & \cdots & -1 & 1
\end{pmatrix}, \quad
\begin{pmatrix}
4 & -2 \\
-2 & 1
\end{pmatrix}
$$

$$
\begin{pmatrix}
2 & 0 & -1 & \cdots & 0 & 0 \\
0 & 2 & -1 & \cdots & 0 & 0 \\
-1 & -1 & 2 & \cdots & 0 & 0 \\
\vdots & \vdots & \vdots & \ddots & \vdots & \vdots \\
0 & 0 & 0 & \cdots & 2 & -1 \\
0 & 0 & 0 & \cdots & -1 & 1
\end{pmatrix}, \quad
\begin{pmatrix}
2 & -1 & 0 \\
-1 & 1 & -1 \\
0 & -1 & 2
\end{pmatrix},
$$

$$
\begin{pmatrix}
1 & -1 & \cdots & 0 & 0 \\
-1 & 2 & \cdots & 0 & 0 \\
\vdots & \vdots & \ddots & \vdots & \vdots \\
0 & 0 & \cdots & 2 & -1 \\
0 & 0 & \cdots & -1 & 1
\end{pmatrix}, \quad
\begin{pmatrix}
1 & -1 \\
-1 & 1
\end{pmatrix},
$$

where $i \in S$ if and only if $a_{ii} = 1$. The last two matrices with $S = \{1\}$ are the Cartan matrices of $A^{(4)}(0, 2n)$, $n \geq 2$, and $A^{(4)}(0, 2)$ respectively.

We next construct an example of an affine Lie superalgebra. It is not a BKM superalgebra but shows that when the odd part is non-trivial, we have to be careful about the base field taken.

Example 2.1.27. The extended affine Lie superalgebra $sl^{(1)}(1, n)$ is constructed in the same way as the extended affine Lie algebra $sl_2^{(1)}$ (see Example 2.1.10) as the central extension of the Loop superalgebra

$$
\mathbf{C}[t, t^{-1}] \otimes_{\mathbf{C}} sl(1, n)
$$

with an added derivation.

Let $E_1, ..., E_n$ and $F_1, ..., F_n$ be the generators of $sl(1, n)$ with $1 \in S$. Then the generators of $sl^{(1)}(1, n)$ are $e_0, ..., e_n$ and $f_0, ..., f_n$, where

$$
e_i = 1 \otimes E_i \quad \text{and} \quad f_i = 1 \otimes F_i \quad \forall i \geq 1.
$$

Let θ be the root of $sl(1, n)$ of maximal height. It has norm 0. Therefore the simple root vectors e_0 and f_0 of $sl^{(1)}(1, n)$ are defined as follows: let $F_0 \in sl(1, n)_\theta$ be such that

$$
(\omega(F_0), F_0) = 1/(\theta, \alpha_1)
$$

and not as $(\omega(F_0), F_0) = 1/(\theta, \theta)$ as in the affine Lie algebra case since θ has norm 0. So calculations show that

$$
F_0 = i[e_1 ... [e_{n-1}, e_n]]
$$

and we have to take $\mathbf{C}$ as our ground field and cannot work over $\mathbf{R}$. Then $E_0 = \omega(F_0)$ and

$$e_0 = t \otimes E_0 \quad \text{and} \quad f_0 = t^{-1} \otimes F_0.$$

Before closing this section, we define two basic automorphisms of BKM superalgebras as they map the subspace N_+ onto the subspace N_- and whose existence follow from Exercise 2.1.5.

Proposition 2.1.28. *There is an automorphism ω of period 4 acting on the BKM superalgebra G, called the Chevalley automorphism given by:*

$$\omega(e_i) = \begin{cases} f_i & \text{if } i \in S, \\ -f_i, & \text{otherwise.} \end{cases};$$

$$\omega(f_i) = -e_i \quad \forall i \in I, \omega(h) = -h \quad \forall h \in H.$$

There is an antilinear automorphism ω_0 of period 4 acting on the BKM superalgebra G, called the compact automorphism given by:

$$\omega_0(e_i) = \begin{cases} f_i & \text{if } i \in S, \\ -f_i, & \text{otherwise.} \end{cases};$$

$$\omega_0(f_i) = -e_i \quad \forall i \in I, \omega_0(h) = -h \quad \forall h \in H_{\mathbf{R}}.$$

We remind the reader that ω_0 is antilinear on the complex vector space G if $\omega_0(cx) = \bar{c}\omega_0(x)$ for all $x \in G$ and $c \in \mathbf{C}$. The need for the antilinear map ω_0 will be seen in Corollary 2.2.6. When the BKM superalgebra G is a Lie algebra, ω (resp. ω_0) is the usual Chevalley (resp. compact) involution. As will be seen in section 2.4, when G is a finite dimensional simple Lie algebra, ω is an inner automorphism, but this is no longer the case in infinite dimension.

Proposition/Definition 2.1.29. *Suppose that G is a BKM algebra. The set $\{x \in G : \omega_0(x) = x\}$ of fixed point of the automorphism ω_0 is a real Lie superalgebra $C(A)$ of the BKM superalgebra G such that $G = \mathbf{C} \otimes_{\mathbf{R}} C(A)$. It is the compact form.*

Proof. The result follows from the fact that $e_i - f_i$, $i(e_i + f_i)$ and ih, for all $h \in H_{\mathbf{R}}$, are vectors in $C(A)$. $\qquad \square$

Exercises 2.1

1. Show that a finite dimensional Lie superalgebra L is semisimple if and only if it has no non-trivial abelian ideal.

Hint: you may use induction on the dimension of L to show that the non-existence of a non-trivial abelian ideal is a sufficient condition.

2. Let L be a finite dimensional Lie superalgebra.

(i) Prove that the Killing form K on L is an invariant bilinear form, i.e. for all $x, y, z \in L$, $K([x, y], z) = K(x, [y, z])$ and K is bilinear.

(ii) Deduce that the kernel of the Killing form is an ideal of L.

(iii) Suppose that L is a Lie algebra. Deduce that the Killing form is non-degenerate if the Lie algebra L is semisimple.

(iv) If the Killing form is non-degenerate on L, prove the Lie superalgebra L is semisimple.

Hint: you may use Exercise 1.

3. Let B be a symmetrizable matrix satisfying conditions $(i') - (iv)$ of Remark 2.1.11. We keep the same notation as in Remark 2.1.11.

(i) Show that for $i, j \in I$, $(h'_i, h'_j) = d_i^{-1} d_j^{-1} a_{ij}$, where d_i is the (i, i)-th entry of the diagonal matrix D.

(ii) Deduce that the Lie superalgebras $G(A, H, S)$ and $G(B, H, S)$ are isomorphic.

We remind the reader that the generalized Cartan matrix is assumed to be indecomposable.

4. Show that the centre of the BKM superalgebra G is the subspace

$$\{h \in H : (h, h_i) = 0 \quad \forall i \in I\}.$$

Deduce that if $H = \langle h_i : i \in I \rangle$ and the form is non-degenerate on H, then the BKM superalgebra G is simple.

5. Consider the Lie superalgebra $\tilde{G}$.

(i) Show that $\tilde{G} = \tilde{N}_- \oplus H \oplus \tilde{N}_+$ as vector spaces, where $\tilde{N}_+$ (resp. $\tilde{N}_-$) is the Lie sub-superalgebra generated by the elements e_i (resp. f_i), $i \in I$, satisfying properties $(1) - (3)$ of Definition 2.1.7.

Hint: Show first that $\tilde{G}$ is the sum of the above three Lie sub-superalgebras. To show that the sum is direct, construct for each $\lambda \in H^*$ (the dual of H), an action of $\tilde{G}$ on the tensor algebra $T(V)$ of a vector space with a basis indexed by I and consider the action of $0 = f + h + e$ on $1 \in T(V)$, where $f \in \tilde{N}_-$, $e \in \tilde{N}_+$, and $[e, f] = h$.

(ii) Deduce that the Lie sub-superalgebra $\tilde{N}_+$ (resp. $\tilde{N}_-$) is freely generated by the vectors e_i (resp. f_i), $i \in I$.

(iii) Deduce that the automorphisms ω and ω_0 defined in Proposition 2.1.28 are well defined.

For a solution see [Kac14, Theorem 1.2] or [Borc6].

6. Let $G = \oplus_{i \in \mathbf{Z}} G_i$ be a $\mathbf{Z}$-graded Lie superalgebra. Show the existence of a minimal Lie superalgebra generated by $G_{-1} + G_0 + G_1$.

For a solution, see [Kac6, Proposition 1.2.2] and [Kac1, Proposition 4]

7. Let $\hat{G}$ be the Lie superalgebra generated by the subspace $\hat{H}$ containing vectors h_{ij}, $i, j \in I$ with $h_{ij} \neq 0$ if and only if $a_{ki} = a_{kj}$ for all $k \in I$, satisfying $(h_{ii}, h_{jj}) = a_{ij}$, and elements e_i, f_i, $i \in I$ satisfying relations $(1) - (6)$ of Definition 2.1.7. and the following extra ones

(7) $[e_i, f_j] = h_{ij}$,

(8) $[h_{ij}, e_k] = 0 = [h_{ij}, f_k]$ if $i \neq j$.

(i) Show that $\hat{G}$ is the universal central extension of the BKM superalgebra G given in Definition 2.1.7.

(ii) Deduce that a Kac-Moody Lie superalgebra is its own universal central extension.

(iii) Deduce the a BKM superalgebra is the semi-direct product of $\hat{G}/Z$, where Z is a central subalgebra of $\hat{G}$ and of an abelian algebra having the generators e_i and f_i of $\hat{G}$ as eigenvectors.

For a solution, [see Borc6]

2.2   Bilinear Forms

In Section 2.1, we saw that BKM superalgebras were defined by generalizing one of the defining properties of finite dimensional semisimple Lie algebras, namely Lemma 2.1.5.b. In this section, we consider the equivalent characterizing property 2.1.5.c. When the Lie algebra G is infinite dimensional, the Killing form obviously makes no sense as the trace is no longer definable. Furthermore, note that when G is a finite dimensional BKM superalgebra, Lemma 2.1.3.c may not hold.

Lemma 2.2.1. *The Killing form on a finite dimensional BKM superalgebra is trivial if and only if it is of type $D(2, 1, \alpha)$.*

To generalize the Killing form to BKM superalgebras, let us consider its essential characteristics. First some more definitions.

Definition 2.2.2. *Let $(.,.)$ be a bilinear form on the Lie superalgebra L.*

(*i*) $(.,.)$ *is said to be supersymmetric if* $(x, y) = (-1)^{d(x)d(y)}(y, x)$ *for all homogeneous elements $x, y \in L$.*

(*ii*) $(.,.)$ *is said to be consistent if* $(G_{\bar{0}}, G_{\bar{1}}) = 0$.

(*iii*) $(.,.)$ *is said to be invariant if* $([g, x], y) = (g, [x, y])$ *for all elements $g, x, y \in L$.*

Lemma 2.2.3. *Let L be a finite dimensional Lie superalgebra and K its Killing form. The form K is bilinear supersymmetric, invariant, and consistent.*

These properties are well known and we leave it for the reader to check. Though Lemma 2.1.3.c. does not hold in general for finite dimensional BKM superalgebras, they have non-degenerate bilinear forms satisfying Lemma 2.2.3.

Lemma 2.2.4. *There is a non-degenerate, invariant, supersymmetric, consistent bilinear form $(.,.)$ on the BKM superalgebra G. Moreover, if the Lie superalgebra G is perfect (i.e. equal to its derived Lie subalgebra), then this form is unique up to multiplication by a non-zero scalar.*

In the rest of this chapter, $(.,.)$ will denote a non-degenerate, invariant, supersymmetric, consistent bilinear form on G.

As a consequence of Lemma 2.2.1, for finite dimensional BKM superalgebras of type $D(2,1,\alpha)$, the non-degenerate forms are not multiples of the Killing form. However these properties of a bilinear form are too weak to characterize BKM algebras, let alone BKM superalgebras. Even in finite dimension, there are nilpotent Lie algebras with non-degenerate invariant symmetric bilinear forms (see Exercise 2.2.2). To find a characterization of BKM algebras via bilinear forms, we need to consider the compact antilinear automorphism.

Lemma 2.2.5. *Let G be a BKM superalgebra. The compact antilinear automorphism keeps the bilinear form invariant.*

We first consider the case when G is a Lie algebra. Usually a Hermitian form on a complex vector space is considered to be linear in the first argument. Here it is more suitable to define them to be linear in the second variable and antilinear in the first. This is due to requirements arising in the more general context of BKM superalgebras (see Remark 2.2.18).

Corollary 2.2.6. *Let G be a BKM algebra. The form $(.,.)_0$ defined by*

$$(x, y)_0 = -(\omega_0(x), y), \quad x, y \in G$$

is Hermitian and contravariant, i.e. $([g, x], y)_0 = -(x, [\omega_0 g, y])_0$ for all $g, x, y \in G$.

Proof. We will only prove that the form $(.,.)_0$ is Hermitian. The contravariance property is left for the reader to check. We apply Proposition 2.1.29. Let $x = x_1 + ix_2 \in G$ and $y = y_1 + iy_2 \in G$, where $x_i, y_i \in C(A)$ for $i = 1, 2$. Then,

$$(y, x)_0 = (y_1 - iy_2, x_1 + ix_2) = \overline{(x_1 - ix_2, y_1 + iy_2)} = \overline{(x, y)_0}.$$

$\square$

In the above result, for the form $(.,.)_0$ to be Hermitian, the map ω_0 has to be assumed to be antilinear.

The following property of the form $(.,.)_0$ for finite dimensional semisimple Lie algebras is classical. For a proof, see Exercise 2.6.6.

Proposition 2.2.7. *If G is a finite dimensional semisimple Lie algebra, then the bilinear form $(.,.)_0$ is positive definite on G.*

This result is clearly false in this strong form for arbitrary BKM algebras since the generalized symmetric Cartan matrix A may have non-positive diagonal entries. As mentioned earlier, Kac-Moody Lie algebras were first constructed in the context of a study of $\mathbf{Z}$-graded Lie algebras and this basic property also needs to be taken into account. Note that the nilpotent Lie algebras carrying non-degenerate invariant bilinear symmetric forms constructed in Exercise 2.2.2 are not $\mathbf{Z}$-graded.

Theorem 2.2.8. *Let G be a BKM algebra. Then:*

(i) *G is graded:* $G = \sum_{i \in \mathbf{Z}} G_i$, $\dim G_i < \infty$ *for all* $i \neq 0$, *and* $[G_i, G_j] \leq G_{i+j}$;

(ii) *there is an antilinear involution ω_0 of G such that $\omega_0(G_i) = G_{-i}$ and ω_0 is multiplication by* -1 *on* $G_{0\mathbf{R}}$;

(iii) *the form $(.,.)_0$ is a contravariant Hermitian form and it is positive definite on all the spaces G_i with $i \neq 0$. Furthermore $(G_i, G_j)_0 = 0$ if $i \neq j$.*

Proof. Set $\deg h = 0$ for all $h \in H$ and $\deg e_i = i = -\deg f_i$ for $i \in I$. This induces a $\mathbf{Z}$-gradation of the Lie algebra G satisfying (i). By definition, the compact involution ω_0 given in Proposition 2.1.28 satisfies (ii) since $h_i \in H_{\mathbf{R}}$ and $G_0 = H$.

(iii) follows from Exercise 2.6.6 as it requires the use of some representation theory. This part proved in [KacP3], is called the Kac-Peterson Theorem. $\square$

It is important to note that the i-th pieces are all finite dimensional except possibly the 0-th piece. As Exercise 2.2.3 shows there are Lie algebras which satisfy all the other conditions but are not BKM algebras.

The above properties characterize BKM algebras. In other words, the converse to Theorem 2.2.8 holds. It is essential in Borcherds' proof of the Moonshine Theorem where it is needed to show that the Monster Lie algebra is a BKM algebra.

Theorem 2.2.9. *Let L be a Lie algebra satisfying the following conditions:*

(i) *L is graded:* $L = \sum_{i \in \mathbf{Z}} L_i$ *with* $\dim L_i < \infty$ *for all* $i \neq 0$ *and* $[L_i, L_j] < L_{i+j}$;

(ii) *there is an antilinear involution ω_0 of L such that $\omega_0(L_i) = L_{-i}$ and ω_0 is multiplication by* -1 *on* $L_{0\mathbf{R}}$;

(iii) *there is a contravariant Hermitian form $(.,.)_0$ on L, invariant under ω_0 and positive definite on all the subspaces L_i for $i \neq 0$. Furthermore $(L_i, L_j)_0 = 0$ if $i \neq j$.*

Then the kernel R of the Hermitian form is contained in the centre of L and L/R is a BKM algebra.

We split the proof of this result into several parts. L will denote a Lie algebra satisfying conditions (i)-(iii) of Theorem 2.2.9.

Lemma 2.2.10. *The Lie subalgebra L_0 is abelian.*

Proof. For any $h_1, h_2 \in (L_0)_{\mathbf{R}}$, condition (i) gives $[h_1, h_2] \in L_0$. Hence by condition (ii), $-[h_1, h_2] = \omega_0([h_1, h_2]) = [\omega_0(h_1), \omega_0(h_2)] = [h_1, h_2]$, and the Lemma follows. $\square$

Lemma 2.2.11. *The kernel R of the Hermitian form is contained in the centre of L.*

Proof. Let $k \in R$. There exists $s \geq 0$ such that $k_j \in L_j$, $-s \leq j \leq s$ and

$$k = \sum_{j=-s}^{s} k_j.$$

So,

$$0 = (k, k_j)_0 = (k_j, k_j)_0,$$

by (iii). This implies that $s = 0$ since the Hermitian form $(.,.)_0$ is non-degenerate on L_i for $i \neq 0$. So $k = k_0 \in L_0$ and thus

$$R \leq L_0.$$

Also, for any $j > 0$ and any $x \in L_j$,

$$([k, x], y)_0 = -(k, [\omega_0(x), y])_0 = 0$$

for all $y \in L_j$. Hence

$$[k, x] = 0$$

as $[k, x] \in L_j$ by (i) and $(.,.)_0$ is non-degenerate on L_j. Therefore as L_0 is abelian, k is contained in the centre of L. $\qquad\square$

Lemma 2.2.12. *Suppose that $R = 0$. For any integer $r > 0$, set M_r to be the Lie algebra generated by the spaces L_i, $|i| < r$ and $V_r = \{v \in L_r : (v, x)_0 = 0 \forall x \in L_r\}$. Then, $L_r = V_r \oplus (M_r \cap L_r)$ as vector spaces and there exists a basis B_r of the vector space V_r consisting of pairwise orthonormal eigenvectors $\{e_i\}$ for L_0.*

Proof. As the Hermitian form $(.,.)_0$ is positive definite on L_r, the direct sum follows.

Now, $[L_0, V_{\pm r}] \leq V_{\pm r}$: indeed, for any $h \in L_0, x \in V_{\pm r}, y \in M_r$,

$$([h, x], y)_0 = -(x, [\omega_0(h), y])_0 = 0$$

by definition of $V_{\pm r}$ since $[\omega_0(h), y] \in M_r$.

Hence there exists a basis of V_r consisting of eigenvectors for L_0. And so, as the Hermitian form is positive definite on V_r, there is an orthonormal set $\{e_j\}$ of elements in V_r consisting of eigenvectors for L_0. $\qquad\square$

We keep the notation of Lemma 2.2.12. For all i, set

$$f_i = -\omega_0(e_i) \quad \text{and} \quad h_i = [e_i, f_i].$$

By (i), $h_i \in L_0$. The following consequence is immediate.

Corollary 2.2.13. *Suppose that $R = 0$. The set $B = (\cup_{r=0}^{\infty} B_r) \cup (\cup_{r=0}^{\infty} \omega_0(B_r))$ together with the Lie subalgebra L_0 generate the Lie algebra L.*

Let $A = (a_{ij})$ be the matrix with entries $a_{ij} = (h_i, h_j)_0$. We next show that the matrix A satisfies conditions (i)-(iii) of a symmetric generalized Cartan matrix, that its entries are real, and that the generators e_i, f_i, and $H = G_0$ satisfy relations (1)-(6) of Definition 2.1.7 of a BKM algebra.

Lemma 2.2.14. *Suppose that $R = 0$. For all i, j, $[e_i, f_j] = 0$ if $i \neq j$ and $[h, e_i] = (h, h_i)_0 e_i$, $[h, f_i] = -(h, h_i)_0 f_i$.*

Proof. Let $i \neq j$. We first show that $[e_i, f_j] = 0$. Without loss of generality, we may assume that $e_i \in L_s$, $f_j \in L_{-t}, s \geq t > 0$. Hence by Lemma 2.2.12, $e_i \in V_s$. So for all $x \in L_{s-t}$,

$$([e_i, f_j], x)_0 = -(e_i, [e_j, x])_0. \tag{1}$$

Since $0 \leq s - t < s$, $x \in M_s$. We show that

$$([e_i, f_j], x)_0 = 0 \tag{2}$$

for all $x \in L_{s-t}$. If $t < s$, then $f_j \in M_s$ and so $[e_j, x] \in M_s$. In this case, (2) follows from (1) since the subspaces M_s and V_s are orthogonal by Lemma 2.2.12. If $t = s$, $x \in L_0$ and so the vector $[e_j, x]$ is a multiple of the vector e_j since e_j is an eigenvector for L_0. Hence (2) again follows from (1) by orthogonality of the generators e_i and e_j (see Lemma 2.2.12).

For $s - t \neq 0$, the form $(.,.)_0$ is positive definite on L_{s-t}, and so equality (2) forces $[e_i, f_j] = 0$.

For $s = t$, because of equality (2), $[e_i, f_j] \in R$ since by condition (iii), $([e_i, f_j], x) = 0$ for all $x \in L_r$, $r \neq 0$. Hence, $[e_i, f_j] = 0$ by the assumption that $R = 0$.

For any $h \in L_0$, $[h, e_i] = ce_i$ for some scalar $c \in \mathbf{C}$ as e_i is an eigenvector for L_0, and
$$c = c(e_i, e_i)_0 = ([h, e_i], e_i)_0 = -(h, [f_i, e_i])_0 = (h, h_i)_0.$$

Similarly $[h, f_i] = -(h, h_i)_0 f_i$.   □

Lemma 2.2.15. $a_{ij} \in \mathbf{R}$ for all i, j and if $i \neq j$ then, $a_{ij} \leq 0$.

Proof. For any i, j,

$$
\begin{aligned}
(h_i, h_j)_0 &= ([e_i, f_i], [e_j, f_j])_0 \\
&= (f_i, [f_i[e_j, f_j]])_0 \\
&= (f_i, [e_j[f_i, f_j]])_0 \quad \text{by Lemma 2.2.14} \\
&= ([f_j, f_i], [f_i, f_j])_0.
\end{aligned}
$$

This is a non-positive real because of condition (iii).   □

Lemma 2.2.16. If $a_{ij} = 0$, then $[e_i, e_j] = 0$ and $[f_i, f_j] = 0$.

Proof. Suppose that $i \neq j$.

$$
\begin{aligned}
([e_i, e_j], [e_i, e_j])_0 &= (e_j, [f_i[e_i, e_j]])_0 \\
&= -(e_j, [h_i, e_j])_0 \quad \text{by Lemma 2.2.14} \\
&= -a_{ij}(e_j, e_j)_0 \quad \text{by Lemma 2.2.14} \\
&= 0.
\end{aligned}
$$

As the Hermitian form is positive definite on each space L_s for $s \neq 0$, $[e_i, e_j] = 0$. Applying the involution ω_0 to this bracket, we get the second equality.   □

Lemma 2.2.17. *If $a_{ii} > 0$ and $i \neq j$, then $\frac{2a_{ij}}{a_{ii}} \in \mathbf{Z}_-$ and*

$$(\text{ad}\,(e_i))^{1 - \frac{2a_{ij}}{a_{ii}}} e_j = 0 = (\text{ad}\,(f_i))^{1 - \frac{2a_{ij}}{a_{ii}}} f_j.$$

Proof. Suppose that $i \neq j$ and $a_{ii} > 0$. The Lie subalgebra $S_i = \langle e_i, h_i, f_i \rangle$ isomorphic to sl_2 acts on L via the adjoint action. If the S_i-module generated by f_j is finite dimensional, then the result follows since by Lemma 2.2.14, $[e_i, f_j] = 0$.

So we only need to show that the S_i-module generated by f_j is finite dimensional. For any positive integer n, as the Hermitian form is contravariant,

$$
\begin{aligned}
((\text{ad}\,f_i)^n f_j, (\text{ad}\,f_i)^n f_j)_0 &= ((\text{ad}\,f_i)^{n-1} f_j, [e_i, (\text{ad}\,f_i)^n f_j])_0 \\
&= -n(\frac{n-1}{2} a_{ii} + a_{ij})((\text{ad}\,f_i)^{n-1} f_j, (\text{ad}\,f_i)^{n-1} f_j)_0.
\end{aligned}
$$

Hence, $(\text{ad}\,f_i)^n f_j \neq 0$ if and only if $\frac{n-1}{2} a_{ii} + a_{ij} \neq 0$ and $((\text{ad}\,f_i)^{n-1} f_j, (\text{ad}\,f_i)^{n-1} f_j)_0 \neq 0$. Therefore the result follows as the Hermitian form is positive definite on the space L_s for $s > 0$. $\square$

This shows that the Lie algebra L/R is the BKM algebra with Cartan matrix A and Cartan subalgebra $H = L_0/R$, proving Theorem 2.2.9.

Theorems 2.2.8 and 2.2.9 assume that G is a Lie algebra. So the immediate question that comes to mind is: Do properties (i)-(iii) of Theorem 2.2.8 characterize BKM superalgebras? Though in many cases a Lie algebra fails to be a BKM algebra because there is no suitable $\mathbf{Z}$-grading with finite dimensional pieces, a most important aspect of the characterization given in Theorem 2.2.8 is the positivity of the Hermitian form $(.,.)_0$. However, positivity is not a natural concept in the context of Lie superalgebras. Indeed, when the odd part is non-trivial, (Hermitian) supersymmetry is the natural concept replacing (Hermitian) symmetry. If the form $(.,.)_0$ on a Lie superalgebra G is Hermitian supersymmetric, then for any $x \in G_{\bar{1}}$,

$$(x, x)_0 = -\overline{(x, x)_0}$$

and so positivity implies that $(x, x)_0$ is purely imaginary (i.e. in $i\mathbf{R}$). Hence, for a characterization of BKM superalgebras in terms of an almost positive definite bilinear form, we need the form to be Hermitian symmetric. As this is rather an artificial construction in the framework of Lie superalgebras, from the outset it appears unlikely that BKM superalgebras can be characterized in this manner. Let us take a closer look at this question.

Remark 2.2.18. When G is a BKM algebra and the base field in taken to be $\mathbf{R}$ instead of $\mathbf{C}$, the form $(.,.)_0$ is defined as $(x, y)_0 = -(x, \omega(y))$ in [Borc4], where $(.,.)$ is the Hermitian form on G induced by the generalized symmetric Cartan matrix A defined in Lemma 2.2.4. As the Chevalley automorphism ω is an involution,

$$(\omega(x), y) = (x, \omega(y)),$$

so it does not matter whether ω acts on the first or second component.

When $S \neq \emptyset$, the above equality no longer holds. Therefore, we have to be careful with the definition of the form $(.,.)_0$. If we keep Borcherds' definition given above, then for $i \in S$

$$(e_i, e_i)_0 = -(e_i, \omega_0(e_i)) = -(e_i, f_i) = -1 < 0$$

by Lemma 2.2.4, and so the form $(.,.)_0$ is not positive definite. This suggests that the definition of the form $(.,.)_0$ should be

$$(x, y)_0 = -(\omega(x), y)$$

over $\mathbf{R}$, and

$$(x, y)_0 = -(\omega_0(x), y)$$

over $\mathbf{C}$, as given in [KacP3] and [Kac14, §11.5] instead of the one given in [Borc4]. It is in fact why the Hermitian form needs to be antilinear in the second factor.

Corollary 2.2.19.

(i) *The form $(.,.)_0$ on the BKM superalgebra G is consistent.*
(ii) $([g, x], y)_0 = (-1)^{d(g)d(x)}(x, [\omega_0(g), y])_0$, *where ω_0 is the compact automorphism.*

Proof. We only prove (ii).

$$
\begin{aligned}
([g, x], y)_0 &= -([\omega_0(g), \omega_0(x)], y) \\
&= -(-1)^{d(g)d(x)}(\omega_0(x), [\omega_0(g), y]) \quad \text{by invariance of the form } (.,.) \\
&= (-1)^{d(g)d(x)}(x, [\omega_0(g), y])_0.
\end{aligned}
$$

$\square$

We will say that a Hermitian form on G satisfying Corollary 2.2.19.(ii) is contravariant. As expected, the next result shows that only a few BKM superalgebras can be characterized in terms of almost positive definite bilinear forms.

Theorem 2.2.20. *Let L be a Lie superalgebra satisfying the following conditions:*

(i) *L is graded: $L = \sum_{i \in \mathbf{Z}} L_i$ with $\dim L_i < \infty$ for all $i \neq 0$ and $[L_i, L_j] < L_{i+j}$;*
(ii) *there is an antilinear automorphism ω_0 of L of period 4 such that $\omega_0(L_i) = L_{-i}$ and ω_0 is multiplication by -1 on $(L_0)_{\mathbf{R}}$;*
(iii) *there is a contravariant Hermitian form, invariant under ω_0, $(.,.)_0$ on L, which is positive definite on all subspaces L_i with $i \neq 0$. Furthermore $(L_i, L_j)_0 = 0$ if $i \neq j$.*

Then the kernel R of the bilinear form is contained in the centre of L and L/R is a sum of BKM algebras and Lie superalgebras of type $A(n,0)$ or $B(n,0)$, $A(\infty,0)$, $B(\infty,0)$, or $B^{(1)}(1,2n)$.

We prove this result in a few steps. In what follows, we keep the notation used in the proof of Theorem 2.2.9 and L will denote a Lie superalgebra satisfying Theorem 2.2.20.

Lemma 2.2.21. *The Lie sub-superalgebra L_0 is abelian and its odd part $(L_0)_{\overline{1}}$ is contained in the centre of L.*

Proof. Replacing ω_0 with ω_0^2 in the proof of Lemma 2.2.10, it follows that the Lie sub-superalgebra L_0 is abelian.

For any integer $s > 0$, consider an arbitrary eigenvector $x \in L_s$ for L_0. For any $h \in L_0 \cap L_{\overline{1}}$, there is some scalar $c \in \mathbf{R}$ such that

$$[h, x] = cx.$$

Now,

$$2[h, [h, x]] = [[h, h]x] = 0$$

as $h \in L_{\overline{1}}$ and L_0 is abelian. However, $[h, [h, x]] = c^2 x$. Therefore, as $x \neq 0$, $c = 0$. It follows that h is a central element, proving the Lemma. $\qquad\square$

Lemma 2.2.11 clearly holds as well. From Lemma 2.2.21, it immediately follows that there is a basis of V_r consisting of homogeneous elements. Hence the form $(.,.)_0$ being positive definite, we can take the elements e_i to be homogeneous elements:

Corollary 2.2.22. *Suppose that $R = 0$. For any integer $r > 0$, $L_r = V_r \oplus (M_r \cap L_r)$ as vector spaces and there exists a basis B_r of the vector space V_r consisting of pairwise orthonormal homogeneous eigenvectors $\{e_i\}$ for L_0.*

As usual, we let I be the indexing set for the subscripts i of the generators $e_i \in L$, and S be the subset giving the odd generators. Without loss of generality, we may assume that the set I is connected. Lemma 2.2.15 and 2.2.16 clearly holds and so does Lemma 2.2.17 for the even generators of the Lie superalgebra L. Moreover,

Lemma 2.2.23. *If $i \in S$, $a_{ii} > 0$ and $i \neq j$, then $\frac{a_{ij}}{a_{ii}} \in \mathbf{Z}_-$ and $(\mathrm{ad}\,(e_i))^{1-\frac{a_{ij}}{a_{ii}}} e_j = 0 = (\mathrm{ad}\,(f_i))^{1-\frac{a_{ij}}{a_{ii}}} f_j$.*

Lemma 2.2.24. *The Lie superalgebras L is a BKM superalgebra with $|S| = 1$. Furthermore, for all $i \in S$, $a_{ii} \geq 0$ and if $a_{ii} = 0$ then $[e_i, e_i] = 0$.*

Proof. Let $i \in S$, $j \in I$.

$$
\begin{aligned}
([e_i, e_j], [e_i, e_j])_0 &= -(-1)^{d(e_j)}(e_j, [f_i[e_i, e_j]])_0 \\
&= \begin{cases} (-1)^{d(e_j)}(e_j, [h_i, e_j])_0 & \text{if } i \neq j \\ 2(e_j, [h_j, e_j])_0 & \text{if } i = j \end{cases} \qquad (*) \\
&= \begin{cases} (-1)^{d(e_j)}a_{ij}(e_j, e_j)_0 & \text{if } i \neq j \\ 2a_{jj}(e_j, e_j)_0 & \text{if } i = j \end{cases}
\end{aligned}
$$

Since the form $(.,.)_0$ is positive definite on the space L_s, if $j \in S$ (resp. $j \notin S$), then $[e_i, e_j] \neq 0$ if and only if $a_{ij} > 0$ (resp. $a_{ij} < 0$). In particular, for all $i \in S$, $a_{ii} \geq 0$ and $a_{ii} = 0$ if and only if $[e_i, e_i] = 0$.

So for $i \in S$, $a_{ii} > 0$, $E_i = [e_i, e_i] \neq 0$ and $\omega(E_i) = F_i = [f_i, f_i] \neq 0$. This gives

$$
\begin{aligned}
([E_i, e_j], [E_i, e_j])_0 &= -(-1)^{d(e_j)}(e_j, [F_i[E_i, e_j]])_0 \\
&= 4a_{ii}(-1)^{d(e_j)}(e_j, [h_i, e_j])_0 \\
&= 4(-1)^{d(e_j)}a_{ii}a_{ij}(e_j, e_j)_0 \\
&= -4a_{ii}a_{ij}(e_j, e_j)_0.
\end{aligned}
$$

Hence since the form $(.,.)_0$ is positive definite on the spaces L_s, $a_{ij} \leq 0$. So from $(*)$, it follows that for all $j \in S$ $a_{ij} = 0$.

Hence, the set I being connected, if there exists $i \in S$ such that $a_{ii} > 0$, then $S = \{i\}$. Also, Lemmas 2.2.15, 2.2.16, 2.2.17, 2.2.23 together with $(*)$ imply that L is BKM superalgebra.

Next suppose that $i \neq j \in S$ and $a_{ii} = 0 = a_{jj}$. If $a_{ij} > 0$, then from $(*)$, the even vectors $E = [e_i, e_j] \neq 0$, $F = [f_i, f_j] \neq 0$. Suppose that there exists $k \in I - \{i, j\}$ such that $a_{jk} \neq 0$. If $k \in S$ (resp $k \notin S$), then $a_{kk} = 0$. Since $[e_j, e_k]$ is an even (resp. odd) vector, from $(*)$, we can deduce that $a_{ik} \geq 0$ (resp. $a_{ik} \leq 0$) and $a_{jk} \geq 0$ (resp. $a_{jk} \leq 0$) and on the other hand $a_{ik} + a_{jk} \leq 0$ (resp. $a_{ik} + a_{jk} \geq 0$). This contradiction and the connectedness of the indexing set I imply that $|S| \leq 2$. Moreover, by Lemmas 2.2.15, 2.2.16, 2.2.17, 2.2.23 and equalities $(*)$, L is a BKM superalgebra and if $|S| = 2$, then L is finite dimensional. The list of finite dimensional BKM superalgebras given in Corollary 2.1.23 together with Theorem 2.4.8 on the uniqueness of the generalized Cartan matrix forces L to be of type $A(1,0)$ when $|S| = 2$. However in this case there is another non-equivalent generalized symmetric Cartan matrix $\begin{pmatrix} 0 & -1 \\ -1 & 2 \end{pmatrix}$ associated to L and the corresponding subset S has cardinality 1 (see Example 2.4.7). $\qquad\qquad\square$

We can now assume that L is a BKM superalgebra with $S = \{1\}$ and $a_{11} \geq 0$. For details on roots, see §2.3.

Lemma 2.2.25. *Let I_1 be a finite connected subset of I such that $1 \in I_1$ and L_1 be the BKM sub-superalgebra of L indexed by I_1. Either $\dim L_1 < \infty$ or L is an affine Kac-Moody Lie superalgebra of type $osp^{(1)}(1, 2n)$. Moreover all roots of L have non-negative norm.*

Proof. Let L_E be the even BKM subalgebra generated by the even generators $e_i, f_i, i \in I \setminus S$. We first show that all the roots of L_E have positive norm. Let α

be a root of L_E such that $(\alpha, \alpha) \leq 0$. Without loss of generality, we may assume that $(\alpha, \alpha_1) < 0$. Then $\beta = \alpha + \alpha_1$ is an odd root and by Lemmas 2.2.23 and 2.2.24, $(\beta, \beta) < 0$. Hence, by Lemma 2.3.15, for any $0 \neq e_\beta \in L_\beta$, $[e_\beta, e_\beta] \neq 0$ and

$$
\begin{aligned}
([e_\beta, e_\beta], [e_\beta, e_\beta])_0 &= 2(e_\beta, [[\omega(e_\beta), e_\beta]e_\beta])_0 \\
&= 2(e_\beta, e_\beta)_0 (e_\beta, \omega(e_\beta))(\beta, \beta) \\
&= 2(e_\beta, e_\beta)_0^2 (\beta, \beta).
\end{aligned}
$$

This contradicts the fact that the form $(., .)_0$ is positive definite and in particular, the Lie superalgebra L has no odd roots of negative norm. If there are roots of L_E of non-positive norm, then there is a root α such that $(\alpha, \alpha) \leq 0$ and $(\alpha, \alpha_1) \neq 0$. Hence, the above shows that if the set I is finite, then $\dim L_E < \infty$. If $a_{11} = 0$, then the result follows from Corollary 2.3.34.

Assume next that $a_{11} > 0$. Lemma 2.2.23 implies that the Lie subalgebra L_p generated by L_E and the elements $[e_1, e_1]$, $[f_1, f_1]$ remains a BKM algebra. If all roots of L_p have positive norm, then again $\dim L_1 < \infty$ as before.

So suppose that the BKM algebra L_p has a root α of non-positive norm. We claim that

$$(\alpha, \alpha) = 0.$$

From what precedes, $1 \in Supp(\alpha)$. By Lemma 2.3.21 (ii), we may assume that $(\alpha, \alpha_j) \leq 0$ for all $j \in I$. Then, $(\alpha, \alpha_1) = 0$ for otherwise the finite dimensional representation theory of sl_2 (see [Kac14, §3.2]) implies that $\alpha + \alpha_1$ is an odd root of L and as its norm is negative, we get a contradiction to the above. Since $a_{1j} \leq 0$ for all $j \in I$, it follows that $a_{1j} = 0$ for all $j \in Supp(\alpha)$, contradicting the connectedness of the support of the root α given in Proposition 2.3.8.

If $(\alpha, \alpha_j) < 0$ for some $j \in I$, then $\alpha + \alpha_j$ is a root of L of negative norm since $a_{jj} > 0$. However, we have just shown that all roots of non-positive norms have norm 0 and so

$$(\alpha, \alpha_j) = 0 \quad \forall j \in I.$$

Moreover as the indexing set I is connected, $I = Supp(\alpha)$. Therefore the Kac-Moody Lie algebra L_p is an affine Lie algebra [Kac14, §4]. As L is a Kac-Moody Lie superalgebra with no roots of negative norm, from the list of affine Kac-Moody Lie superalgebras given [Kac7] we can deduce that L is of type $osp^{(1)}(1, 2n)$. □

We can now finish the proof of Theorem 2.2.20.

Proof of Theorem 2.2.20. It only remains to show that when $S = \{1\}$ and the BKM superalgebra L has finite dimension, it is of type $sl(1, n)$ or $osp(1, 2n)$. Since all roots have non-negative norm, the list given in Corollary 2.1.23 shows that these are the only possibility. □

The converse to Theorem 2.2.20 holds:

Lemma 2.2.26. *The Lie superalgebras of type $A(m, 0)$, $B(0, n)(= osp(1, 2n))$, $A(\infty, 0)$, $B(0, \infty)$, and $B^{(1)}(0, n)$ satisfy conditions (i)-(iii) of Theorem 2.2.20.*

Proof. Let G be a BKM superalgebra a type stated in the Lemma. In each case, there is a unique simple odd root α_1. For the automorphism ω_0, we can take the compact automorphism. The Hermitian form $(.,.)_0$ is defined as in Remark 2.2.18. We only show that the Hermitian form is positive definite on root spaces, the other properties being easily verified. As the Lie subalgebra G_p generated by e_i, f_i for $i > 2$ and $[e_1, e_1]$, $[f_1, f_1]$ is a BKM algebra, Theorem 2.2.8 shows that the Hermitian form $(.,.)_0$ is positive definite on the root spaces of this BKM subalgebra. Hence, we only need to show that $(.,.)_0$ is positive definite on root spaces G_α not contained in G_p. This is the case for a positive root α only when there is an even positive root β such that $\beta + \alpha_1 = \alpha$ and for any $x \in G_\alpha$, there exists $y \in G_\beta$ such that $x = [e_1, y] \neq 0$. Therefore,

$$(x, x)_0 = -(y, y)_0 (\beta, \alpha_1).$$

Since $(\beta, \alpha_1) \leq 0$, it follows that $(x, x)_0 > 0$, proving the result. $\square$

Remark 2.2.27. The finite dimensional and affine Lie superalgebras that are not BKM superalgebras do not have an almost positive definite contravariant form since any Lie superalgebra satisfying Theorem 2.2.20 must be a BKM superalgebra. In particular, non-twisted affine Lie superalgebras $sl^{(1)}(1, n)$ do not have such a form even though the finite dimensional Lie superalgebras $sl(1, n)$ do.

It is important to notice that we want the Hermitian form $(.,.)_0$ to be positive definite on the root spaces and not on the whole of G. The difference between $sl(1, n)$ and $osp(1, 2n)$ is that in the first case $(.,.)_0$ is not positive definite on G, whereas it is in the second case. Indeed $sl(1, n)$ has roots of norm 0, i.e. there is an element $h \in H$ such that $(h, h)_0 = 0$, whereas $osp(1, 2n)$ does not. This is the reason why there is a Hermitian form satisfying Theorem 2.2.20 on the affine Lie superalgebras $osp^1(1, 2n)$ but not on $sl^{(1)}(1, n)$.

Example 2.2.28. The Lie superalgebra $sl^{(1)}(1, 2)$ (see Example 2.1.27 and for details see [Kac7]) is generated by vectors e_i, f_i, $i = 0, 1, 2$, where e_i, f_i are odd vectors for $i = 0, 2$. Setting $h_i = [e_i, f_i]$, the matrix giving the bilinear form (h_i, h_j) is $\begin{pmatrix} 0 & -1 & 1 \\ -1 & 2 & -1 \\ 1 & -1 & 0 \end{pmatrix}$. Hence for the vector $x = [[e_0, e_1][e_1, e_2]]$,

$$\begin{aligned}(x, x)_0 &= -([e_1, e_2], [[[f_0, f_1][e_0, e_1]][e_1, e_2]])_0 \\ &= -([e_1, e_2], [h_1 - h_0, [e_1, e_2]])_0 \\ &= -([e_1, e_2], [e_1, e_2])_0\end{aligned}$$

since

$$[[f_0, f_1], [e_0, e_1]] = -[[h_0, f_1]e_1] + [e_0[f_0, h_1]] = h_1 - h_0.$$

Therefore the form $(.,.)_0$ cannot be positive definite on the root spaces.

From Theorems 2.2.9 and 2.2.20 together with [Kac14, §4], we can deduce the following result.

Corollary 2.2.29.

(i) *A finitely generated Lie superalgebra L satisfies Theorem 2.2.20 and the bilinear form $(.,.)_0$ is positive definite on L if and only if L is the sum of finite dimensional simple Lie algebras and copies of $sl(1,n)$.*

(ii) *A finitely generated Lie superalgebra L satisfies Theorem 2.2.20 and the bilinear form $(.,.)_0$ is semi-positive definite on L_0 if and only if L is the sum of finite dimensional simple Lie algebras, copies of $sl(1,n)$, affine Lie algebras and copies of $osp^{(1)}(1,2n)$.*

Exercises 2.2

1. Let $G = G(A, H, S)$ be a BKM superalgebra.

(i) Construct a non-degenerate, invariant, bilinear, consistent, supersymmetric form $(.,.)$ on G such that $(e_i, f_i) = 1$ for all $i \in I$ and $(h_i, h_j) = a_{ij}$.

Hint: Find a proper $\mathbf{Z}$-grading for G and use induction. For a solution, see [Kac14, Theorem 2.2].

(ii) Show that for any root $\alpha, \beta \in \Delta$, $(G_\alpha, G_\beta) = 0$ unless $\alpha + \beta = 0$.

(iii) Show that the form $(.,.)$ is left invariant by the Chevalley automorphism.

(iv) Show that when $H = \langle h_i : i \in I \rangle$, any other invariant symmetric bilinear form on G is a multiple of the form $(.,.)$.

2. Let $\mathcal{L}_n$ be the filiform Lie algebra spanned by the elements $e_0, e_1, \cdots, e_n$ with only non-trivial brackets $[e_0, e_i] = e_{i+1}$ for all $1 \leq i \leq n - 1$.

(i) Construct a non-degenerate bilinear symmetric form on the Lie algebra $\mathcal{L}_n$.

(ii) Show that the Lie algebra $\mathcal{L}_n$ cannot be $\mathbf{Z}$-graded.

3. Let L be an extended affine Lie algebra and consider the Lie algebra: $G = \mathbf{C}[t^{-1}, t] \otimes L$ with bracket given by $[f(t)x, g(t)y] = f(t)g(t)[x, y]$ for all $f(t), g(t) \in R[t^{-1}, t]$ and $x, y \in L$.

(i) Show that there is a $\mathbf{Z}$-grading of G and an antilinear involution ω such that $\omega(G_i) = G_{-i}$ and $\omega(h) = -h$ for all $h \in G_{0\mathbf{R}}$.

(ii) Show that there is a contravariant Hermitian form on G (with respect to ω), positive definite on each piece G_i for $i \neq 0$.

(iii) Show that G is not a BKM algebra.

4. Let G be a Kac-Moody Lie algebra. Let D be a finite group of diagram automorphisms of G, i.e. for all $d \in D$, there is a bijection $\bar{d}$ of the indexing set I satisfying

$$a_{\bar{d}(i)\bar{d}(j)} = a_{ij} \quad \forall \ i, j \in I,$$

such that for $i \in I$, $d(e_i) = e_{\bar{d}(i)}$ and $d(f_i) = f_{\bar{d}(i)}$. Let G_D be the Lie subalgebra of elements fixed by D:

$$G_D = \{x \in G : d(x) = x \quad \forall d \in D\}.$$

Show that the Lie algebra G_D is a BKM algebra but need not be a Kac-Moody Lie algebra. *For a solution, see [Borc4]*.

(ii) Does (i) remain true for BKM superalgebras?

Hint: Consider affine BKM superalgebras.

5. Show that the BKM algebra G is finite dimensional if and only if the Hermitian form $(.,.)_0$ is positive definite on G. Show that the BKM algebra G is an affine Lie algebra if and only if it is semi-positive definite on G.

2.3 The Root System

The generalized Cartan subalgebra H acts semisimply on the BKM superalgebra G via the adjoint action. Hence to know the structure of G, it is worth finding its eigenvalues and eigenspaces. It is most natural to take the eigenvalues in the dual space H^*. In order to do this without any complications, the elements h_i are assumed in [Kac14] to be linearly independent. However, in the most well known and important examples of BKM superalgebras, namely the Monster Lie algebra (see Example 2.3.11.) and the Fake Monster Lie algebra (see Example 2.6.40), that are not Kac-Moody Lie algebras, the elements h_i are linearly dependent. So in the general case, one has to be very careful about the definition of roots or more precisely what set we take them to be elements of.

Definition 2.3.1. *The formal root lattice Q is defined to be the free abelian group generated by the elements α_i, $i \in I$ with a real valued bilinear form given by $(\alpha_i, \alpha_j) = a_{ij}$. The elements α_i $i \in I$ are called the simple roots.*

Note that Q may not strictly speaking be an integral lattice (see Definition 3.1.1) since in general the indexing set I of a BKM superalgebra is countably infinite in which case the rank of Q is not finite (see Example 2.3.11 of the Monster Lie algebra).

For reasons of simplicity, we shall keep the same notation $(.,.)$ for the bilinear form on Q, H or H^*. From the triangular decomposition of the BKM superalgebra G given in Corollary 2.1.19, the following definition of eigenspaces makes sense.

Definition 2.3.2. *For $\alpha = \sum_{k=1}^j \alpha_{i_k} \in Q$, the root space G_α (resp. G_α) is the subspace of G generated by the elements $[e_{i_j}[...[e_{i_2}, e_{i_1}]]]$ (resp. $[f_{i_j}[...[f_{i_2}, f_{i_1}]]]$). A non zero element α of the formal root lattice Q is said to be a root of G if the subspace G_α is non trivial. The dimension of the root space G_α is called the multiplicity of the root α.*

The simple root spaces are easily described. Moreover, there is a concept of positive and negative roots.

Proposition 2.3.3.

(i) *For all $i \in I$, $G_{\alpha_i} = \mathbf{C}e_i$ and $G_{-\alpha_i} = \mathbf{C}f_i$. In particular,*

$$\dim G_{\alpha_i} = 1 = \dim G_{-\alpha_i}.$$

(ii) *If the element $\alpha \in Q$ is a root, then either α or $-\alpha$ is a sum of simple roots.*

(iii) *A root space G_α is either contained in the even part $G_{\bar{0}}$ or odd part of $G_{\bar{1}}$ of the BKM superalgebra G.*

Definition 2.3.4.

(i) *A root α is said to be positive (resp. negative) if α (resp. $-\alpha$) is a sum of simple roots.*

(ii) *A root α is said to be even (resp. odd) if $G_\alpha \leq G_{\bar{0}}$ (resp. $G_{\bar{1}}$). We then write $d(\alpha) = \bar{0}$ (resp. $d(\alpha) = \bar{1}$).*

(iii) *The height of a root $\alpha = \sum k_i \alpha_i$ is defined to be $\sum k_i$ and is written $\mathrm{ht}\,(\alpha)$.*

(iv) *The support of α is the set $\{i \in I : k_i \neq 0\}$ and is written $\mathrm{supp}(\alpha)$.*

(v) *A base of the set of roots Δ is a linearly independent subset Π such that for any $\alpha \in \Delta$, $\alpha = \sum_{\beta \in \Pi} k_\beta \beta$, where for all $\beta \in \Pi$, either all the scalars $k_\beta \in \mathbf{Z}_+$ or all $k_\beta \in \mathbf{Z}_-$.*

Proposition 2.3.5. *For any root $\alpha \in \Delta$,*
$$\mathrm{mult}\,\alpha = \mathrm{mult}(-\alpha).$$

Proof. From Proposition 2.3.3 we get $\omega(G_\alpha) = G_{-\alpha}$ for any root $\alpha \in \Delta$, where ω is the Chevalley automorphism defined in Proposition 2.1.28. This proves the result. $\qquad\square$

We can immediately deduce the following.

Corollary 2.3.6. *A root α in Δ is negative if and only if $-\alpha$ is a positive root. Hence the set Δ decomposes into the set Δ_+ of positive roots and the set $-\Delta_+$ of negative roots:*
$$\Delta = \Delta_+ \cup (-\Delta_+).$$

This together with Corollary 2.1.19 shows that the Cartan decomposition holds for BKM superalgebras:

Corollary 2.3.7. *The BKM superalgebra $G = G(A, H, S)$ is a triangular direct sum:*
$$G = (\oplus_{\alpha \in \Delta_+} G_{-\alpha}) \oplus H \oplus (\oplus_{\alpha \in \Delta_+} G_\alpha).$$

This is called the generalized Cartan decomposition of the BKM superalgebra G.

The following is a basic property of roots, well known to hold in the Kac-Moody case. The proof remains the same and is a direct consequence of condition (5) in Definition 2.1.7 of a BKM superalgebra. Hence we leave it for the reader to check.

Proposition 2.3.8. *The support of every root α is connected.*

The indexing set I of most of the BKM superalgebras with a "nice" construction that are not Kac-Moody Lie superalgebras and that we are primarily interested in this book (see chapter 5) is countably infinite. Therefore their formal root lattice has countably infinite rank. However, the $\mathbf{Z}$-lattice generated by the elements h_i of their Cartan subalgebra has an interesting structure. So, instead of artificially making the roots linearly independent by taking them as elements of the root lattice Q, it is more natural to consider them as vectors in the Cartan subalgebra H or its dual H^*, and hence to consider the integral root lattice as being embedded in H or H^*. This lattice has in many cases very interesting properties. This is the case for the class of BKM superalgebras that we want to classify (see Chapter 5). Note that as the bilinear form in non-degenerate on the Cartan subalgebra H, it makes no difference whether one considers the roots in H or its dual H^*. If the roots are taken to be in the Cartan subalgebra H, then the elements h_i are called the simple roots.

Set f_Q (resp. g_Q) to be the natural map from Q to H (resp. H^*) taking α_i to h_i (resp. the linear functional β_i in H^* mapping h to (h, h_i)). These maps extend to linear maps from $\mathbf{C} \otimes_{\mathbf{Z}} Q$ to H and H^*.

One has to be very careful when roots are taken to be in H or H^* for these maps from Q to H or its dual H^* are not in general injective as the elements h_i are not necessarily linearly independent. This is what happens in the most well known examples of BKM algebras that are not Kac-Moody Lie algebras such as the Monster Lie algebra (see Example 2.3.11). There may be $n > 1$ simple roots in Q of non-positive norm with the same image in H (resp. H^*) if the generalized Cartan matrix A has equal distinct columns.

Proposition 2.3.9.

   (i) $f_Q(\alpha_i) = f_Q(\alpha_j)$ if and only if $g_Q(\alpha_i) = g_Q(\alpha_j)$.
   (ii) $f_Q(\alpha_i) = f_Q(\alpha_j)$ implies that the i-th and j-th columns of A are equal.

If the roots are taken to be in H, then the root space of the root $h \in H$ is

$$G_h = \{x \in G : [h', x] = (h', h)x, \quad h' \in H\}.$$

It is important to note that as a consequence of Proposition 2.3.9 when roots are considered to be in the Cartan subalgebra H or its dual H^* instead of the formal root lattice Q, the simple root spaces are no longer necessarily of dimension 1.

Corollary 2.3.10. *For all* $i \in I$, $G_{\alpha_i} \leq G_{h_i}$.

When a root is considered in the generalized Cartan subalgebra H, its height can be defined in a way similar to that of Definition 2.3.4 (ii). It clearly has the same value as its pre-image in the root lattice Q.

Example 2.3.11. 1. The Monster Lie algebra.

Let $c(n)$ be the coefficients of the q-expansion of the normalized modular invariant J (see section 3.2), i.e. $c(-1) = 1$ and $J(q) = q^{-1} + \sum_{n \geq 1} c(n)q^n$. Let I be the countably infinite set

$$\{-1, 1_1, 1_2, ..., 1_{c(1)}, 2_1, ..., 2_{c(2)}, ...\},$$

where each integer i occurs $c(i)$ times. Let $II_{1,1}$ be the unique (up to isomorphism) even unimodular Lorentzian lattice of rank 2 (see section 3.1). In other words, it is the free abelian group $\mathbf{Z}^2$ with bilinear form given by the matrix

$$\begin{pmatrix} 0 & -1 \\ -1 & 0 \end{pmatrix}$$

with respect to the standard basis $(1,0), (0,1)$. Let $H = II_{1,1} \otimes_{\mathbf{Z}} \mathbf{C}$ be the complex space given by the lattice $II_{1,1}$. To avoid confusion, the bilinear form on H will be written ".". For $1 \leq j \leq c(i)$, set

$$h_{i_j} = (1, i).$$

The Monster Lie algebra M is the BKM algebra with generalized Cartan subalgebra H and generalized symmetric Cartan matrix A indexed by the set I and with (i_j, k_l)–th entry equal to $h_{i_j}.h_{k_l} = -i - k$. Thus A is the following symmetric matrix with a countably infinite number of columns:

$$\begin{pmatrix}
2 & 0 & \cdots & 0 & -1 & \cdots & -1 & -2 & \cdots \\
0 & -2 & \cdots & -2 & -3 & \cdots & -3 & -4 & \cdots \\
\vdots & \vdots & \vdots & \vdots & \vdots & \vdots & \vdots & \vdots & \vdots \\
0 & -2 & \cdots & -2 & -3 & \cdots & -3 & -4 & \cdots \\
-1 & -3 & \cdots & -3 & -4 & \cdots & -4 & -5 & \cdots \\
\vdots & \vdots & \vdots & \vdots & \vdots & \vdots & \vdots & \vdots & \vdots \\
-1 & -3 & \cdots & -3 & -4 & \cdots & -4 & -5 & \cdots \\
-2 & -4 & \cdots & -4 & -5 & \cdots & -5 & -6 & \cdots \\
\vdots & \vdots & \vdots & \vdots & \vdots & \vdots & \vdots & \vdots & \vdots
\end{pmatrix}.$$

We will show later that the multiplicity of the root (m, n) is $c(mn)$ (see Exercise 5.2.2).

2. Consider $I = \{1, 2, 3\}$, $S = \{1\}$, $G = G(A, H, S)$ with $H = \mathbf{R}h_1 \oplus \mathbf{R}h_2$, $h_3 = h_1$ and

$$A = \begin{pmatrix} 0 & -1 & 0 \\ -1 & 2 & -1 \\ 0 & -1 & 0 \end{pmatrix}.$$

Then both $e_1, e_3 \in G_{h_1}$ since $h_1 = h_3$. Therefore, G_{h_1} is not contained in either the even nor the odd part of the BKM superalgebra G.

The latter example shows that the concept of an odd or even root may not make sense when roots are no longer considered to be in the formal root lattice Q. However the aspect to be particularly careful about when the roots are taken to be elements of the generalized Cartan subalgebra H or its dual is the problematic notion of positivity for the simple roots $h_i \in H$ are in general linearly dependent. This is in particular the case when the indexing set I is infinite and the generalized Cartan subalgebra is finite as in the BKM algebras we will study in Chapter 5 (see for example the above Example 2.3.11.1 of the Monster Lie algebra).

The situation may be more complicated than in the previous example for the non-positive norm root spaces. Part of a non-positive norm root space – but not necessarily the full space – may be generated by simple root spaces. Hence we may need to differentiate between the multiplicity of a root – i.e. the dimension of the root space – and the codimension of its maximal subspace generated (as a Lie superalgebra) by simple root spaces.

Definition 2.3.12. *The element $h \in H$ is said to have root multiplicity n if $n = \dim G_h$ and restricted simple root multiplicity m if there are m simple roots $\alpha_i \in Q$ such that the simple root spaces G_{h_i} (or the spaces G_{-h_i}) generate (as a Lie superalgebra) the root space G_h.*

The above result can be rewritten in terms of roots in the dual space H^* by taking the image of h in H^*.

Proposition 2.3.13.

(i) *The restricted simple root multiplicity of a root $h \in H$ is the codimension of the maximal subspace of G_h that can be generated (as a Lie superalgebra) by root spaces G_{h_i} (or G_{-h_i}).*

(ii) *Suppose that the restricted root multiplicity of h is strictly less than its root multiplicity, then the maps f_Q and g_Q are not injective.*

This distinction between the multiplicity of a root and its restricted simple root multiplicity is essential for a natural class of BKM subalgebras of the Monster Lie algebra ([Ray2]).

Examples 2.3.14. 1. For $r \in \mathbf{N}$, the Lie subalgebra M_r of the monster Lie algebra M generated by the subspaces $M_{(m,rn)}$, $m, n \in \mathbf{Z}$ is a BKM algebra with the same Cartan subalgebra $H = II_{1,1} \otimes_{\mathbf{Z}} \mathbf{R}$ as M. It is hard to show this from the definition of a BKM algebra given in Definition 2.1.7 but it follows easily from Theorem 2.2.9.

Assume that $r > 1$, in other words that M_r is a proper Lie subalgebra of M and consider its root space $M_{(2,2r)}$. As stated in Example 2.3.11.1, its dimension is $c(4r)$. We calculate the restricted simple root multiplicity of the root $(2, 2r)$ of the BKM algebra M_r.

We first need to find the positive roots (m_i, rn_i), $1 \le i \le s$ distinct from $(2, 2r)$ such that $\sum_{i=1}^{s}(m_i, rn_i) = (2, 2r)$. As the root (m_i, rn_i) is positive, the integer $m_i \ge 1$. This follows from the description of simple roots given in Example 2.3.11.1 And so $s = 2$,

$$m_i = 1$$

and $n_1 + n_2 = 2$. Since $r > 1$ and $c(l) = 0$ for $l < -1$, we get $n_i > 0$. So

$$n_i = 1.$$

Clearly $(1, r)$ is a simple root of the BKM algebra M_r. Thus the restricted simple root multiplicity of the root $(2, 2r)$ is equal to the codimension of the subspace

$$[M_{(1,r)}, M_{(1,r)}]$$

in the root space $M_{(2,2r)}$. So we now compute the dimension of this subspace.

Let e_j, $1 \leq j \leq c(r)$, be a basis for the subspace $M_{(1,r)}$. Then $[e_j, e_k]$, $j < k$, are linearly independent vectors in $M_{(2,2r)}$. Indeed if $c_{jk} \in \mathbf{R}$, for scalars

$$\sum_{j,k,j<k} c_{jk}[e_j, e_k] = 0,$$

then for the smallest integer $1 \leq l \leq r$ for which some $c_{lk} \neq 0$,

$$\sum_{j,k,j<k} c_{jk}[f_l[e_j, e_k]] = 0$$

implies that

$$-\sum_{k,l<k} c_{lk}[h_l, e_k] = 0.$$

Thus,

$$-\sum_{k,l<k} c_{lk}h_l.h_k e_k = 0.$$

As $h_l = h_k = (1, r)$, $h_l.h_k = -2r$. This implies that

$$\sum_{k,l<k} c_{lk}e_k = 0,$$

forcing $c_{lk} = 0$ for all $k > l$ as the vectors e_k are linearly independent, contradicting the above assumption.

Hence the dimension of the subspace $[M_{(1,r)}, M_{(1,r)}]$ is the cardinal of the set $\{(a, b) : 1 \leq a < b \leq c(r)\}$, i.e.

$$\dim[M_{(1,r)}, M_{(1,r)}] = \sum_{i=1}^{c(r)-1} i = \frac{c(r)(c(r) - 1)}{2}.$$

Therefore $(2, rn)$ has root multiplicity $c(4r)$ and restricted simple root multiplicity

$$c(4r) - \frac{c(r)(c(r) - 1)}{2}.$$

This is a non-trivial integer strictly less than $c(4r)$ (see [Rad]). For example $c(4) = 20245856256$ and $c(1) = 196884$.

2. Let $G = G(A, H)$ be the BKM algebra with indexing set $I = \{1, 2, 3\}$, a two dimensional generalized Cartan subalgebra H with basis h_1, h_2 and such that $h_3 = h_1 + h_2$ and non-degenerate bilinear form given by the matrix

$$\begin{pmatrix} 2 & -3 \\ -3 & 2 \end{pmatrix}.$$

Thus the generalized symmetric Cartan matrix of G is

$$A = \begin{pmatrix} 2 & -3 & -1 \\ -3 & 2 & -1 \\ -1 & -1 & -2 \end{pmatrix}.$$

Then the restricted simple root multiplicity of h_3 is 1, whereas its multiplicity is 2. Indeed the root space G_{h_3} has dimension 2 since

$$G_{h_3} = \mathbf{R}[e_1, e_2] \oplus \mathbf{R}e_3.$$

In this last example, $h_2 = h_3 - h_1$. This shows that one may have to be careful about the concept of positive and negative root when the roots are taken to be in H or its dual.

When we will be speaking of roots without mentioning the space they belong to, it will mean that the statements hold whichever they are taken to be elements of. When we will talk of positive or negative or even or odd roots, we will mean that either the roots are considered to be elements in the formal root space Q or else if they belong to H or H^*, these concepts have no ambiguity.

Before going further, we give a very useful result involving the bilinear form and Chevalley involution.

Lemma 2.3.15. *Let $\alpha \in \Delta$ and $h_\alpha \in H$ be such that for all $x \in G_\alpha$ and $h \in H$, $[h, x] = (h_\alpha, h)x$. Then, $[x, y] = (x, y)h_\alpha$ for all $y \in G_{-\alpha}$.*

Proof. For any $h \in H$, $([x, y] - (x, y)h_\alpha, h) = -(x, [h, y]) - (x, y)(h_\alpha, h) = 0$ by definition of h_α. The result follows from the non-singularity of the bilinear form (see Exercise 2.2.1).

Roots of Finite and Infinitety Type

We first make some elementary observations.

Lemma 2.3.16. *Let L be a Lie superalgebra. For any $x \in L$, $[x[x, x]] = 0$ for all $x \in L$.*

Proof. Let x be a homogeneous element.

$$[x[x, x]] = [[x, x]x] + (-1)^{d(x)}[x[x, x]] = -(-1)^{d(x)}[x[x, x]] + (-1)^{d(x)}[x[x, x]] = 0.$$

$\square$

There are two types of roots:

Definition 2.3.17. *A root α is said to be of finite type if all elements in the root space G_α act locally nilpotently on G, i.e. for any $x \in G_\alpha$, $y \in G$, there is a non-negative integer n (depending on x and y) such that $(\operatorname{ad} x)^n y = 0$. Otherwise it is said to be of infinite type.*

The reader may be more familiar with the usual notions of real and imaginary roots. It makes sense in the context of BKM algebras but as we shall see below this is less so in the larger framework of BKM superalgebras. The next result about odd roots though elementary is very helpful.

Lemma 2.3.18. *If $\alpha \in \Delta_1$, then for all $x \in G_\alpha$, $[x, x] \neq 0$, in particular $2\alpha \in \Delta_0$, unless α is a root of norm 0 and finite type.*

Proof. Let $\alpha \in \Delta_1$ and $0 \neq x \in G_\alpha$. Suppose that $[x, x] = 0$. Let $y \in G_{-\alpha}$ be such that $(x, y) = 1$ (by Exercise 2.2.1 such an element y exits). Then, by Lemma 2.3.15, $0 = [y[x, x]] = 2(x, y)(\alpha, \alpha)x$. Hence, $(\alpha, \alpha) = 0$ and for all $z \in G$, $[x[x, z]] = \frac{1}{2}[[x, x]z] = 0$, proving the result. $\qquad\square$

When a root has non-zero norm, there is a nicer definition of finite type, given by the length of the root chains they give rise to.

Lemma 2.3.19. *Let α be a root of non-zero norm and finite type and $x \in G_\alpha$ be locally nilpotent on G. Then, the Lie sub-superalgebra $S_x = \langle x, \omega(x) \rangle$ (generated by the vectors x and $\omega(x)$ as a Lie superalgebra) is isomorphic to the 3-dimensional simple Lie algebra sl_2 if $d(x) = \overline{0}$ and to the 5-dimensional simple Lie superalgebra $osp(1, 2)$. Moreover, as a S_x-module G is a direct sum of finite dimensional S_x-modules.*

Proof. We only consider the case when $x \in G_{\overline{1}}$ as the even case is well known (see [Kac14, Proposition 3.6]). In this case, $[x, x] \neq 0$ by Lemma 2.3.18. Hence, by Lemma 2.3.16, S_x has basis $x, [x, x], \omega(x), \omega([x, x]), [x, \omega(x)]$ and as $a_{ii} \neq 0$, it is therefore isomorphic to $osp(1, 2)$ (see Corollary 2.1.23). To prove the second part, it suffices to consider the case $x \in G_{\overline{0}}$. The result then follows from Exercise 2.3.2 $\qquad\square$

Together with Lemma 2.3.15 and the finite dimensional representation theory of sl_2, Lemma 2.3.15 yields the following result.

Proposition 2.3.20. *A root α of non-zero norm is of finite type if and only if for any $\beta \in \Delta$, $n\alpha + \beta \in \Delta$ only for finitely many integers n, in which case, $n \in [-a, b] \cap \mathbf{Z}$, where*

$$a - b = \begin{cases} 2(\alpha, \beta)/(\alpha, \alpha) & \text{if } d(x) = \overline{0} \\ (\alpha, \beta)/(\alpha, \alpha) & \text{if } d(x) = \overline{1} \end{cases}.$$

Proof. Let $\alpha \in \Delta$ be a root of finite type such that $(\alpha, \alpha) \neq 0$. We only check that for $e_\alpha \in G_\alpha \cap G_{\overline{1}}$ and $e_\beta \in G_\beta$ for which $[e_{-\alpha}, e_\beta] = 0$, where $e_{-\alpha} \in G_{-\alpha}$ satisfies $(e_\alpha, e_{-\alpha}) = 1$, $[e_\alpha, e_\beta] \neq 0$ implies that $[[e_\alpha, e_\alpha]e_\beta] \neq 0$ for then the result follows from Lemma 2.3.19 and the finite dimensional representation theory of sl_2. Suppose that $[[e_\alpha, e_\alpha]e_\beta] = 0$. Then, $[e_\alpha[e_\alpha, e_\beta]] = 0$. From Lemma 2.3.15, we get

$$0 = [e_{-\alpha}[e_\alpha[e_\alpha, e_\beta]]] = ((-\alpha, \alpha + \beta) - (-\alpha, \beta))[e_\alpha, e_\beta] = -(\alpha, \alpha)[e_\alpha, e_\beta].$$

Hence, by assumption on the root α, $[e_\alpha, e_\beta] = 0$. $\qquad\square$

Simple odd roots α_i with norm 0 behave strangely since the action of the corresponding root vector on a root space G_α does not depend on the inner product

(α_i, α). This is why the structure of BKM superalgebras is more complex than that of a BKM algebra.

Lemma 2.3.21. *If $a_{ii} = 0$ and $e_i \in G_{\bar{1}}$, then $(\operatorname{ad} e_i)^2 x = 0$ for all $x \in G$.*

Proof. Let $a_{ii} = 0$ and $e_i \in G_{\bar{1}}$. Because of Condition (6) of Definition 2.1.7 of a BKM superalgebra, $[e_i, e_i] = 0$. Hence, using the Jacobi identity, we then get $[e_i[e_i, x]] = \frac{1}{2}[[e_i, e_i]x] = 0$ for all $x \in G$. $\qquad\square$

In spite of this result, simple odd roots α_i of norm 0 may not satisfy Proposition 2.3.20 for if $a_{jj} \leq 0$, then there may well be simple roots $\alpha_{k,n} \in Q$ such that $f(\alpha_{k,n}) = f(n\alpha_i + \alpha_j)$ for countably infinite integers $n \geq 0$. Furthermore this may not happen for consecutive integers n and so the root α_i would neither satisfy Theorem 2.3.33, characterizing infinite type roots as roots with unbroken root chains in at least one direction. This is why we do not use Proposition 2.3.20 as the definition of roots of finite type.

Applying Proposition 2.3.20, Lemma 2.3.21 and the representation theories of sl_2 and of the 3-dimensional Heisenberg algebra, the type of a simple root can easily be deduced from its norm and parity.

Proposition 2.3.22.

(i) *Suppose that $a_{ii} > 0$. Then the simple root α_i is of finite type. If moreover, $i \in S$, then $2\alpha_i \in \Delta$ is an even root of finite type.*

(ii) *Suppose that $a_{ii} = 0$ and $i \notin S$. Then the simple root α_i is of infinite type.*

(iii) *Suppose that $a_{ii} = 0$ and $i \in S$. Then the simple root $\alpha_i \in Q$ is of finite type and so is $h_i \in H$ if and only if $G_{h_i} \cap G_{\bar{0}} = 0$.*

(iv) *Suppose that $a_{ii} < 0$. Then the simple root α_i is of infinite type.*

To study arbitrary roots, we need to introduce a most important object in the theory of BKM superalgebras, namely their (even) Weyl group, which controls the essential of their structure. It makes sense to define the reflection r_α along a hyperplane perpendicular to α when α is an even (resp. odd) root of finite type and non-zero norm. It is well known that for all weights $\beta \in H$ or $\mathbb{C} \otimes_{\mathbb{Z}} Q$,

$$r_\alpha(\beta) = \begin{cases} \beta - \frac{2(\beta,\alpha)}{(\alpha,\alpha)}\alpha & \text{if } d(\alpha) = \bar{0} \\ \beta - \frac{(\beta,\alpha)}{(\alpha,\alpha)}2\alpha & \text{if } d(\alpha) = \bar{1} \end{cases}.$$

For $i \in I$, the reflection r_{α_i} will be written r_i.

Definition 2.3.23. *The even Weyl group W_E is defined to be the group generated by the reflections r_i, $i \in I$ such that $a_{ii} > 0$. The Weyl group W is generated by all the reflections r_α, where $\alpha \in \Delta^+$ is a root of finite type with non-zero norm.*

Note that the even Weyl group of the BKM superalgebra G is the Weyl group of the Kac-Moody Lie sub-algebra of G generated by the vectors e_i and f_i for $i \in I \setminus S$ and $[e_i, e_i]$, $[f_i, f_i]$ for $i \in S$ such that $a_{ii} > 0$. Hence it is a Coxeter group [Kac14, §3.13].

It is well known that in the case of Kac-Moody Lie superalgebras, the groups W_E and W are the same. This remains true for all infinite dimensional BKM superalgebras but not necessarily for finite dimensional ones. Indeed a finite dimensional BKM superalgebra may contain roots of negative norm and they are necessarily of finite type since the dimension of the BKM superalgebra is finite.

Example 2.3.24. The exceptional Lie superalgebra $F(4)$ (for details, see [Kac3]; for its generalized Cartan matrix see Corollary 2.1.23) has four simple roots, $S = \{1\}$ and $a_{11} = 0$. The set of even simple roots $\alpha_2, \alpha_3, \alpha_4$ generates the set of positive roots of the simple finite dimensional Lie algebra F_4. However, there is one positive even root not generated by them, namely $2\alpha_1 + 3\alpha_2 + 2\alpha_3 + \alpha_4$, containing $1 \in S$ in its support and with negative norm.

Theorem 2.3.25. *If the derived BKM superalgebra $[G, G]$ is isomorphic to $A(m, 1)$, $B(m, 1)$, $G(3)$, $F(4)$ or $D(2, 1, \alpha)$ for $\alpha > 1$, then the Weyl group W is isomorphic to $W_E \times \mathbf{Z}_2$. Otherwise, $W_E = W$. In particular, $W_E = W$ for all BKM superalgebras with infinite dimensional derived Lie sub-superalgebras.*

We leave the proof of this result for later as it requires further properties of the root system. It is a direct consequence of Theorem 2.3.44.

The next technical result is well known in the Kac-Moody setup, and will be often used.

Lemma 2.3.26.

(i) $(w(\alpha), w(\beta)) = (\alpha, \beta)$, for all roots α, β.

(ii) Let α be a positive root of non-positive norm and minimal height in $W_E(\alpha)$. Then $(\alpha, \alpha_i) \leq 0$ for all $i \in I$. Furthermore if $(\alpha, \alpha) = 0$, then either the root α is simple or for all $i \in Supp(\alpha)$, $(\alpha, \alpha_i) = 0$ and $a_{ii} > 0$.

(iii) If α and β are positive roots of non-positive norm, then $(\alpha, \beta) \leq 0$.

Proof. (i) follows from Proposition 2.3.20.
Suppose that $\alpha \in W_E(\alpha)$ is of minimal height. Let $i \in \text{supp}(\alpha)$. If $a_{ii} \leq 0$, then $(\alpha, \alpha_i) \leq 0$ since in this case $a_{ij} \leq 0$ for all $j \in I$. Hence, $(\alpha, \alpha_i) > 0$ implies that $a_{ii} > 0$. By Lemma 2.3.22 (i), α_i is therefore a root of finite type and so from the finite dimensional representation theory of sl_2, $r_i(\alpha)$ is a root, contradicting the minimality of α. Hence $(\alpha, \alpha_i) \leq 0$ for all $i \in I$. If $(\alpha, \alpha) = 0$, then this implies that for all $i \in \text{supp}(\alpha)$ such that $a_{ii} \leq 0$, $(\alpha_i, \alpha_j) = 0$ for all $j \in \text{supp}(\alpha)$. So (ii) now follows from the connectedness of root supports (see Lemma 2.3.8).
(iii) follows from (ii). □

We next show that roots of finite type are conjugate under the action of the even Weyl group W_E to simple roots or "nearly" simple ones.

Proposition 2.3.27.

(i) For any $\alpha \in \Delta^+$, if α has positive norm, then there exits $w \in W_E$, such that $w(\alpha)$ or $\frac{1}{2}w(\alpha)$ is a simple root.

(ii) *If α is an odd root of norm 0 of finite type, then there exits $w \in W_E$ such that $w(\alpha)$ is a simple root.*

Proof. We only prove the second part. Let α be a positive root of finite type and norm 0 of minimal possible height in $W_E\alpha$. Suppose that the root α is not simple. Then by Lemma 2.3.27, supp(α) gives rise to a Kac-Moody Lie sub-superalgebra; and so, from Exercise 2.3.5, the root α is of infinite type, contradicting assumption. $\qquad\square$

In part (i), the half factor is needed for some even roots 2α such that α is an odd root. For example if $i \in S$ and $a_{ii} > 0$, then $\beta = 2\alpha_i$ is a root that is not necessarily conjugate to any simple root, but $\frac{1}{2}\beta$ is simple.

Proposition 2.3.27 has several interesting consequences. The above example $F(4)$ (see Example 2.3.24) is an illustration of the following result.

Corollary 2.3.28. *Suppose that G is a finite dimensional BKM superalgebra and that α is an even root containing an index $i \in S$ in its support such that $a_{ii} = 0$. Then the norm of α is negative.*

Proof. Let $\alpha \in \Delta^+$ be an even root containing in its support an index $i \in S$ such that $a_{ii} = 0$. The root α is not conjugate to a simple root under the action of the even Weyl group W_E, for otherwise it would necessarily be conjugate to α_i, which is not possible since the root α is even, whereas α_i is odd. Hence, Proposition 2.3.27 (i) implies that $(\alpha, \alpha) \leq 0$. If $(\alpha, \alpha) = 0$, then by Lemma 2.3.26 (ii) and Proposition 2.3.27 (ii), we may assume that for all $i \in$ supp(α), $a_{ii} > 0$, contradicting assumption. Hence, $(\alpha, \alpha) < 0$. $\qquad\square$

Whether roots of non-negative norm are of finite or infinite type can now be easily deduced.

Corollary 2.3.29.

  (i) *All roots of positive norm are of finite type.*
  (ii) *If the norm of the even root α is positive, then $k\alpha$ is a root if and only if $k = \pm 1$.*
 (iii) *All roots of norm 0 and finite type are odd.*

We next study roots of infinite type.

Corollary 2.3.30. *If α is a root of infinite type, then $(\alpha, \alpha) \leq 0$ and for all $w \in W_E$, $w(\alpha) > 0$ if and only if $\alpha > 0$.*

Proof. The first part follows from Corollary 2.3.29.(i). Suppose that α is a positive root having non-positive norm. If $r_i(\alpha) < 0$, then $\alpha - c\alpha_i < 0$ for some scalar $c \in \mathbf{Z}$. Hence, $\alpha < c\alpha_i$ and so its norm is positive. This contradiction proves the Corollary. $\qquad\square$

Lemma 2.3.31. *Let $\alpha, \beta \in \Delta^+$ be positive roots with non-positive norm and such that $(\alpha, \beta) \neq 0$. If the root α or β has norm 0, then assume that it is of infinite type. Then, for all non-trivial vectors $e_\alpha \in G_\alpha$, $e_\beta \in G_\beta$, $[e_\alpha, e_\beta] \neq 0$*

unless $\alpha = \beta$ and e_β is a multiple of e_α or $\alpha = 2\beta$ (resp. $\beta = 2\alpha$) and $[e_\beta, e_\beta]$ (resp. $[e_\alpha, e_\alpha]$) is a non-trivial multiple of e_α (resp. e_β).

Proof. Assume that the result does not hold for $e_\alpha \in G_\alpha$, and $e_\beta \in G_\beta$, where $\alpha, \beta \in \Delta^+$, with $(\alpha, \alpha) \le 0$, $(\beta, \beta) \le 0$ and

$$(\alpha, \beta) \ne 0. \tag{1}$$

So

$$[e_\alpha, e_\beta] = 0. \tag{2}$$

When $\alpha = \beta$, e_β is not a multiple of e_α by assumption. Since neither root is of finite type and with norm 0, by Lemma 2.3.18, we may assume that both α, β are even roots. Let $e_{-\alpha} \in G_{-\alpha}$ (resp. $e_{-\beta} \in G_{-\beta}$) be such that $(e_\alpha, e_{-\alpha}) = 1$ (resp. $(e_\beta, e_{-\beta}) = 1$).

Case 1: $\alpha = \beta$.

Equality (2) and Lemma 2.3.15 imply that

$$\alpha([e_{-\alpha}, e_\beta])e_\alpha = [[e_{-\alpha}, e_\beta]e_\alpha] = -(\alpha, \alpha)e_\beta.$$

Hence, as e_α is not a multiple of e_β, $(\alpha, \alpha) = 0$. Without loss of generality, we may assume that α is of minimal height in $W_E(\alpha)$, Hence, by Lemma 2.3.26 (ii), $a_{ii} > 0$ for all $i \in \text{supp}(\alpha)$ and the elements e_i, f_i with $i \in \text{supp}(\alpha)$ generate an affine Kac-Moody Lie superalgebra (see [Kac14, Chapter 4] and [Kac7]). Their structure contradicts (2). Hence the result holds when $\alpha = \beta$.

Case 2: $\alpha \ne \beta$.

Since $(\alpha, \beta) \ne 0$, equality (2) implies that $[e_\alpha, e_{-\beta}] \ne 0$. Hence $\alpha - \beta$ is a root. Without loss of generality, we may assume that the root $\alpha - \beta$ is positive. Lemma 2.3.26 (ii) and 2.3.30 allow us to assume that

$$(\beta, \alpha_i) \le 0 \quad \forall \ i \in I \tag{3}$$

since $(\alpha, \alpha) \le 0$.

Claim: $(ad e_\beta)^n(e_{-\alpha}) \ne 0 \quad \forall n \in \mathbf{Z}_+$.

Suppose this equality does not hold. Let $n \in \mathbf{Z}_+$ be the minimal integer satisfying

$$(ad e_\beta)^{n+1}(e_{-\alpha}) = 0. \tag{4}$$

Making the operator $ad(e_\beta)$ act on this equation and using (2), we get

$$(\alpha, \beta) = \frac{n}{2}(\beta, \beta). \tag{5}$$

So the assumption that $(\alpha, \beta) \ne 0$ gives

$$(\alpha, \beta) < 0 \quad \text{and} \quad (\beta, \beta) < 0. \tag{6}$$

By minimality of n, $n\beta - \alpha \in \Delta \cup \{0\}$. Since

$$(n\beta - \alpha, \beta) = \frac{n}{2}(\beta, \beta) < 0, \tag{7}$$

$$n\beta - \alpha > 0 \tag{8}$$

because of (3). Hence

$$n > 1 \tag{9}$$

since otherwise (8) becomes $\beta - \alpha > 0$, contradicting assumption. Also

$$(n\beta - \alpha, \alpha) \neq 0. \tag{10}$$

Otherwise,

$$(n\beta - \alpha, \alpha - \beta) > 0 \quad \text{and} \quad (n\beta - \alpha, n\beta - \alpha) = n(n\beta - \alpha, \beta) < 0$$

from (7). And so by Lemma 2.3.26 (iii),

$$0 < (\alpha - \beta, \alpha - \beta) = (\alpha, \alpha) - 2(\alpha, \beta) + (\beta, \beta) = (n - 2)(\alpha, \beta) + (\beta, \beta),$$

which is false because of (9) and (7). Now $[e_\alpha, (\operatorname{ad} e_\beta)^n e_{-\alpha}] = 0$ by (9). Therefore,

$$n\beta - 2\alpha \in \Delta \cup \{0\}.$$

Moreover, (5) can be written as $(n\beta - 2\alpha, \beta) = 0$. Suppose first that $n\beta - 2\alpha$ is a root. So from (3) and Lemma 2.3.26 (ii) we can deduce that $(\beta, \alpha_i) = 0$ for all $i \in \operatorname{supp}(n\beta - 2\alpha)$. Hence, for all $j \in \operatorname{supp}(\beta)$ such that $j \notin \operatorname{supp}(n\beta - 2\alpha)$, $(\alpha_j, \alpha_i) = 0$. As $\operatorname{supp}(\beta)$ is connected (see Proposition 2.3.8), either $\operatorname{supp}(\beta) = \operatorname{supp}(n\beta - 2\alpha)$ or $\operatorname{supp}(\beta) \cap \operatorname{supp}(n\beta - 2\alpha) = \emptyset$. If the former holds, then $(\beta, \alpha_i) = 0$ for all $i \in \operatorname{supp}(\beta)$, and so $(\beta, \beta) = 0$, contradicting (7). The latter cannot hold since $\alpha > \beta$ and $n\beta - \alpha > 0$ imply that $\operatorname{supp}(\alpha) = \operatorname{supp}(\beta)$. So

$$n\beta = 2\alpha.$$

Suppose that n is even. Then, $h = (\operatorname{ad} e_\beta)^{\frac{n}{2}} e_{-\alpha} \in H$. By minimality of n, $[h, e_\beta] \neq 0$. From Lemma 2.3.15 we know that $[h, e_\beta] = (h, h_\beta) e_\beta$, where $h_\beta = [e_\beta, e_{-\beta}]$. However

$$(h_\beta, h) = (\beta, \beta)(e_\beta, (\operatorname{ad} e_\beta)^{\frac{n}{2} - 1} e_{-\alpha}) = 0$$

unless $n = 2$, in which case $\alpha = \beta$, contradicting assumption. Hence, the integer n is odd and

$$e_{\frac{\beta}{2}} = (\operatorname{ad} e_\beta)^{\frac{n+1}{2}} e_{-\alpha} \neq 0.$$

Then, $h = (\operatorname{ad} e_{\frac{\beta}{2}})^n e_{-\alpha} \neq 0$ for otherwise same arguments as before imply that $\frac{1}{2}(\alpha, \beta) = \frac{l}{8}(\beta, \beta)$ for some integer $l \leq n - 1$, whereas $\frac{1}{2}(\alpha, \beta) = \frac{n}{4}(\beta, \beta)$ since $n\beta = 2\alpha$, and so $l = 2n$ contradicting the assumption on l. Then, the same argument as in the even case applied to $\beta/2$ instead of β implies a contradiction since $n > 1$ by (9). This proves the Claim.

Hence for n large enough,

$$n\beta - \alpha \in \Delta^+, \ (n\beta - \alpha, n\beta - \alpha) < 0, \ (n\beta - \alpha, \alpha) < 0, \ [e_\alpha, (\operatorname{ad} e_\beta)^n e_{-\alpha}] = 0.$$

So, $n\beta - 2\alpha \in \Delta \cup \{0\}$. For $n \gg 0$, $n\beta - 2\alpha > 0$. Therefore applying the above Claim with $n\beta - \alpha$ replacing α and α replacing β, it follows that for $n \gg 0$,

$$n\beta - m\alpha \in \Delta \cup \{0\}$$

for all integers $m > 0$. So for all α_i, α_j in the support of α, $nl_i - mk_i \leq 0$ if and only if $nl_j - mk_j \leq 0$. Therefore $k_i l_j = l_i k_j$, and $\beta = \frac{k_i}{l_i} \sum_{j \in I} l_j \alpha_j = \frac{k_i}{l_i} \alpha$ for some $l_i \neq 0$. It follows that for any integer $r > 0$, if $m = k_i r$ and $n = l_i r$, then $n\beta - m\alpha = 0$ and so $0 \neq (\operatorname{ad} e_{-\alpha})^{m-1}(\operatorname{ad} e_\beta)^n e_{-\alpha} = h \in H$. For $r \gg 0$, $[e_{-\alpha}, h] \neq 0$ and $m \gg 0$ and hence we get a contradiction as above. This proves the Lemma. $\qquad \square$

Lemma 2.3.32. *Let α be a positive root with non-positive norm and of minimal height in $W_E(\alpha)$. If $(\alpha, \alpha) = 0$, assume that α is of infinite type. Let β be a positive root such that $(\alpha, \beta) \neq 0$ and $(\beta, \beta) > 0$. Then, $\alpha + \beta \in \Delta$.*

Proof. By Lemma 2.3.26 (ii), $(\alpha, \beta) < 0$ since α and β are positive roots. Hence $\alpha + \beta \in \Delta$ by the finite dimensional representation theory of sl_2. $\qquad \square$

The next two result on the length of root chains induced by roots of infinite type and justifying the term "infinite type" is immediate from the previous two Lemmas.

Theorem 2.3.33. *Let α be a positive root with non-positive norm. If $(\alpha, \alpha) = 0$, assume that α is of infinite type. Let β be any positive root. Suppose that if β has norm 0, then it is of infinite type. Then, $n\alpha + \beta \in \Delta$ for all $n \in \mathbf{Z}_+$, more precisely $(\operatorname{ad} e_\alpha)^n(e_\beta) \neq 0$ for any linearly independent vectors $e_\alpha \in G_\alpha$, $e_\beta \in G_\beta$ or $-n\alpha + \beta \in \Delta$ for all $n \in \mathbf{Z}_+$, more precisely $(\operatorname{ad} e_{-\alpha})^n(e_\beta) \neq 0$ for any non-trivial vectors $e_{-\alpha} \in G_{-\alpha}$, $e_\beta \in G_\beta$. If the latter holds, then $(\alpha, \beta) > 0$ and $(\beta, \beta) > 0$.*

Corollary 2.3.34. *Consider the roots as elements of the lattice Q. Let α be a root of infinite type. Then, there is a root β such that for any non-trivial vector $x \in G_\alpha$, there exists a vector $y \in G_\beta$ such that $(\operatorname{ad} x)^n y \neq 0$ for all integers $n \geq 0$. In particular, $n\alpha + \beta$ is a root for all integers $n \geq 0$.*

Proof. Without loss of generality, we may assume that the infinite type root α is a positive even root. Suppose that there exists a positive root β such that $(\alpha, \beta) \neq 0$ and $\beta \notin \{\alpha, 2\alpha, \frac{1}{2}\alpha\}$ and if $(\beta, \beta) = 0$ then β is of infinite type. Then the result follows from Theorem 2.3.33.

So suppose that no such root β. We show that this leads to a contradiction.

Since the generalized Cartan matrix A is indecomposable by assumption, it then follows that $\operatorname{supp}(\alpha) = I$. Since the root α is of infinite type, it is not of the type α_i, $2\alpha_i$, $\frac{1}{2}\alpha_i$ for then $I = \{\alpha_i\}$ and the derived Lie superalgebra G' is finite dimensional by Lemma 2.3.16. In particular the root α is of finite type, contradicting assumption. Hence,

$$(\alpha, \alpha_i) = 0 \tag{1}$$

unless $i \in S$ and $a_{ii} = 0$. Suppose that $j \in I$ and $a_{jj} < 0$. Then, unless $j \in S$ and $a_{jj} = 0$, (1) implies that $a_{ij} = 0$ for all $i \in I$, contradicting the connectedness of supp(α) (see Proposition 2.3.8). Hence, for all $i \in I$, $a_{ii} > 0$ or $i \in S$ and $a_{ii} = 0$. Let $x \in G_\alpha$ be a non-trivial vector.

Case 1: for all $i \in I$, $a_{ii} > 0$.

In this case, $(\alpha, \alpha_i) = 0$ for all $i \in I$ and for all $i \in I$, e_i, f_i generate an affine Kac-Moody Lie superalgebra and their structure (see [Kac14, §4,6,7,8] and [Kac7]) leads to a contradiction of the above assumption.

Case 2: There exists $i \in S$, $a_{ii} = 0$.

Without loss of generality, assume that $a_{11} = 0$.

Claim 1: $[x, e_i] = 0$ for all $i \in I$.

Suppose this statement is false and $[x, e_i] \neq 0$. Then, $\alpha + \alpha_i \in \Delta$. If $a_{ii} = 0$, then, $(\alpha, \alpha + \alpha_i) < 0$ and $(\alpha + \alpha_i, \alpha + \alpha_i) < 0$ and so by Theorem 2.3.33, our assumption is again contradicted. Thus $a_{ii} > 0$. As the generalized Cartan matrix is assumed to be indecomposable, it follows that there exists a positive root $\beta \in \Delta$ with $1 \in$ supp(β) such that $\alpha + \beta \in \Delta$. Since $1 \in$ supp(β), $(\beta, \beta) \leq 0$ by Proposition 2.3.27 (i). Hence, the previous argument applied to β instead of α_i leads once more to a contradiction. As a result, our Claim holds.

Claim 2: The root α is the root of maximal height in Δ.

Suppose that there is a positive root β such that $\beta \not\leq \alpha$. If for all $i \in$ supp(β), $a_{ii} > 0$, then as the matrix A is indecomposable, there is a root $\gamma > \beta$ such that $1 \in$ supp(γ). So we may suppose that there exists $1 \in$ supp(β). Then, $(\alpha, \beta) < 0$. From Claim 1, we can deduce that $\alpha + \beta$ is not a root. Hence, $\alpha - \beta$ is a root and so $\alpha \leq \beta$. However, Claim 1 also implies that there are no such roots, proving Claim 2.

Hence there are only finitely many roots and as $\dim G_\beta < \infty$ for all $\beta \in \Delta$, the derived Lie subalgebra G' is finite dimensional. In particular the root α is of finite type, contradicting assumption and proving the result. $\qquad\square$

In the above Corollary, the statement can be easily rewritten in terms of roots belonging to the subspaces H or H^* but it is slightly more cumbersome. The next example illustrates why the length of the root chains through infinite type roots may be sometimes finite or more precisely why in Theorem 2.3.33, we have to exclude the case when β is a root of finite type and norm 0.

Example 2.3.35. Let $A = \begin{pmatrix} 0 & -1 & -1 \\ -1 & 1 & 0 \\ -1 & 0 & -1 \end{pmatrix}$, where $S = \{1, 2\}$. The root $\alpha = \alpha_1 + \alpha_2$ is a root of infinite type since $n\alpha + \alpha_3$ is a root for all positive integers n. This follows from the representation theory of sl_2. However, $(\alpha, \alpha_1) < 0$ but $\alpha + \alpha_1$ is not a root. Indeed,

$$[[e_1, e_2]e_1] = -[[e_1, e_1]e_2] + [e_1[e_1, e_2]] = [e_1[e_1, e_2]]$$

by Lemma 2.3.21; and so

$$[[e_1, e_2]e_1] = 0.$$

We can now explain the reason why the terminology real and imaginary roots makes sense if one restricts oneself to BKM algebras. It is due to the following equivalences.

Proposition 2.3.36. *Suppose that G is a BKM algebra. Assume that the derived Lie algebra G' is not the 3-dimensional Heisenberg algebra. Then the following three properties are equivalent:*

(a) For any $\beta \in \Delta$, $\beta + n\alpha \in \Delta$ only for finitely many integers n;

(b) $(\alpha, \alpha) > 0$;

(c) There exists an element $w \in W$ such that $w(\alpha)$ is a simple root of positive norm.

Proof. (b) implies (c) by Proposition 2.3.27 (i) and (c) implies from (a) by Corollary 2.3.29. We check that (a) implies (b): Suppose that α is a positive root satisfying (a) and $(\alpha, \alpha) \leq 0$. Since G is a Lie algebra, by Theorem 2.3.33, $(\alpha, \beta) = 0$ for all $\beta \in \Delta$. In particular, $(\alpha, \alpha_i) = 0$ for all $i \in \operatorname{supp}(\alpha)$. Therefore from [Kac14, Chapter 4], either $\operatorname{supp}(\alpha)$ generates an affine Lie algebra and α is a root of infinite type, or $I = \{1\}$, $a_{11} = 0$ and $\alpha = \alpha_1$. This contradicts assumptions. $\qquad\square$

However, as Example 2.3.24 or more generally Corollary 2.3.28 show, Proposition 2.3.36 does not hold for all finite dimensional BKM superalgebras. Example 2.3.24 goes further and illustrates the fact that a finite dimensional BKM superalgebra with $S \neq \emptyset$ may have roots of negative norm and finite type. For infinite dimensional BKM superalgebras, a slightly altered version holds.

Corollary 2.3.37. *Suppose that the G is a BKM superalgebra such that the derived sub-superalgebra G' has infinite dimension. Then the following three properties are equivalent:*

(a) For any $\beta \in \Delta$, $\beta + n\alpha \in \Delta$ only for finitely many integers n;

(b) $(\alpha, \alpha) \geq 0$ and if $(\alpha, \alpha) = 0$ then the root α is odd and $2\alpha \notin \Delta$;

(c) There exists an element $w \in W$ such that $w(\alpha)$ is a simple root of non-negative norm.

To prove this result, we only need to show that for infinite dimensional BKM superalgebras, roots having negative norm are necessarily of infinite type (see Theorem 2.3.44). Some prerequisites are required for this. First notice that for BKM algebras, all real roots give rise to reflections and so contribute to the Weyl group. However this is false both for finite or infinite dimensional BKM superalgebras as they may have odd roots of norm 0 and finite type. Hence, in the larger context of BKM superalgebras, it is best to differentiate between roots by using the essential characteristic of real roots, namely the lengths of the root chains they give rise to. Since the generalized Cartan matrix is assumed to be indecomposable, Theorem 2.3.33 can be rewritten in terms of roots of finite type having negative norm.

Corollary 2.3.38. *Let α be a root of finite type and negative norm. Then,*

(i) $I = Supp(\alpha)$ and

(ii) $\forall \beta \in \Delta$, $(\alpha, \beta) = 0$ unless $\beta \in \{\frac{1}{2}\alpha, \alpha, 2\alpha\}$ or β is of finite type and has norm 0.

Therefore there are very few roots of finite type having negative norm.

Corollary 2.3.39. *There are at most two positive roots of finite type having negative norm in the set of roots* Δ, *one even root* α *and one odd root* $\frac{1}{2}\alpha$. *Moreover, the corresponding root spaces have dimension 1.*

Proof. Let α and β be positive roots of finite type having negative norm. Then by Corollary 2.3.38, $\text{supp}(\alpha) = \text{supp}(\beta)$ contains $i \in S$ such that $a_{ii} = 0$ and for all $i \in I$ such that $a_{ii} \neq 0$, $(\alpha, \alpha_i) = 0$. Hence, $(\alpha, \beta) < 0$ and so $\beta = \frac{1}{2}\alpha$ or α or 2α as desired and the dimensions are stated (see Theorem 2.3.33). □

For a more in depth study of the influence of these negative norm roots of finite type on the structure of BKM superalgebras, we need to consider automorphisms on G induced by the Weyl group. As in the Kac-Moody setup, there is an inner automorphism (see section 2.4) of the BKM superalgebra corresponding to a reflection r_i when $a_{ii} > 0$. For reasons of simplicity, we will keep the same notation for the reflection and the inner automorphism.

Definition 2.3.40. *For* $a_{ii} > 0$, *the inner automorphism corresponding to the reflection* r_i *is defined to be*

$$r_i := \begin{cases} (\exp \text{ad} \, (f_i))(\exp \text{ad} \, (-e_i))(\exp \text{ad} \, (f_i)), & \text{if } i \in I \setminus S \\ (\exp \text{ad} \, ([f_i, f_i]))(\exp \text{ad} \, (-[e_i, e_i]))(\exp \text{ad} \, ([f_i, f_i])), & \text{if } i \in S \end{cases}.$$

The next well known result in the context of Kac-Moody Lie algebras clearly still holds.

Proposition 2.3.41. *For all* $i \in I$, $r_i(G_\alpha) = G_{r_i(\alpha)}$ *for any root* α.

Corollary 2.3.42. *If* α *is a root of positive norm, then the root space* G_α *has dimension 1. This is also the case for roots of norm 0 and finite type when the roots are assumed to be in the formal root lattice* Q.

Proof. Let α be a root of positive norm. By Propositions 2.3.27 and 2.3.41 we may assume that the root α is simple. Since all the non-diagonal entries of the matrix A are non-positive, if α_i and α_j are simple roots in Q of positive norm, then $f(\alpha_i) = f(\alpha_j)$ implies that $i = j$. The result follows since the same arguments clearly hold when α is a root of norm 0 and finite type and the roots are taken to be in Q. □

Corollary 2.3.43. *Let* G_p *be the (even) Lie subalgebra generated by the elements*

$$e_i, f_i, i \in I \setminus S, \quad [e_i, e_i], [f_i, f_i], i \in S.$$

Then W_E *is the Weyl group of* G_p. *Moreover, if* G_p *is finite dimensional and if the set* $\{i \in S : a_{ii} = 0\}$ *is finite, then the derived Lie superalgebra* $[G, G]$ *is finite dimensional.*

Proof. Suppose that $\dim G_p < \infty$. Since for all $i \in S$ such that $a_{ii} \neq 0$, $[e_i, e_i] \neq 0$, all the simple roots of G are of finite type and there are finitely many roots of non-zero norm. Furthermore, by Exercise 2.3.9, W_E is a finite group. Hence, by Corollary 2.3.27 (ii), there are only finitely many roots of norm 0. The result now follows from Corollary 2.3.39 since the latter implies that there are at most two positive roots of negative norm, α, $\frac{1}{2}\alpha$, and $\dim G_\alpha = \dim G_{\frac{1}{2}\alpha} \leq 1$.

$\square$

We are now ready to prove the main result on roots of finite type having negative norm.

Theorem 2.3.44. *Suppose that the BKM superalgebra G contains roots of finite type and negative norm. Then, the Lie superalgebra $G' = [GG]$ is a finite dimensional Lie superalgebra of type $A(m,1)$, $B(m,1)$, $G(3)$, $F(4)$ or $D(2,1,\alpha)$ for $\alpha > 1$;*

Proof. Let α be a positive even root of finite type having negative norm.

Claim: All roots of G are of finite type.

Let β be a root of infinite type. By Corollary 2.3.29 (i),

$$(\beta, \beta) \leq 0.$$

Without loss of generality, we may assume that the root β is of minimal height in $W_E\beta$. From Lemma 2.3.38 (ii),

$$(\alpha, \beta) = 0.$$

From Corollary 2.3.39 and Lemma 2.3.26 (ii), $(\alpha, \alpha_i) \leq 0$ for all $i \in I$ and so

$$(\alpha, \alpha_i) = 0 \tag{1}$$

for all $i \in \operatorname{supp}(\beta)$. Since $I = \operatorname{supp}(\alpha)$ by Lemma 2.3.39 (i), it then follows that for $i \in I - \operatorname{supp}(\beta)$, $a_{ij} = 0$ for all $j \in \operatorname{supp}(\beta)$. Since $\operatorname{supp}(\alpha)$ is connected by Proposition 2.3.8, this forces $I = \operatorname{supp}(\beta)$ and so equalities (1) give $(\alpha, \alpha) = 0$, contradicting the definition of the root α. This proves our Claim.

Consider the even Lie subalgebra G_p of G generated by the elements e_i, f_i, $i \in I \setminus S$ and $[e_i, e_i], [f_i, f_i], i \in S$. The above Claim implies that G_p is a Kac-Moody Lie algebra (i.e. its simple roots have positive norm) all of whose roots are of finite type. Hence by Proposition 2.3.36, all its roots have positive norm. Furthermore as $I = \operatorname{supp}(\alpha)$, the indexing set I is finite. Therefore the Lie algebra G_p is finite dimensional [Kac14, Proposition 4.9]. Therefore by Corollary 2.3.43, $\dim[G, G] < \infty$,. Then the list given in Corollary 2.1.23 together with the uniqueness of the generalized Cartan matrix (see Theorem 2.4.8) proves the Theorem.

$\square$

Theorem 2.3.44 together with Corollary 2.3.37 allows us to deduce Theorem 2.3.25, i..e when the group $W_E = W$ and when it is a proper subgroup. Another consequence of Theorem 2.3.44 is that infinite dimensional BKM superalgebras contain roots of infinite type:

Corollary 2.3.45. *Suppose that the indexing set I is finite and that all roots of the BKM superalgebra G are of finite type. Then, the derived Lie superalgebra $[G, G]$ has finite dimension.*

Proof. If there is a root of negative norm and finite type, then the result follows from Theorem 2.3.44. So suppose that all roots have non-negative norm. Consider the BKM subalgebra generated by the elements e_i, f_i for $i \in I \setminus S$ and $[e_i, e_i]$, $[f_i, f_i]$ for $i \in S$. Then, it is finite dimensional and the Weyl group W of G is finite by Exercise 2.3.9. Therefore by Proposition 2.3.27, the number of roots of norm 0 is finite and so the result follows from Corollary 2.3.41.   □

In Lemma 2.3.31, we considered root chains induced by non-positive norm roots through roots that are not in its orthogonal hyperplane. We next study what happens when the root α has non-positive norm and the root β is orthogonal to α. When $(\beta, \beta) > 0$, applying the finite dimensional representation theory of sl_2, one can easily conclude.

Proposition 2.3.46. *Let α and β be orthogonal roots of non-positive norm, both positive or both negative. Then either $\beta = n\alpha$ for some integer $n > 0$ or* $\operatorname{supp}(\alpha) \cap \operatorname{supp}(\beta) = \emptyset$, $(\alpha, \alpha_i) = 0$ *for all $i \in \operatorname{supp}(\beta)$ and $(\beta, \alpha_i) = 0$ for all $i \in \operatorname{supp}(\alpha)$. Assume further that when both $\alpha = \beta$ has norm 0 and is of infinite type, α is an even root. Then the corresponding root spaces commute.*

Proof. Let α and β be positive orthogonal roots of non-positive norm. Without loss of generality, by Corollary 2.3.31, we may assume that the root α is a root of minimal height in $W_E(\alpha)$ and so by Lemma 2.3.26 (ii), $(\alpha, \alpha_i) \leq 0$ for all $i \in I$. As a consequence

$$(\alpha, \alpha_i) = 0 \quad \forall i \in \operatorname{supp}(\beta), \tag{1}$$

and so due to the connectedness of the support of a root (Proposition 2.3.8), either $\operatorname{supp}(\alpha) \cap \operatorname{supp}(\beta) = \emptyset$ and $\alpha + \beta$ is not a root or $\operatorname{supp}(\alpha) = \operatorname{supp}(\beta)$ and $\operatorname{supp}(\alpha)$ generates an affine Kac-Moody Lie superalgebra (see [Kac14, Chapter 4] and Kac7]). The result then follows from the theory of affine Kac-Moody Lie algebras [Kac14, Chapters 6–8].   □

Before concluding this section, we define the Weyl vector which, as will be seen later sections and chapters, plays a fundamental role in the theory of BKM superalgebras.

Definition 2.3.47. *A Weyl vector is defined to be a vector ρ either in the dual of the space $\mathbf{C} \otimes_{\mathbf{Z}} Q$ or in H or in its dual H^* satisfying*

$$(\rho, \alpha_i) = \frac{1}{2}(\alpha_i, \alpha_i) \quad \text{for all} \quad i \in I$$

for all $i \in I$.

Remark 2.3.48. There may be no Weyl vector in the space $\mathbf{C} \otimes_{\mathbf{Z}} Q$ since the bilinear form may not be non-degenerate on the formal root lattice. This is for example the case for affine Lie algebras (see Example 2.1.10). It exists in both H and its dual since the bilinear form on the generalized Cartan subalgebra H is assumed to be non-degenerate by definition. It is therefore also uniquely defined in H.

Note also that the *fundamental Weyl chamber* in the BKM superalgebra setup is defined to be the subset $\{h \in H_{\mathbf{R}} : (h, h_i) \geq 0, \forall i \in I, a_{ii} > 0\}$ of the real subspace $H_{\mathbf{R}}$. Hence it only depends on the simple roots of positive norm and is equal to the fundamental Weyl chamber of the Lie subalgebra G_p (see Corollary 2.3.43 for its definition). The Weyl chambers are the conjugates of this set under the action of the Weyl group W.

Example 2.3.49. For the monster Lie algebra M (see example 2.2.11.1), $(1, 0)$ is the Weyl vector in $H = \mathbf{C} \otimes_{\mathbf{Z}} II_{1,1}$ since

$$(1, 0).(1, n) = -n = (1, n).(1, n), \quad \forall n \geq -1.$$

Exercises 2.3

1. Prove that the support of a root is always connected.

2. Show that for $i \in I$ such that $a_{ii} > 0$, considered as a S_i-module, where S_i is the Lie sub-superalgebra of the BKM superalgebra generated by the elements e_i and f_i, G is a direct sum of S_i-modules.

3. For $r \in \mathbf{N}$, show that the Lie subalgebra M_r of the monster Lie algebra M generated by the subspaces $M_{(m,rn)}$, $m, n \in \mathbf{Z}$ is a BKM algebra with the same generalized Cartan subalgebra $H = \mathbf{C} \otimes II_{1,1}$, as M.

4. Suppose that $G = G(A, H, S)$ is a Kac-Moody Lie superalgebra. Prove that if the restricted simple root multiplicity of the root $h \in H$ is non-trivial, then it is equal to the multiplicity of h and $\text{mult}(h) = 1$.

Show that there are Kac-Moody Lie superalgebras for which the map f_Q is not injective. However, if the elements h_i are assumed to be linearly independent then show that the map f_Q is injective.

5. Let G be a BKM superalgebra and the root $\alpha \in \Delta$ satisfies $I = \text{supp}(\alpha)$ and $(\alpha, \alpha_i) = 0$ for all $i \in I$.

   (i) Show that the matrix with entries given by the bilinear form (α_i, α_j) with $i, j \in \text{supp}(\alpha)$ is the generalized Cartan matrix of an affine Lie superalgebra.
   (ii) Show that in affine BKM superalgebras the roots of norm 0 are of infinite type.
   (iii) Deduce that the root α is of infinite type unless $\alpha = \alpha_i$ for some $i \in S$.

Hint: For a solution see [Kac14, Chapters 4,6] and [Kac7].

6. Prove that all roots of positive norm are conjugate to a simple root of positive norm under the action of the group W_E. Deduce that they are always of finite type.

7. Enumerate the roots of the BKM superalgebra $F(4)$ and calculate their norms.

8. Let $\alpha \in Q$ be a root and $h_\alpha = f_Q(\alpha)$. Show that $[x, y] = (x, y)h_\alpha$ for all $x \in G_\alpha$, $y \in G_{-\alpha}$. Deduce that for all $0 \neq x \in G_\alpha$, $[x, \omega(x)] \neq 0$.

9. Show that the derived BKM superalgebra $[G, G]$ is finite dimensional if and only if its Weyl group is finite.

Hint: for a partial solution, see [Kac14, Proposition 4.9].

10. Show that if all roots of the BKM superalgebra G have non-negative norm, then the derived Lie superalgebra G' is either finite dimensional simple or affine. In particular, if the norms are all positive then deduce that the Lie superalgebra G is finite dimensional and simple.

11. Set

$$K = \{\alpha \in Q_+ : (\alpha, \alpha_i) \leq 0, \forall i \in I, \text{supp}(\alpha) \quad \text{connected}\},$$

where $Q_+ = \{\alpha \in Q : \alpha = \sum_{i \in I} k_i \alpha_i, k_i \geq 0 \, \forall i \in I\}$.

 (i) Let $\alpha \in K$ be such that $\text{supp}(\alpha)$ does not contain any index $i \in S$ with $a_{ii} = 0$. Show that α is either a root or $\alpha = j\alpha_i$ for some $i \in I$ such that $a_{ii} \leq 0$ and some integer $j \geq 2$.

Hint: Consider the positive root β of maximal height such that $\beta \leq \alpha$. Show that $\beta = \alpha$ by applying Theorem 2.3.33 and using the theory of affine Kac-Moody Lie superalgebras.

 (ii) Does (i) hold for elements $\alpha \in K$ containing an index $i \in S$ such that $a_{ii} = 0$? Either prove this statement or find a counterexample.

 (iii) Deduce that when $S = \emptyset$, the set of positive roots of infinite type is

$$\bigcup_{w \in W} w(K) - \cup_{j \geq 2} j\Pi_m,$$

where $\Pi_m = \{\alpha_i : i \in I, a_{ii} \leq 0\}$ but that this is not necessarily true for BKM superalgebras.

12. Let G be a finite dimensional semisimple Lie algebra.

 (i) Show that G is the direct sum of a maximal abelian Lie subalgebra H and its eigenspaces (with respect to the adjoint action).

 (ii) Studying the roots of G with respect to H, show that G is generated by Chevalley generators satisfying the Serre-Chevalley relations.

For a solution, see [Serr1].

2.4 Uniqueness of the Generalized Cartan Matrix

Obviously a BKM superalgebra can have distinct generalized symmetric Cartan matrices. So let us start by defining clearing what we mean by this.

Definition 2.4.1. *Two symmetric matrices A and B are said to be equivalent if they can be obtained from one other by a permutation of rows and by positive scalar multiplication.*

It is fundamental to know that different Cartan decompositions give rise to the same generalized symmetric Cartan matrix (up to equivalence). This is well known in the context of Kac-Moody Lie algebras and as expected still holds in the more general setup of BKM superalgebras. For this, the conjugacy of Cartan subalgebras under the action of inner automorphisms, also a classical result in the context of Kac-Moody Lie algebras [KacP6] must continue to hold in our larger framework. First let us define the group $Inn(G)$ of inner automorphisms. For details, see [KacP4] and [Hum]. We keep the notation used in [KacP6].

Consider the root spaces G_α as abelian groups. Let $\mathcal{G}^*$ be the free product of the groups G_α for all $\alpha \in \Delta$ with $(\alpha, \alpha) > 0$ with inclusion maps $i_\alpha : G_\alpha \to \mathcal{G}^*$, i.e. the group $\mathcal{G}^*$ is the disjoint union of the groups G_α and the only relations are those in the subgroups G_α. Let V be an integrable G'-module (see Definition 2.6.29) and $\pi : G' \to gl(V)$ the corresponding representation. Let $\pi^* : \mathcal{G}^* \to GL(V)$ be the homomorphism given by $\pi^*(i_\alpha x) = \exp(\pi(x))$ for $x \in G_\alpha$. Set $N = \ker \pi^*$, where the intersection is taken over all integrable G'-modules.

Definition 2.4.2. *The Lie group associated to the BKM superalgebra G is the group $\mathcal{G} = \mathcal{G}^*/N$. For $x \in G_\alpha$ with $(\alpha, \alpha) > 0$, write $\exp(\alpha) := N + (i_\alpha(x)) \in \mathcal{G}$. The group of inner automorphisms of G, $Inn(G)$ is the image of the group $\mathcal{G}$ under the homomorphism $\mathrm{Ad} : \mathcal{G} \to GL(G)$ given by $\mathrm{Ad}\,(\exp(x)) = \exp(\mathrm{ad}\,(x))$ for $x \in G_\alpha$.*

Theorem 2.4.3. *All generalized Cartan subalgebras of the BKM superalgebra G are conjugate under the action of inner automorphisms.*

To prove this, a little preparation is necessary. First we need some more notation and then we state an obvious fact.

Set
$$J = \{i \in I : a_{ii} > 0\}$$
and, for any subset K of the indexing set I, set G_K to be the Kac-Moody Lie superalgebra generated by the elements e_i and f_i for $i \in K$. Define U^+ (resp. U^-) as the Lie sub-superalgebra generated by the root spaces G_α where the support of the root $\alpha \in \Delta^+$ (resp. $-\alpha \in \Delta^+$) contains at least an index $i \in I$ such that $a_{ii} \leq 0$.

The next result is immediate from the generalized Cartan decomposition given in Corollary 2.3.7.

Lemma 2.4.4.
$$G = U^+ \oplus (G_J + H) \oplus U^-$$

as a direct sum of vector spaces.

Proposition 2.4.5. *If the generalized symmetric Cartan matrix A is indecomposable and the derived BKM superalgebra $[G, G]$ is infinite dimensional, then*

(i)   *The adjoint action on G of an element of the generalized Cartan subalgebra H is locally finite;*
(ii) *every element of G with locally finite action on G via the adjoint action is contained in the Lie sub-superalgebra $G_J + H$;*
(iii) *all generalized Cartan subalgebras of G are contained in the Lie sub-superalgebra $G_J + H$.*

Proof. (i) is an immediate consequence of the fact that the Lie superalgebra G is generated by eigenvectors for H. So we prove (ii) and (iii). Suppose that K is a generalized Cartan subalgebra of G. By definition, K is an even Lie subalgebra of G.

Since K is a generalized Cartan subalgebra, it acts on G in a locally finite way, i.e. for any element $x \in K$ and $y \in G$, there is a finite dimensional subspace of G containing $(adx)^n y$ for all $n \in \mathbf{Z}_+$. Let x be an element of K not contained in $G_J + H$. By Lemma 2.4.4, there are elements $u^+ \in U^+$, $u^- \in U^-$ and $v \in G_J + H$ such that

$$x = u^+ + v + u^-.$$

We first assume that $u^+ \neq 0$. So there exists positive roots $\beta_1, ..., \beta_r \in \Delta$ such that for all $1 \leq i \leq r$, $\exists i_j \in \operatorname{supp}(\beta_j)$ satisfying $a_{i_j i_j} \leq 0$, and root vectors $u_i \in G_{\beta_i}$ such that

$$u^+ = \sum_{i=1}^{r} u_i.$$

From Proposition 2.3.27 we can infer that $(\beta_i, \beta_i) \leq 0$ for all $1 \leq i \leq r$. Since the Lie subalgebra K is even, the roots β_i are even for all $1 \leq i \leq r$. Hence the roots β_i must be of infinite type by Proposition 2.3.27 (ii) and Corollary 2.3.30. Therefore, applying Corollary 2.3.34, there exists a root β such that $n\beta_1 + \beta$ is a root for all integers $n \geq 0$ and for some vector $y \in G_\beta$, $(\operatorname{ad} u_1)^n y \neq 0$ for all integers $n \geq 0$.

We claim that infinitely vectors in the set $\{(\operatorname{ad} x)^n(y) : n \in \mathbf{Z}_+\}$ are linearly independent. We can choose $j \in \{1, ..., r\}$ to be such that β_j is of maximum height with the property that $\beta_j + \beta$ is a root. Then for each positive integer n, $(adu^+)^n(y)$ is the sum of root vectors, exactly one of which corresponds to the root $n\beta_j + \beta$. Hence there are infinitely many linearly independent vectors $(adx)^n(y)$, $n \in \mathbf{Z}_+$. Indeed for all $n \in \mathbf{Z}_+$, the component of $(adx)^n(y)$ in the $n\beta_k + \beta$-root space cannot be a linear combination of components of $(adx)^r(y)$, $r \leq n - 1$ since all the components of u^- belong to negative root spaces and those of v belong to root spaces generated by simple roots of positive norm and to the generalized Cartan subalgebra H. It follows that the claim holds. This contradicts the fact that K acts in a locally finite way. So

$$u^+ = 0.$$

Similarly,

$$u^- = 0,$$

and so

$$x \in G_J + H,$$

proving (ii). This forces the Cartan subalgebra K to be contained in $G_J + H$ as desired in (i). $\qquad\square$

We are now ready to prove Theorem 2.4.3.

Proof of Theorem 2.4.3. We first suppose that the derived Lie superalgebra $[G, G]$ is infinite dimensional. Let H_1 be a generalized Cartan subalgebra of G. Then,

$$H_1 \leq H + G_J$$

by Proposition 2.4.5 (iii). Let K be a finite subset of the indexing set J and $H_K = H_1 \cap (H + G_K)$. The Lie subalgebra H_K is a generalized Cartan subalgebra of the BKM superalgebra $H + G_K$. Write Z for the centre of the Lie superalgebra $H + G_K$. By Corollary 2.1.18, $Z \leq H_K$.

Claim: The vector space H_K/Z has finite dimension.

For $i \in K$, let e_i and f_i together with H_K be the usual generators of $H + G_K$. In particular the vectors e_i and f_i are eigenvectors for the Lie subalgebra H_K. By Definition 2.1.7, for $h \in H_K$, $[h, e_i] = 0 = [h, f_i]$ if and only if $(h, h_i) = 0$, where $(.,.)$ is the bilinear form on H_K. So,

$$Z = \cap_{i \in K} Z_i,$$

where $Z_i = \{h \in H_K, (h, h_i) = 0\}$. Elementary arguments from linear algebra imply that the subspaces Z_i are hyperplanes in H_K, i.e. subspaces of codimension 1. From this, we can deduce that the subspace Z has finite codimension in H_K since the set K is finite.

Let the vector space L be a complement of Z in the vector space H_K. As L is finite dimensional, with obvious modifications for the superalgebra case, the Conjugacy Theorems of Peterson-Kac for Kac-Moody Lie algebras [KacP6] can directly be extended to $L + G_K$. Thus, the subalgebra L of $H + G_J$ is conjugate to a subalgebra of H under the action of some inner automorphism g of the BKM superalgebra G. Since Z is central in $H + G_K$ and contained in H, it follows that H_K is conjugate to a subalgebra of H under the action of this automorphism g. A symmetric argument applied to H implies that H_K is conjugate to H under the action of g. Similarly H_K is conjugate to H_1 under the action of some inner automorphism. Hence $H_K = H_1$ since $H_K \leq H_1$, and H and H_1 are conjugate under the action of some inner automorphism.

When the derived Lie superalgebra $[G, G]$ is finite dimensional, the argument above shows that we may assume G to be finite dimensional. The result is then a well known result in the Lie algebra case (see [Serr1]) and has been proved in [Serg1] in the more general context of Lie superalgebras. $\qquad\square$

The proof of Theorem 4.2.3 by Kac and Peterson for Kac-Moody Lie algebras (used in the above proof) relies on the representation theory of Kac-Moody Lie algebras, and so we leave it as an exercise in section 2.6. The uniqueness of the Cartan matrix is a classical result in the framework of finite dimensional simple Lie algebras. However as the next example shows, it does not necessarily hold in the larger context of BKM superalgebras.

Example 2.4.6. Both

$$\begin{pmatrix} 0 & -1 \\ -1 & 0 \end{pmatrix} \quad \text{with} \quad S = I$$

and

$$\begin{pmatrix} 0 & -1 \\ -1 & 2 \end{pmatrix} \quad \text{with} \quad S = \{1\}$$

are generalized Cartan matrices of $A(1,0)$ (see Corollary 2.1.23).

In general, when the odd part is non-trivial, a finite dimensional BKM super-algebra may also be decomposable as the sum of eigenspaces for an abelian Lie subalgebra H in such a way that it is generated by H and eigenvectors e_i, f_i satisfying Serre-Chevalley relations (though these may not be sufficient for a presentation) and the matrix of the bilinear form with respect to this decomposition is not a generalized Cartan matrix according to definition 2.1.7. For example,

$$\begin{pmatrix} 0 & -1 & 0 \\ -1 & 2 & -1 \\ 0 & -1 & 0 \end{pmatrix}$$

is a generalized Cartan matrix for the Lie superalgebra $A(1,1)$; and the latter also has a decomposition with respect to which the bilinear form is given by the matrix

$$\begin{pmatrix} 2 & -1 & 0 \\ -1 & 0 & 1 \\ 0 & 1 & -2 \end{pmatrix}.$$

Before stating the uniqueness result, we need to study the conjugacy of bases under the action of the Weyl group.

Theorem 2.4.7. *If* $\dim H < \infty$ *and* $|\Delta| = \infty$, *then all bases of* Δ *are conjugate to the base* Π *or* $-\Pi$ *under the action of the Weyl group* W, *where* $\Pi = \{\alpha_i, i \in I\}$.

Proof. By Corollary 2.3.45, $W_E = W$ since $|\Delta| = \infty$. Let Π_1 be another set of simple roots in Δ with respect to which the generalized symmetric Cartan matrix is $B = (b_{ij})$, and Δ_1^+ the set of positive roots with respect to Π_1. Since the matrix A is indecomposable, so is B by Lemma 2.1.15. By Corollary 2.3.46, the set of roots Δ contains roots of infinite type. We consider two cases.

Case A: All roots of infinite type have norm 0.

Let α be a root of infinite type and assume that α is of minimal height in $W(\alpha)$. Hence $(\alpha, \alpha_i) = 0$ for all $i \in \text{supp}(\alpha)$ and so

$$\forall i \in \text{supp}(\alpha), \ a_{ii} > 0 \tag{1}$$

(see Exercise 2.3.5). Suppose that there is an index $i \in I - \text{supp}(\alpha)$. Then as the set I has been assumed to be connected, we may assume that $(\alpha, \alpha_i) < 0$. Hence $\alpha + \alpha_i$ is a root. If $a_{ii} \leq 0$, then the root $\alpha + \alpha_i$ has negative norm and so by Theorem 2.3.44, it is of infinite type, contradicting assumption. Therefore $a_{ii} > 0$ and

$$(\alpha + \alpha_i, \alpha + \alpha_i) = (\frac{2(\alpha, \alpha_i)}{(\alpha_i, \alpha_i)} + 1)(\alpha_i, \alpha_i) \leq 0$$

since $\frac{2(\alpha, \alpha_i)}{(\alpha_i, \alpha_i)}$ is a non-positive integer by above assumption on the simple root α_i and definition of a generalized symmetric Cartan matrix. The previous argument thus forces

$$(\alpha + \alpha_i, \alpha + \alpha_i) = 0. \tag{2}$$

Since $j \in \text{supp}(\alpha)$, $a_{jj} > 0$ by condition (1), Proposition 2.3.27 (ii) implies that the root $\alpha + \alpha_i$ is not of finite type. Therefore as $(\alpha, \alpha + \alpha_i) < 0$, from Theorem 2.3.33 we can deduce that $2\alpha + \alpha_i$ is a root. Equality (2) shows that this root has negative norm and thus this again contradicts assumption because of Theorem 2.3.44. As a consequence,

$$I = \text{supp}(\alpha).$$

By Exercise 2.3.5, the matrix A is thus the generalized Cartan matrix of an affine BKM superalgebra. So the result holds by Exercise 2.4.2.

Case B: There exists a root of infinite type having non-zero norm.

Claim 1: The roots of non-positive norm in Δ^+ are either all in Δ_1^+ or all in $-\Delta_1^+$.

Let $\alpha \in \Delta^+$ be a root having negative norm. By Theorem 2.3.44 it is of infinite type. Without loss of generality, assume that $\alpha \in \Delta_1^+$. Let $\beta \in \Delta^+$ be a root of non-positive norm. Lemma 2.3.26 (iii) gives $(\alpha, \beta) \leq 0$. It also shows that if $(\alpha, \beta) < 0$ then, $\beta \in \Delta_1^+$ since $\alpha \in \Delta_1^+$. Suppose that $(\alpha, \beta) = 0$. By Proposition 2.3.47, either β is a positive multiple of α and so is in Δ_1^+, or else the supports of α and β are disconnected and $(\alpha, \alpha_i) = 0$ (resp. $(\beta, \alpha_i) = 0$) for all $i \in \text{supp}(\beta)$ (resp. $i \in \text{supp}(\alpha)$). In the latter case, as the indexing set I is connected there exists a root of infinite type and negative norm $\gamma = \alpha + \alpha_{i_1} + ... \alpha_{i_s}$ such that $(\alpha, \gamma) < 0$ and $(\gamma, \beta) < 0$. So the above arguments applied to the roots α and γ imply that $\gamma \in \Delta_1^+$ and then applying the same reasoning to the roots γ and β, we can again deduce that $\beta \in \Delta_1^+$.

Without loss of generality, we may assume that all roots of non-positive norm are in Δ_1^+. Suppose that $i \in J$ and $\alpha_i \in (-\Delta_1^+)$. Then, $a_{ii} > 0$.

Claim 2: $|\Pi \cap (-\Delta_1^+)| < \infty$

Claim 1 implies that any submatrix of the generalized Cartan matrix A induced by finitely many simple roots in the set $\Pi \cap (-\Delta_1^+)$ is positive definite. Therefore, the elements h_i in $\Pi \cap (-\Delta_1^+)$ are linearly independent, and so Claim 2 follows since $\dim H < \infty$.

$|\Pi \cap r_i(-\Delta_1^+)| < |\Pi \cap (-\Delta_1^+)|$ since for $j \neq i$, $\alpha_j \in \Delta_1^+$ if and only if $r_i(\alpha_j) \in \Delta_1^+$. Hence Claim 2 implies that there is some $w \in W$ such that $w(\Pi) \leq \Delta_1^+$. We can then conclude from Definition 2.3.4 (v) of a base that $w(\Pi) = \Pi_1$. $\qquad \square$

The uniqueness of the generalized (symmetric) Cartan matrix now follows.

Theorem 2.4.8. *Assume that* $\dim H < \infty$. *Let* $G = G(A, H_A, S_A)$ *be a BKM superalgebra. If the derived sub-superalgebra* $[G, G]$ *is finite dimensional, then assume that it is not of type* $A(1, 0)$. *If* B *is a generalized symmetric Cartan matrix and* $G = G(B, H_B, S_B)$, *then the matrices* A *and* B *are equivalent.*

Proof. By Exercise 1, for any $h \in H$, $i \in I$ and $x \in G_{\pm\alpha_i}$,

$$[\mathrm{Ad}\,(g)h, \mathrm{Ad}\,(g)(x)] = \mathrm{Ad}\,(g)\mathrm{ad}\,(h)(\mathrm{Ad}\,g)^{-1}(\mathrm{Ad}\,g)(x) = \pm(h_i, h)\mathrm{Ad}\,(g)(x).$$

Hence the result is an immediate consequence of Theorems 2.4.3 and 2.4.7. $\square$

Remark 2.4.9. It is important to note that for a finite dimensional BKM superalgebra $G = G(A, H, S)$, there may well be bases with respect to which the "Cartan" matrix B is not equivalent to the generalized Cartan matrix A. However in this case the matrix B would not be a generalized symmetric Cartan matrix, i.e. would not satisfy conditions (i)-(iv) given section 2.1. Example 2.4.6 illustrates this subtlety.

Theorems 2.4.3 and 2.4.7 also imply the following classical property of Borel sub-superalgebras.

Theorem 2.4.10. *Suppose that* $\dim H < \infty$ *and* $\dim G = \infty$. *The number of conjugacy classes is precisely two.*

It is well known that all Borel subalgebras of finite dimensional Lie algebras are conjugate under the action of inner automorphisms. For infinite dimensional Kac-Moody Lie algebras, there are two conjugacy classes of Borel subalgebras as shown in [KacP6]. The above result generalizes this property to the context of infinite dimensional BKM superalgebras. Most finite dimensional BKM superalgebras with non-trivial odd part have more than two conjugacy classes of Borel sub-superalgebras [Serg1] except $B(0, n)$ that has one as all its roots have positive norm and $B(1, 1)$ that has two. For example, if G is of type $A(1, 0)$, then there are three conjugacy classes of Borel subalgebras [Exercise 2.4.2].

Hence for finite dimensional simple Lie superalgebras, the situation differs much from the Lie algebra one. This suggests that the structure of infinite dimensional BKM superalgebras is closer to that of Kac-Moody Lie algebras than to finite dimensional Lie superalgebras with non-trivial odd part.

As a consequence of Theorem 2.4.8, the rank and the Kac-Moody rank are well defined concepts.

Definition 2.4.11. *Suppose that* $\dim H < \infty$. *The rank (resp. Kac-Moody rank) of the BKM superalgebra* $G = G(A, H, S)$ *is defined to be the dimension of the subspace generated by the elements* h_i, $i \in I$ *(resp.* $i \in J$).

Corollary 2.4.12. *If* $\dim H < \infty$, *then the rank and Kac-Moody rank are well defined.*

Automorphisms of BKM superalgebras

The above results allow us to give the decomposition of an arbitrary automorphism of the BKM superalgebra G. We suppose that $\dim H < \infty$.

Note that the Weyl group W acts on the subspace of H generated by the elements h_i, $i \in I$, in a natural way. For each $i \in I$, set

$$I_i = \{j \in I : a_{kj} = a_{ki}, \quad \forall k \in I\}.$$

This set labels all the columns equal to the i-th one.

We choose a set of representatives from each class of indices I_i:

$$\hat{I} = \{i \in I : i \leq j, \quad \forall j \in I_i\}.$$

The vectors e_j (resp. f_j), $j \in I_i$ form a basis of G_{h_i} (resp G_{-h_i}). For each $i \in \hat{I}$, for each bijective linear map μ_i of the vector space $G_{h_i} \oplus G_{-h_i}$ satisfying $\mu_i(G_{h_i}) = G_{h_i}$, $\mu_i(G_{-h_i}) = G_{-h_i}$, and $[\mu_i(e_j), \mu_i(f_j)] = \mu_i([e_j, f_j])$ for all $j \in I_i$, let $\hat{\mu}_i$ be the automorphism of G given by

$$\hat{\mu}_i(e_j) = \begin{cases} \mu_i(e_j), & \text{if } j \in I_i \\ e_j, & \text{otherwise} \end{cases}$$

$$\hat{\mu}_i(f_j) = \begin{cases} \mu_i(f_j), & \text{if } j \in I_i \\ f_j, & \text{otherwise} \end{cases}$$

For each $i \in I$, we choose an ordering of $I_i = \{j_1, ..., j_{n_i}\}$, where $n_i = \dim G_{h_i}$, and write $e_i^{(s)} = e_{j_s}$ and $f_i^{(s)} = f_{j_s}$. Each bijection $\dot{d}$ of the indexing set $\hat{I}$ satisfying $a_{\dot{d}(i)\dot{d}(j)} = a_{ij}$ for all $i, j \in \hat{I}$ and $\dot{d}(S) = S$ gives rise to an automorphism d of G called a *diagram automorphism* and satisfying $d(e_i^{(s)}) = e_{\dot{d}(i)}^{(s)}$ and $d(f_i^{(s)}) = f_{\dot{d}(i)}^{(s)}$ for all $i \in \hat{I}$ and $1 \leq s \leq n_i$.

We write $AutG$ for the group of automorphisms of the BKM superalgebra G.

Theorem 2.4.13. *Suppose that* $\dim H < \infty$ *and* $|\Delta| = \infty$.

(i) *Let* ϕ *be an automorphism of the BKM superalgebra* G. *Then there is a diagram automorphism* ϕ_1, *an inner automorphism* ϕ_2 *of* G, *and for each* $i \in \hat{I} - J$, *automorphisms* μ_i *of* G *such that*

$$\phi = (\prod_{i \in \hat{I} - J} \mu_i) \omega^i \phi_1 \phi_2 \quad \text{for some integer} \quad 0 \leq i \leq 3.$$

This decomposition (in the given order) is unique.

(ii) *Let* $A_i = \{\hat{\mu}_i : \mu_i \in GL(G_{h_i})\}$. *This is a group isomorphic to the general linear group* $GL(n_i)$, *where* $n_i = \dim G_{h_i}$; *and*

$$AutG = <\omega> \times (D \ltimes (\bigsqcup A_i \times In(G))).$$

The proof follows from Exercise 2.4.3.

Remark 2.4.14. In the above definition of diagram automorphisms, we could have as is more usual included the maps μ_i by considering the entire indexing set I rather than $\hat{I}$. However, roots are more naturally considered as elements of the Cartan subalgebra H than of the abstract root lattice Q (see Example 2.3.11.1 of the Monster Lie algebra) and so the above statement is more convenient.

Exercises 2.4.

1. (i) Show that $Inn(G) \cong \mathcal{G}/Z$, where Z is the centre of $\mathcal{G}$.

(ii) Let V be an integrable G'-module and $\pi : G \to gl(V)$ the corresponding representation. Write also π for the representation of $\mathcal{G}$ given by $\pi(N+g) = \pi^*(g)$ for $g \in \mathcal{G}^*$. Show that $\pi(\mathrm{Ad}\,(g)x) = \pi(g)\pi(x)\pi(g)^{-1}$ for $x \in G$ and $g \in \mathcal{G}$.

2. Suppose that G is a finite dimensional BKM superalgebra and that $S \neq \emptyset$. In each case, find the number of conjugacy classes of Borel sub-superalgebras under the action of inner automorphisms.

3. Suppose that $\dim H < \infty$. Let $\phi \in Aut(G)$.

(i) Show that there exists an automorphism $\tau \in Inn(G)$ such that $\psi_1 = \phi\tau^{-1}$ fixes the subalgebra H.

(ii) Deduce that there are automorphisms $\phi_1 \in Inn(G)$, $d \in D$ and an integer $0 \leq l \leq 3$ such that $\phi\phi_1\omega^l d(G_\alpha) = G_\alpha$ for all roots $\alpha \in H$.

(iii) Deduce the decomposition of the automorphism ϕ given in Theorem 2.4.13.

(iv) Show that the group of $Inn(G)$ is normal in $Aut(G)$.

(v) Deduce the decomposition of the group $Aut(G)$ given in Theorem 2.4.13.

(vi) Show that the Chevalley automorphism is an inner automorphism if and only if G is a finite dimensional simple Lie algebra or is of type $B(0,n)$.

4. Show that Theorem 2.4.7 holds for affine BKM superalgebras.

2.5 A Characterization of BKM Superalgebras

It is usually very hard to apply Definition 2.1.7 in terms of generators and relations to a given Lie superalgebra in order to find whether it is a BKM superalgebra or not. Hence it is useful to find different characterizations of BKM superalgebras. Theorem 2.2.9 that gives a characterization in terms of an almost positive definite contravariant bilinear form is very powerful as it is needed in the proof of the Moonshine Theorem to show that the Monster Lie algebra [Example 2.3.11.1] is a BKM algebra. However, as shown in Theorem 2.2.20, this result applies to all BKM algebras but only to a restricted class of BKM superalgebras. Even in the case of a given Lie algebra, though there might well exist a bilinear form and an involution with the adequate properties, it may not be easy to construct them. Note that even if one finds a contravariant bilinear form with respect to some involution, the said involution may not be an adequate one. Hence it is important to find different characterizations. Here we give one based on properties of the root system and spaces. It is helpful when a given Lie superalgebra is constructed from a root lattice to find whether it is a BKM superalgebra or not. And it so happens that so far, most of the known concrete examples of BKM superalgebras that are not Kac-Moody Lie algebras are constructed in this manner, via vertex algebras (see section 5.4).

We start by selecting the typical properties of root systems and spaces of BKM superalgebras.

Lemma 2.5.1.
(i) *The Cartan subalgebra H is self-centralizing.*
(ii) *There is an element $h \in H$ such that $C_G(h) = H$, where $C_G(h)$ is the centralizer of h in G.*

Proof. (i): Let $x \in G - H$ be such that $[x, h] = 0$ for all $h \in H$. Then, $([\omega(x), x], h) = (\omega(x), [x, h]) = 0$ for all $h \in H$, where ω is the Chevalley automorphism. Hence $[x, \omega(x)] = 0$ since $[x, \omega(x)] \in H$ and the bilinear form in non-degenerate on H. However, for any $\alpha \in \Delta$, if $x \in G_\alpha$, then $[x, \omega(x)] \neq 0$ by Exercise 2.3.8. This proves the first part.

(ii): Consider the simple roots as elements of the dual space H^*, i.e. $\alpha_i(h) = (h, h_i)$ for $h \in H$. As the bilinear form is non-degenerate on H, for any $i \in I$, $Ker(\alpha_i) \neq H$. Therefore, since the indexing set is either finite or countably infinite, $\cup_{i \in I} Ker(\alpha_i) \neq H$. Any element in $H - \cup_{i \in I} Ker(\alpha_i)$ has the required property. □

Definition 2.5.2. *Any element of the Cartan subalgebra satisfying condition (ii) of Lemma 2.5.1 is said to be regular.*

The following is an immediate consequence of Lemma 2.2.9.

Lemma 2.5.3. *Suppose that there are only finitely many indices $i \in I$ such that $a_{ii} > 0$. The norms of the roots of a BKM superalgebra are bounded above.*

Lemmas 2.5.1-2.5.3 together with Lemma 2.3.26 (iii) and Proposition 2.3.46 and the existence of a non-degenerate supersymmetric consistent bilinear form essentially characterize BKM superalgebras [Borc8, Ray3].

Theorem 2.5.4. *Any Lie superalgebra L satisfying the following conditions is a BKM superalgebra.*

1. *L has a self centralizing even subalgebra H with the property that L is the direct sum of eigenspaces of H, and all the eigenspaces are finite dimensional. A root of G is defined to be a nonzero eigenvalue of H (roots belong to the dual space of H).*
2. *There is a nondegenerate invariant supersymmetric bilinear form $(.,.)$ defined on L.*
3. *There is an element $h \in H$ such that $C_G(h) = H$ and there only exist finitely many roots α of L with $|\alpha(h)| < r$ for any $r \in \mathbf{R}$. If $\alpha(h) > 0$ (resp. $\alpha(h) < 0$), α is called a positive (resp. negative) root.*
4. *The norms of roots are bounded above.*
5. *Let α and β be both positive or both negative roots of non-positive norm. Then, $(\alpha, \beta) \leq 0$. Moreover, if $(\alpha, \beta) = 0$ and if $x \in G_\alpha$ and $[x, G_{-\gamma}] = 0$ for all roots γ such that $0 < \gamma(h) < \alpha(h)$, then $[x, G_\beta] = 0$.*

We prove Theorem 2.5.4 in several steps. L will denote a Lie superalgebra satisfying conditions $(1) - (5)$ of the preceding Theorem. For any subspace U

of the Lie superalgebra L, we will write $U_{\bar{i}}$ for $U \cap L_{\bar{i}}$ ($\bar{i} = \bar{0}$ or $\bar{1}$), i.e. for the even and odd parts of U. We write L_α for the α-root space and Δ for the set of roots. We take $\alpha \in H^*$.

The existence of regular elements in H allows the use induction. The method of proof is then similar to that of Theorem 2.2.9.

Let $h \in H$ be a fixed regular element. If x is a homogeneous element of L in the root space of $\alpha \in \Delta$, then we define the degree of x to be: $\deg x := \alpha(h)$. This gives a gradation of L:

$$L = \oplus_{r \in \mathbf{R}} L_r.$$

Let $r \in \mathbf{R}$ be a given real number, and P and M be the Lie sub-superalgebra of L generated by the subspaces L_s for $|s| < r$ and $|s| \le r$ respectively. Assume that P is a BKM superalgebra. To prove Theorem 2.5.4 it suffices to show that M is also a BKM superalgebra. We start with basic properties of the bilinear form.

Lemma 2.5.5.

   (i) If $\alpha, \beta \in \Delta \cup \{0\}$, then $(L_\alpha, L_\beta) = 0$ unless $\alpha + \beta = 0$. In particular the bilinear form $(.,.)$ is non-degenerate on H.
   (ii) The bilinear form $(.,.)$ is consistent.

Proof. Let $\alpha, \beta \in \Delta \cup \{0\}$, $x \in L_\alpha$, $y \in L_\beta$. Then,

$$\alpha(h)(x,y) = ([h,x],y) = (x,[h,y]) = \begin{cases} 0 & \text{if } y \in H \\ -\beta(h)(x,y) & \text{if } y \in L_\beta \end{cases}$$

and so $(x,y) = 0$ unless $\alpha + \beta = 0$ since by definition of the regular element h, $\alpha(h) \ne 0$. Hence condition 1 implies that $(.,.)$ is nondegenerate on H.

If $x \in L_\alpha \cap L_{\bar{0}}$ and $y \in L_{-\alpha} \cap L_{\bar{1}}$, then $[x,y] \in H$ and so $[x,y] = 0$ since H is an even subspace by condition 1. Therefore,

$$0 = ([x,y],h) = (x,[y,h]) = \alpha(h)(x,y),$$

giving $(x,y) = 0$. And so the form is consistent.

Lemma 2.5.6. *There exists a set of linearly independent elements $\{e_i\}_{i \in I}$ and $\{f_i\}_{i \in I}$ in M, satisfying $(e_i, f_j) = \delta_{ij}$, and $(e_i, e_j) = 0 = (f_i, f_j)$ for all i, j, such that together with a basis of H, they generate M as a Lie superalgebra.*

Proof.   Since H is an even subalgebra, the homogeneous subspaces $(L_r)_{\bar{i}}$, $i = 0, 1$ are eigenspaces for H. Let $(U_r)_{\bar{i}} = P \cap (G_r)_{\bar{i}}$, $i = 0, 1$. Similarly for U_{-r}. Set

$$(V_r)_{\bar{i}} = \{x \in (L_r)_{\bar{i}} : (x,y) = 0, \forall y \in P\}.$$

Similarly for V_{-r}. Let K be the kernel of the restriction of the bilinear form to P. By Lemma 2.5.5 (i) and as the Lie superalgebra P is a BKM superalgebra, by Theorem 2.1.17, $K = 0$ since it intersects H trivially; and so

$$(L_r)_{\bar{i}} = (U_r)_{\bar{i}} \oplus (V_r)_{\bar{i}}.$$

Moreover, H leaves $(V_r)_{\bar{i}}$ invariant since for any $x \in (V_r)_{\bar{i}}$, $y \in P$, $z \in H$, $([z, x], y) = -(x, [z, y]) = 0$ since $[z, y] \in L$.

Hence, for $i = 0, 1$, there exists a linearly independent set of elements in $(V_r)_{\bar{i}}$ of eigenvectors for H, which are orthogonal to P. Let $\{e_j\}$ be this set (they form a basis for $(V_r)_{\bar{i}}$). Furthermore, since the form is consistent by Lemma 2.5.5 (ii), there exists a linearly independent set of elements in $(L_{-r})_{\bar{i}} - L$ of eigenvectors of H, which is dual to the set $\{e_j\}$ with respect to $(.,.)$. We denote this set $\{f_j\}$. We then add these elements to the set of elements e_i, f_i already chosen as generators of the derived Lie superalgebra $[P, P]$.   □

In the rest of the Proof, $\{e_i\}_{i \in I}$, and $\{f_i\}_{i \in I}$ will denote fixed sets of generators, as described in Lemma 2.5.6. We denote by $\alpha_i \in H^*$ the root corresponding to e_i. Set

$$h_{ij} = [e_i, f_j], \quad h_i = h_{ii},$$

and let A be the real matrix with entries $a_{ij} = (h_i, h_j)$. We next show that the matrix A and the above generators satisfy Definition 2.1.7 of a BKM superalgebra.

Lemma 2.5.7. *The matrix A is symmetric.*   □

Proof. This result follows since H is an even subalgebra and the form is supersymmetric.   □

Lemma 2.5.8.

(i) $[h_i, e_j] = a_{ij} e_j$, and $[h_i, f_j] = -a_{ij} f_j$.
(ii) If $i \neq j$ then $h_{ij} = [e_i, f_j] = 0$.

Proof. (i): Since e_j is an eigenvector of H, there is some scalar $c \in \mathbf{R}$ such that $[h_i, e_j] = ce_j$. Hence

$$c = (ce_j, f_j) = ([h_i, e_j], f_j) = (h_i, [e_j, f_j]) = (h_i, h_j) = a_{ij}.$$

Similarly $[h_i, f_j] = -a_{ij} f_j$, proving (i).

(ii): If $e_i, f_j \in P$, then this is true by assumption. So, assume that $e_i \in V_r$.

If $f_j \in P$, then $[e_i, f_j] \in L_s$ where $s < r$. Let $x \in L_{-s}$, so that

$$([e_i, f_j], x) = (e_i, [f_j, x]) = 0$$

as $[f_j, x] \in P$ and e_i is orthogonal to P by definition. So $[e_i, f_j] = 0$ since $(.,.)$ is nondegenerate on $L_s \oplus L_{-s}$.

Now suppose that $f_j \in V_{-r}$. Hence $[h, [e_i, f_j]] = 0$. Thus since H is the centralizer of h in L, $[e_i, f_j] \in H$. So since $(e_i, f_j) = 0$ by duality of the sets $\{e_i\}$ and $\{f_i\}$, for any $z \in H$,

$$([e_i, f_j], z) = (e_i, [f_j, z]) = 0.$$

Therefore as $(.,.)$ is nondegenerate on H, $h_{ij} = 0$.

The case $f_j \in V_{-r}$ and $e_i \in L$ is dealt with in a similar manner.   □

We next consider roots of positive norm.

Lemma 2.5.9. *Suppose that $a_{ii} > 0$.*

(i) *If, $e_i, f_i \in L_{\bar{0}}$, then $2\frac{a_{ij}}{a_{ii}}$ is a non-positive integer. Moreover, $(\mathrm{ad}\,(e_i))^{1-\frac{a_{ij}}{a_{ii}}} e_j = 0$; and $(\mathrm{ad}\,(f_i))^{1-\frac{a_{ij}}{a_{ii}}} f_j = 0$.*

(ii) *If $e_i, f_i \in L_{\bar{1}}$, then $\frac{a_{ij}}{a_{ii}}$ is a non-positive integer. Moreover, $(\mathrm{ad}\,(e_i))^{1-2\frac{a_{ij}}{a_{ii}}} e_j = 0$; and $(\mathrm{ad}\,(f_i))^{1-2\frac{a_{ij}}{a_{ii}}} f_j = 0$.*

Proof. (i): The Lie superalgebra S_i generated by the elements e_i and f_i is isomorphic to sl_2. Consider the S_i-module generated by the element e_j. If the result does not hold, then the representation theory of sl_2 implies that $n\alpha_i + \alpha_j$ is a root for all integers $n \geq 0$, contradicting condition 4. The second part of (i) can proved in a similar fashion.

(ii): The element $E_i = [e_i, e_i] \neq 0$, for otherwise

$$0 = [f_i[e_i, e_i]] = [[f_i, e_i]e_i] - [e_i[f_i, e_i]] = 2a_{ii}e_i.$$

Similarly $F_i = [f_i, f_i] \neq 0$. The Lie superalgebra S_i generated by the elements E_i and F_i is isomorphic to sl_2 and now (ii) follows from arguments similar to those in (i). □

Lemma 2.5.10. *If $a_{ij} = 0$, then $[e_i, e_j] = 0 = [f_i, f_j]$.*

Proof. If either $a_{ii} > 0$ or $a_{jj} > 0$, then the result follows from Lemma 2.5.9. Otherwise it is a direct consequence of condition 5. □

This proves Theorem 2.5.4.

When there are infinitely many simple roots of positive norm, there may not necessarily be an upper bound for the norms of the roots. We can replace Conditions 4 and 5 of Theorem 2.5.4 by root type conditions so as to be sure to include all BKM superalgebras.

Corollary 2.5.11. *Any Lie superalgebra L satisfying the following conditions is a BKM superalgebra.*

1. *L has a self centralizing even subalgebra H, L is the direct sum of eigenspaces of H, and all the eigenspaces are finite dimensional. A root of G is defined to be a nonzero eigenvalue of H (roots belong to the dual space of H).*

2. *There is a nondegenerate invariant supersymmetric bilinear form $(.,.)$ defined on L.*

3. *There is an element $h \in H$ such that $C_G(h) = H$ and there only exist finitely many roots α of L with $|\alpha(h)| < r$ for any $r \in \mathbf{R}$. If $\alpha(h) > 0$ (resp. $\alpha(h) < 0$), α is called a positive (resp. negative) root.*

4. *A root is either of finite type (according to Definition 2.3.17) or else it is said to be of infinite type.*

5. *Let α and β be either infinite type roots or of norm 0, both positive or both negative. Then, $(\alpha, \beta) \leq 0$. Moreover, if $(\alpha, \beta) = 0$, then for all $x \in G_\alpha$ satisfying $[x, G_{-\gamma}] = 0$ for all roots γ such that $0 < |\gamma(h)| < |\alpha(h)|$, $[x, G_\beta] = 0$.*

The proof of this result is similar to that of Theorem 2.5.4.

Remark 2.5.12. 1. If the first part of condition 5 of Corollary 2.5.11 only holds for roots of infinite type, then L is either a BKM superalgebra or a finite dimensional simple classical Lie superalgebra or an affine Lie superalgebra. The latter are not necessarily BKM superalgebras since they may not have "Cartan" matrices with all non diagonal entries non-positive.

2. As the next example shows, even though it would be more satisfying, characterizing a Lie superalgebra solely in terms of root chain lengths would include too large a class of Lie superalgebras: when the simple root α_i is of infinite type, it would not follow that $a_{ii} \leq 0$. Therefore, Theorem 2.5.4 and Corollary 2.5.11 are better characterizations.

Example 2.5.13. Let $G(A, H)$ be the Lie algebra satisfying properties (1) and (2) of Definition 2.1.7 and with $A = \begin{pmatrix} 2 & -1 \\ -1 & 3 \end{pmatrix}$. Then, the root α_1 is of finite type, and the root α_2 is of infinite type in the sense of Theorem 2.3.33, though it has positive norm.

Open Problem. The above two characterizations given in §2.2 and 2.4 are useful in some contexts but not in all. So, it would be worthwhile to find other characterizations of BKM superalgebras in order to be able to tell when a given Lie superalgebra is a BKM superalgebra.

Exercises 2.5.

1. Show that the even part $G_{\bar{0}}$ of a BKM superalgebra is a BKM algebra.

Hint: Show that the conditions of Corollary 2.5.11 hold for the Lie algebra $G_{\bar{0}}$.

2. Suppose that L is a Lie superalgebra satisfying conditions 1-4 of Corollary 2.5.11. Let α and β be either infinite type roots or of norm 0. If $x \in G_\alpha$ satisfying $[x, G_{-\gamma}] = 0$ for all roots γ such that $0 < |\gamma(h)| < |\alpha(h)|$, $[x, G_\beta] = 0$. Let the Lie superalgebra M and the reals a_{ij} be defined as above.

  (i) Show that if $a_{ii} = 0$ implies that $a_{ij}a_{ik} \geq 0$ for all j, k then M is a BKM superalgebra.
  (ii) Suppose that there is some i such that $a_{ii} = 0$, $a_{ij} < 0$ and $a_{ik} > 0$.
    (a) Show that the roots α_j of M with non-zero norm are of finite type and that if $a_{jj} = 0$, then $[e_j, e_j] = 0$.
    (b) Show that all non-zero norm root spaces of M have dimension 1.
    (c) Show that all norm 0 root spaces have finite dimension.
    (d) Deduce that the Lie superalgebra M is contragredient and has finite growth.
  (iii) Deduce that the Lie superalgebra M is a direct sum of a BKM superalgebra and of finite dimensional classical Lie superalgebras with symmetrizable Cartan matrix (see [Kac3]).

For a solution, see [Ray3].

3. Let G be a BKM superalgebra with a finite dimensional Cartan subalgebra H. Let Ω be a finite group of diagram automorphisms of G. Assume that when the Cartan matrix of G has a principal submatrix of affine type $C^{(2)}(n+1)$, $n \geq 1$, all elements of Ω act trivially on the corresponding sub-superalgebra. Show that the sub-superalgebra G^Ω of G of elements fixed by Ω is a BKM superalgebra.

Hint: Use Exercise 2. For a solution, see [Ray3].

2.6 Character and Denominator Formulas

Highest weight representations form an important class of representations of BKM superalgebras. They lead to the denominator formula, which is central to the theory of BKM superalgebras because it not only contains the essential information about the corresponding Lie superalgebra but also because it provides the link with automorphic forms (see Chapter 5). This denominator formula plays an essential role in the proof of the Moonshine Theorem.

We start with universal enveloping algebras as they are basic objects in representation theory but do not give proofs as they are well known standard ones (for details see [Jac]). The first result defines the structure of a Lie superalgebra on a $\mathbf{Z}_2$-graded associative algebra.

Lemma 2.6.1. *Let L be a $\mathbf{Z}_2$-graded associative algebra. If x, y are homogeneous elements,*

$$[x, y] = xy - (-1)^{d(x)d(y)}yx$$

gives the structure of a Lie superalgebra to L.

This allows us to now define the tensor superalgebra of a Lie superalgebra.

Definition 2.6.2. *Let L be a Lie superalgebra. The $n-th$-tensor product $T_n(L)$ of L is the vector space spanned by the elements $x_1 \otimes \cdots \otimes x_n$, $x_i \in L$. It is an induced $\mathbf{Z}_2$-grading. The tensor Lie superalgebra is the Lie superalgebra $T(L)$ with underlying vector space $\bigsqcup_{n=0}^{\infty} T_n(L)$ with associative structure given by*

$$(x_1 \otimes \cdots \otimes x_n)(y_1 \otimes \cdots \otimes y_m) = x_1 \otimes \cdots \otimes x_n \otimes y_1 \otimes \cdots \otimes y_m$$

and Lie superalgebra structure given by Lemma 2.6.1.

We next give the definition of the universal enveloping superalgebra and show how it can be constructed from the tensor Lie superalgebra.

Definition 2.6.3. *A universal enveloping superalgebra of the Lie superalgebra L is an associative $\mathbf{Z}_2$-graded algebra $U(L)$ with 1 together with a linear map: $p : L \longrightarrow U(L)$ satisfying*

$$p([x, y]) = xy - (-1)^{d(x)d(y)}yx \qquad (i)$$

and such that for any other associative algebra $V(L)$ with 1 and any linear map
$q : L \longrightarrow V(L)$ satisfying (i), there is a unique homomorphism of $\mathbf{Z}_2$-algebras
$f : U(L) \longrightarrow V(L)$ such that $f \circ p = q$ and $f(1) = 1$.

Proposition 2.6.4. *The universal $\mathbf{Z}_2$-graded algebra is uniquely defined. Let*
K be the two-sided ideal of $T(L)$ generated by the elements

$$[x, y] - x \otimes y - (-1)^{d(x)d(y)} y \otimes x,$$

where $x, y \in L$ are homogeneous elements. Then, $U(L) = T(L)/K$.

Lemma 2.6.5. *The map $p : L \to U(L)$ is injective.*

This last result allows us to identify the Lie superalgebra L with its image
$p(L)$ in $U(L)$. The Poincaré-Birkhoff-Witt Theorem holds [Jac, pp. 159–160].
As the proof remains the same as in the Lie algebra case, we leave it for the
reader to check. Note simply that if y is an odd element of the Lie superalgebra
L, then $y^3 = \frac{1}{2}[y, y]y$ in $U(G)$ and that $[y, y]$ is an even element.

Theorem 2.6.6 [PBW Theorem]. *Let L be a Lie superalgebra and $x_1, ... x_m$*
(resp. $y_1, ..., y_n$) be an ordered basis for the even part $L_{\bar{0}}$ (resp. odd part $L_{\bar{1}}$),
then the vectors

$$x_1^{i_1} ... x_m^{i_m} y_1^{j_1} ... y_n^{j_n}, \quad i_k \in \mathbf{Z}_+ \quad \forall \ 1 \le k \le m, \qquad 0 \le j_l \le 1 \quad \forall \ 1 \le l \le n$$

form a basis for $U(L)$.

Together with Corollary 2.1.19, the PBW Theorem allows us to give the
structure of the enveloping $\mathbf{Z}_2$-graded universal algebra of the BKM superalgebra G.

Corollary 2.6.7. *The universal enveloping algebra of the BKM superalgebra G*
decomposes as follows:

$$U(G) = U(N_-) \otimes U(H) \otimes U(N_+).$$

We next remind the reader of the elementary notion of a module for a Lie
superalgebra.

Definition 2.6.8. *A representation of a Lie superalgebra L is a homomorphism*

$$G \to gl(V),$$

where V is a $\mathbf{Z}_2$-graded vector space. The space V is called a L-module.

For the BKM superalgebra G, we want to study the class of G-modules that
generalizes finite dimensional modules. Suppose that V is a finite dimensional
G-module and that G is finite dimensional. Since the Cartan subalgebra H is
abelian, H acts semisimply on V. The eigenvalues are elements in the dual of H
and are called weights. As the module V is finite dimensional, there is a highest
weight, namely there is weight $\Lambda \in H^*$ such that $V_{\Lambda + \alpha_i} = 0$ for all $i \in I$. Here
V_λ is the eigenspace or weight space with eigenvalue $\lambda \in H^*$. The G-module V

has a unique highest weight if and only if it is irreducible, in which case V is said to be a highest weight module. This notion of highest weight modules can be generalized to infinite dimensional BKM superalgebras. We do this now.

Remark 2.6.9. In section 2.6, roots can be considered to be elements of either Q, H or its dual H^*. We will use the notation α_i for the simple roots.

Definition 2.6.10. *The G-module $V(\Lambda)$ is said to be a highest weight module with highest weight $\Lambda \in H$ if $V(\Lambda)$ is generated by a v satisfying:*

$$xv = 0, x \in G_\alpha, \alpha \in \Delta^+, \quad hv = (\Lambda, h)v, h \in H.$$

A vector $v \in V(L)$ with this property is called a highest weight vector.

A fundamental property of highest weight modules is that they are sums of finite dimensional eigenspaces of the generalized Cartan subalgebra H.

Definition 2.6.11. The eigenspaces

$$V_\lambda = \{v \in V(\Lambda)|hv = (\lambda, h)v, h \in H\}, \lambda \in H$$

are called *weight spaces*.

Lemma 2.6.12.

(i) *The generalized Cartan subalgebra H acts semisimply on $V(\Lambda)$. So*

$$V(\Lambda) = \oplus_{\lambda \in H} V_\lambda$$

(ii) *The weight spaces are finite dimensional. In particular $\dim V(\Lambda)_\Lambda = 1$.*

Proof. The first part is an immediate consequence of Definition 2.6.10. By Corollary 2.6.7, $V(\Lambda) = U(N_-)v$, where v is a highest weight vector of $V(\Lambda)$. This leads to the second part since the root spaces are finite dimensional. $\quad\square$

As we saw above, irreducible finite dimensional G-modules are highest weight modules. The converse also holds in finite dimension, namely that a finite dimensional highest weight module is irreducible. Furthermore it is well known that when G is a finite dimensional simple Lie algebra, its finite dimensional modules are completely reducible. As the next result shows, highest weight modules are always indecomposable.

Lemma 2.6.13. *All highest weight G-modules are indecomposable.*

Proof. Let $V(\Lambda)$ be a highest weight G-module and v a highest weight vector of V. Suppose that $V = V_1 \oplus V_2$, where the V_i's are G-submodules of V. Then, $v = v_1 + v_2$ for some $v_i \in V_i$. Now, for $h \in H$,

$$hv_1 + hv_2 = hv = \Lambda(h)v = \Lambda(h)v_1 + \Lambda(h)v_2$$

implies that $hv_i = \Lambda(h)v_i$. Hence v_1 and v_2 are highest weight and so by Lemma 2.6.12 (ii), v_1 and v_2 are multiple of v, giving a contradiction and proving the result. $\quad\square$

There are infinite dimensional highest weight modules even when the Lie superalgebra G is finite dimensional. Whatever the dimension of G, they are not generally irreducible, though by the previous Lemma they are indecomposable. This is well illustrated by the next example.

Example 2.6.14. Consider the G-module $M(0) = U(G)/K$, where K is the left ideal of $U(G)$ generated by the positive Borel sub-superalgebra B_+ of G. It is a highest weight G-module with highest weight 0 and highest weight vector $K + 1$. It is infinite dimensional since $M_\lambda \neq 0$ for all $\lambda \in H^*$ such that $\lambda \leq 0$. The submodule $N(0)$ generated by the vectors $v_i = K + f_i$, $i \in I$ is a proper G-submodule of $M(0)$. So $M(0)$ is not irreducible.

Assume that $|I| \geq 2$. We then show that $N(0)$ is not a direct sum of highest weight modules. The vectors v_i, $i \in I$, are highest weight vectors of $N(0)$ of weight $-\alpha_i$ and

$$N(0) = \{N + x : x \in U(N_-)N_-\}.$$

Hence the vectors v_i are the only highest weight vectors of $N(0)$ (up to non-trivial multiplication by a scalar) and if $a_{ij} \neq 0$, then the vector $N + [f_i, f_j] \in N(0)$ is not contained in $\oplus_{i \in I} U(G)(N + f_i)$. Moreover, the highest weight spaces being 1-dimensional, this also shows that the module $N(0)$ is not completely reducible.

The above example shows that complete reducibility is restricted to finite dimension even when the BKM superalgebra G is simple (see Exercise 2.1.4). Before going further, we need to define the category of modules that highest weight modules belong to. The previous example gives an indication what this category should be like. It should include all highest weight modules and their submodules.

Definition 2.6.15. *Let $\mathcal{O}$ be the category of $\mathbf{Z}_2$-graded G-modules V on which the Cartan subalgebra H acts semisimply with finite dimensional weight spaces, i.e.*

$$V = \oplus_{\lambda \in H} V_\lambda, \tag{1}$$

for all $\lambda \in H$, $\dim V_\lambda < \infty$, and there exist finitely many elements $\Lambda_i \in H$, $i = 1, ..., m$ such that the weight spaces V_λ are non-trivial only if $\lambda \leq \Lambda_i$ for some $1 \leq i \leq m$.

Finite sums of modules in the category $\mathcal{O}$ are also in $\mathcal{O}$. The same holds for quotients, submodules, and tensor products.

A first natural question is to find an expression for the dimensions of the weight spaces of a G-module $V \in \mathcal{O}$ and to be able to differentiate its odd and even weight spaces, in other words to find its character and its super-character. To describe the dimensions of the weight spaces V_λ, equality (1) in Definition 2.6.15 cannot be translated as $\sum_{\lambda \in H} \dim V_\lambda \lambda$ since this would lead to a confusion between the dimension of the weight space $\lambda + \mu$ and the sum of the dimensions of the λ-weight space and μ-weight space, i.e. between $V_{\lambda+\mu}$ and $V_\lambda \oplus V_\mu$. Hence the need to introduce formal exponentials $e(\lambda)$.

Definition 2.6.16. *Let $\mathcal{E}$ be the commutative associative algebra of formal series*

$$\sum_{\lambda \in H} x_\lambda e(\lambda)$$

for which there exist finitely many elements $\lambda_i \in H$, $i = 1, ..., m$ such that the coefficients x_λ are non-trivial only if $\lambda \leq \lambda_i$ for some $1 \leq i \leq m$. Multiplication is defined by $e(\lambda)e(\mu) = e(\lambda + \mu)$. The character and super-character of the G-module $V = V_{\bar{0}} \oplus V_{\bar{1}} \in \mathcal{O}$ are the elements of the algebra $\mathcal{E}$ defined respectively to be the formal sums:

$$\mathrm{ch}V = \sum_{\lambda \in H} \dim V_\lambda e(\lambda) \quad and \quad \mathrm{sch}\, V = \sum_{\lambda \in H} (\dim V_{\bar{0}\lambda} - \dim V_{\bar{1}\lambda}) e(\lambda).$$

When $V(\Lambda) \in \mathcal{O}$ is a highest weight module of highest weight Λ, it is assumed that $d(\Lambda) = \bar{0}$.

Before dealing with more general modules in $\mathcal{O}$, the first step is obviously to compute the character and super-character of irreducible highest weight modules. We will follow the method given in [Kac14, Chapters 9,10,11] in the Kac-Moody setup. The proofs that do not require any changes in the more general context of BKM superalgebras are left as exercises. It was first proved for Kac-Moody Lie algebras in [Kac2], for BKM algebras in [Borc4], for Kac-Moody Lie superalgebras in [Kac7], and for BKM superalgebras in [Ray1,6]. There is an equivalent proof that uses the cohomology theory of BKM superalgebras. For finite dimensional simple Lie algebras, see [Kos]. For Kac-Moody Lie algebras this is done in [GarL].

For the calculation of the character and super-character formulae we need at the start some means of telling when two irreducible highest weight modules are "the same" or "different". In other words, a basic requirement is to find the isomorphism classes of irreducible highest weight G-modules. This can be done by considering Verma modules.

Lemma 2.6.17. *For each element $\Lambda \in H$, there is, up to isomorphism, a unique G-module $M(\Lambda)$ of highest weight Λ, called the Verma module, such that any highest weight G-module is a homomorphic image of it. It is isomorphic to $U(G)/K$, where K is the left ideal in $U(G)$ generated by N_+ and the elements $h - \Lambda(h)$ for $h \in H$. As a $U(N_-)$-module, it is a free module generated by a highest weight vector. Furthermore, the Verma module has a unique maximal G-submodule $N(\Lambda)$.*

Note that as is shown in Example 2.6.14, the submodule $N(\Lambda)$ is not usually a highest weight module. The classification of irreducible highest weight modules is now an immediate consequence.

Theorem 2.6.18. *For each element $\Lambda \in H$, there is, up to isomorphism, a unique irreducible G-module $L(\Lambda)$ of highest weight Λ. It is isomorphic to $M(\Lambda)/N(\Lambda)$.*

We next fix some more notation that will be used throughout this section.

Notation. For $\alpha \in \Delta^+$, let $e_{\alpha,i}$, $1 \leq i \leq \dim G_\alpha$ be a basis for the root space G_α and $f_{\alpha,i}$, $1 \leq i \leq \dim G_\alpha$ the dual basis for $G_{-\alpha}$. Since the bilinear form is consistent, $f_{\alpha,i} \in G_{\bar{0}}$ if and only if $e_{\alpha,i} \in G_{\bar{0}}$.
Let $m_0(\alpha) = \dim G_\alpha \cap G_{\bar{0}}$ and $m_1(\alpha) = \dim G_\alpha \cap G_{\bar{1}} = \text{mult}(\alpha) - m_0(\alpha)$.
Set

$$R = \frac{\prod_{\alpha \in \Delta_0^+}(1 - e(-\alpha))^{m_0(\alpha)}}{\prod_{\alpha \in \Delta_1^+}(1 + e(-\alpha))^{m_1(\alpha)}} \quad \text{and} \quad R' = \frac{\prod_{\alpha \in \Delta_0^+}(1 - e(-\alpha))^{m_0(\alpha)}}{\prod_{\alpha \in \Delta_1^+}(1 - e(-\alpha))^{m_1(\alpha)}}.$$

For $V \in \mathcal{O}$, set $P(V) = \{\lambda \in H : V_\lambda \neq 0\}$.
Set $P^+ = \{\Lambda \in H : \forall i \in I, (\alpha, \alpha_i) \geq 0, \frac{2(\Lambda, \alpha_i)}{(\alpha_i, \alpha_i)} \text{ (resp. } \frac{(\Lambda, \alpha_i)}{(\alpha_i, \alpha_i)}) \in \mathbf{Z}_+ \text{ if } a_{ii} > 0$
and $i \in I \backslash S$ (resp. $i \in S$)$\}$.
For $\lambda = \sum_{i \in I} x_i \alpha_i \in H$, $x_i \in \mathbf{C}$, let $\text{ht}(\lambda) = \sum_{i \in I} x_i$ be the height of λ.

In view of Theorem 2.6.18, it seems sensible to first try and compute the character and super-character of the Verma module $M(\Lambda)$ and to use this information to find the character and super-character of the irreducible module $L(\Lambda)$.

Lemma 2.6.19. *For $\Lambda \in H$, the character and super-character of the Verma G-module $M(\Lambda)$ are as follows:*

$$\text{ch}(M(\Lambda)) = e(\Lambda)R^{-1}, \quad \text{sch}(M(\Lambda)) = e(\Lambda)R'^{-1}$$

Proof. Let v_Λ be a highest weight vector of the module $M(\Lambda)$.
The set of positive roots may be ordered. For example by setting

$$\deg \alpha_i = i,$$

for each degree we get finitely many roots. Let β_i (resp γ_i), $i = 1, 2, \ldots$ be a list of the positive even (resp. odd) roots. Then Lemma 2.6.17 implies that the vectors

$$f_{\beta_{1,1}}^{n_{1,1}} \ldots f_{\beta_{1,i_1}}^{n_{1,i_1}} f_{\beta_{2,1}}^{n_{2,1}} \ldots f_{\beta_{2,i_2}}^{n_{2,i_2}} \ldots f_{\gamma_{1,1}}^{k_{1,1}} \ldots f_{\gamma_{1,j_1}}^{k_{1,j_1}} f_{\gamma_{2,1}}^{k_{2,1}} \ldots f_{\gamma_{2,j_2}}^{k_{2,j_2}} (v_\Lambda)$$

such that

$$(n_{1,1} + \ldots n_{1,i_1})\beta_1 + (n_{2,1} + \ldots n_{2,i_2})\beta_2 + \ldots + (k_{1,1} + \ldots k_{1,j_1})\gamma_1$$
$$+ (k_{2,1} + \ldots k_{2,j_1})\gamma_2 + \ldots = \mu$$

and $n_{1,l} \in \mathbf{Z}^+$, $k_{i,l} = 0$ or 1, form a basis of the weight space $M(\Lambda)_{\Lambda-\mu}$. So

$$\text{ch}M(\Lambda) = e(\Lambda) \prod_{\alpha \in \Delta_0^+} (1 + e(-\alpha) + e(-2\alpha) + \ldots)^{m_0(\alpha)} \prod_{\alpha \in \Delta_1^+} (1 + e(-\alpha))^{m_1(\alpha)},$$

and

$$\text{sch}\, M(\Lambda) = e(\Lambda) \prod_{\alpha \in \Delta_0^+} (1 + e(-\alpha) + e(-2\alpha) + ...)^{m_0(\alpha)} \prod_{\alpha \in \Delta_1^+} (1 - e(-\alpha))^{m_1(\alpha)},$$

which imply the desired answers.   □

Our aim now is to find an expression for the character and super-character of the irreducible highest weight G-module $L(\Lambda)$ in terms of the character and super-character of the corresponding Verma module.

A finite dimensional module V has a composition series and so its character and super-character can be given in terms of the characters and super-characters of its irreducible factors. In infinite dimension, a G-module V in $\mathcal{O}$ may not necessarily have a composition series. So it requires a little work to deduce that its character (resp. super-character) is still the sum of the characters (resp. super-characters) of irreducible $L(\lambda)$, where λ is a primitive weight. For any $\gamma \in H$, there exist finitely many G-submodules of V

$$V = V_0 \supset V_1 \supset ... \supset V_{m_\gamma - 1} \supset V_{m_\gamma} = 0$$

such that either

$$V_{i-1}/V_i \cong L(\mu), \quad \text{for some} \quad \mu \geq \gamma \quad \text{or} \quad (V_{i-1}/V_i)_\mu = 0 \quad \forall \mu \geq \gamma.$$

Proposition 2.6.20. *We keep the above notation and assume that no root is both odd and even. Let $V \in \mathcal{O}$ be a G-module. For all $\lambda \in H$, let x_λ be the number of indices $1 \leq i \leq m_\gamma$ such that $V_{i-1}/V_i \cong L(\lambda)$, where $\gamma \leq \lambda$. The character and super-character of the G-module V satisfy the following:*

$$\text{ch}V = \sum_{\lambda \in H} x_\lambda \text{ch}L(\lambda) \quad \text{and} \quad \text{sch}\, V = \sum_{\lambda \in H} (-1)^{d(\lambda)} x_\lambda \text{sch}\, L(\lambda),$$

where $d(\lambda)$ is the parity of the weight λ in V. In particular, if $V = V(\Lambda)$ is a highest weight module with highest weight Λ, then $x_\Lambda = 1$.

Proof. The proof follows from the fact that x_λ does not depend on the weight $\gamma \in H$ chosen such that $\gamma \leq \lambda$, nor on the series.   □

The case when roots are elements in H and can be both odd and even can be extrapolated from the above in a straightforward manner but is cumbersome to write.

The weights $\lambda \in H$ for which $x_\lambda \neq 0$ have a special property.

Definition 2.6.21. *Let $V \in \mathcal{O}$. The vector $v \in V$ is said to be primitive if there exists a G-submodule U of V such that $v \notin U$ but $xv \in U$ for all $x \in N_+$. If moreover, $v \in V_\lambda$, then λ is called a primitive weight of V.*

Lemma 2.6.22. *$x_\lambda \neq 0$ for the module $V \in \mathcal{O}$ if and only if $\lambda \in H$ is a primitive weight of V.*

Proof. Suppose that $\lambda \in H$ is a primitive weight of V. There is a maximal index $1 \leq i \leq m_\lambda - 1$ such that V_i contains a primitive vector v. Let U be a submodule of V such that $v \notin U$ but $N_+ v \subseteq U$. Since $(V_i/V_{i+1})_\lambda \neq 0$, by definition of the submodules V_j, $V_i/V_{i+1} \cong L(\mu)$ for some $\mu \geq \lambda$. Since the module $L(\mu)$ is irreducible, the vector $V_{i+1} + v$ generates the module V_i/V_{i+1}. Suppose that

$$N_+ v \subseteq V_{i+1}. \tag{1}$$

Then, $\mu = \lambda$, proving the result. So we only need to show that (1) holds.

Otherwise, there is an index $j \in I$ for which $V_{i+1} + e_j v$ is another generator of V_i/V_{i+1} and so $V_i/V_{i+1} \leq (U + V_{i+1})/V_{i+1}$. In particular, $v = v_1 + u$ for some $v_1 \in (V_{i+1})_\lambda$ and $u \in U_\lambda$. Hence, $N_+ v_1 \subseteq U$. By definition of v and U, $v_1 \notin U$ and so v_1 is a primitive vector of V of weight λ contained in V_{i+1}. This contradicts the maximality of i.
The converse is left for the reader to check. $\qquad\square$

Given a weight $\Lambda \in H^*$, we want to find an algebraic way of describing the set of primitive weights λ of a highest weight module $V(\Lambda)$ with highest weight Λ and in particular of showing that it is a finite set. Then, we can apply Proposition 2.6.20 to the corresponding Verma modules $M(\lambda)$ and a "matrix inversion technique" leads to an expression for the character (resp. super-character) of the irreducible module $L(\Lambda)$ in terms of the characters (resp. super-characters) of these Verma modules.

Primitive weights of $V(\Lambda)$ satisfy the following algebraic condition.

Proposition 2.6.23. *Let $V = V(\Lambda)$ be a highest weight G-module and λ a primitive weight of V. Then,*

$$\left|\Lambda + \rho\right|^2 = \left|\lambda + \rho\right|^2.$$

In order to prove Proposition 2.6.23, we need to use the action of the generalized Casimir operator introduced by Kac in [Kac5].

Definition 2.6.24. *Let V be a G-module in the category $\mathcal{O}$ and v be a vector in the λ-weight space V_λ of V. The operator Ω_0 on V is defined to be as follows:*

$$\Omega_0(v) = 2 \sum_{\alpha \in \Delta^+} \sum_i f_{\alpha,i} e_{\alpha,i}(v).$$

The generalized Casimir operator Ω is the operator acting on V as follows:

$$\Omega(v) = \Omega_0(v) + (2\rho + \lambda, \lambda)v.$$

Remark 2.6.25. The operator Ω_0 is well defined: By definition of the category $\mathcal{O}$, there exist finitely many weights $\lambda_i \in H$ such that for any given weight $\lambda \in P(V)$ of V, $\alpha + \lambda \in P(V)$, $\alpha \in \Delta^+$, only if $\alpha < \lambda_i - \lambda$. So for all $v \in V_\lambda$, $\Omega_0(v)$ has only finitely many nonzero terms. Note that both the operator Ω_0 and the generalized Casimir operator are even elements in the $\mathbf{Z}_2$-graded universal enveloping algebra $U(G)$.

We next show that the action of the generalized Casimir operator on highest weight modules commutes with the action of G; or equivalently that the element Γ belongs to the centre of $U(G)$.

Lemma 2.6.26. *Let V be a G-module in the category $\mathcal{O}$. Then, $[\Omega, x]v = 0$ for all $x \in G$, $v \in V$.*

Proof. We only need to show that $[\Omega, e_i]v = 0$ and $[\Omega, f_i]v = 0$, for all $i \in I$, $v \in V_\lambda$. Now

$$[\Omega_0, e_i]v = 2 \sum_{\alpha \in \Delta^+} \sum_s (f_{\alpha,s}[e_{\alpha,s}, e_i] + (-1)^{d(e_\alpha)d(e_i)}[f_{\alpha,s}, e_i]e_{\alpha,s})v.$$

Since $[e_{\alpha,s}, e_i] \in G_{\alpha+\alpha_i}$, $[e_{\alpha,s}, e_i] = \sum_t k_t e_{\alpha+\alpha_i,t}$ and by definition of a dual basis, we get $k_t = ([e_{\alpha,s}, e_i], f_{\alpha+\alpha_i,t})$. Hence,

$$\sum_{\alpha \in \Delta^+} \sum_s f_{\alpha,s}[e_{\alpha,s}, e_i]v$$

$$= \sum_{\alpha \in \Delta^+} \sum_s \sum_t f_{\alpha,s}([e_{\alpha,s}, e_i], f_{\alpha+\alpha_i,t})e_{\alpha+\alpha_i,t}v$$

$$= \sum_{\alpha \in \Delta^+} \sum_s \sum_t (e_{\alpha,s}, [e_i, f_{\alpha+\alpha_i,t}])f_{\alpha,s}e_{\alpha+\alpha_i,t}v$$

$$= \sum_{\alpha \in \Delta^+} \sum_t [e_i, f_{\alpha+\alpha_i,j}]e_{\alpha+\alpha_i,t}v$$

$$= \sum_{\alpha \in \Delta^+ - \{\alpha_i\}} \sum_t [e_i, f_{\alpha,t}]e_{\alpha,t}v$$

$$= -\sum_{\alpha \in \Delta^+ - \{\alpha_i\}} \sum_t (-1)^{d(e_\alpha)d(e_i)}[f_{\alpha,t}, e_i]e_{\alpha,t}v.$$

Thus

$$\begin{aligned}[\Omega_0, e_i]v &= 2(-1)^{d(e_i)d(e_i)}[f_i, e_i]e_i v \\ &= -2h_i e_i v \\ &= -2(\lambda + \alpha_i, \alpha_i)e_i v,\end{aligned}$$

which implies that $[\Omega, e_i]v = 0$. Similar calculations give $[\Omega, f_i]v = 0$. $\square$

In particular, this gives the action of the generalized Casimir operator on a highest weight module.

Corollary 2.6.27. *If $V = V(\Lambda)$ is a highest weight G-module with highest weight $\Lambda \in H$, then for any $v \in V$,*

$$\Omega(v) = (|\rho|^2 - |\Lambda + \rho|^2)v.$$

Proposition 2.6.23 now follows from the definition of primitive weights.

The character (resp. super-character) of irreducible highest weight modules can now be expressed in terms of characters (resp. super-characters) of Verma modules.

Corollary 2.6.28. *For any weight $\Lambda \in H$, the character and super-character of the highest weight G-module $V(\Lambda)$ with highest weight Λ are given by:*

$$\mathrm{ch}V(\Lambda) = \sum_{\substack{\lambda \leq \Lambda \\ |\lambda+\rho|^2 = |\Lambda+\rho|^2}} c_\lambda \mathrm{ch}M(\lambda) \quad and$$

$$\mathrm{sch}\,V(\Lambda) = \sum_{\substack{\lambda \leq \Lambda \\ |\mu-\rho|^2 = |\lambda-\rho|^2}} c'_\lambda \mathrm{sch}\,M(\lambda),$$

where $c_\lambda, c'_\lambda \in \mathbf{Z}$ and $c_\Lambda = 1 = c'_\Lambda$.

As we pointed out at the start of this chapter, we are primarily interested in modules generalizing finite dimensional ones. When the BKM superalgebra G is finite dimensional, if V is finite dimensional G-module, then all the root vectors of G must necessarily act in a locally nilpotent way on V. As we have seen in section 2.3, the roots of finite type and non-zero norm are the roots that behave like the roots of a finite dimensional simple Lie algebra. It is therefore natural to consider G-modules on which the action of the finite type root vectors is locally nilpotent.

Definition 2.6.29. *A G-module V is said to be integrable if for any root α of finite type, and any element $x \in G_\alpha$, x acts in a locally nilpotent way, i.e. for any $v \in V$, there exits $n \in \mathbf{N}$ such that $x^n v = 0$ (n depends both on v and x)*

Lemma 2.6.30. *Let $V \in \mathcal{O}$ be a G-module. Suppose that $\alpha \in \Delta$ is conjugate to a simple root under the action of the Weyl group and that $(\alpha, \alpha) = 0$. Then, for any $x \in G_\alpha \cap G_{\bar{1}}$, $x^2 v = 0$. In particular, x act in a locally nilpotent way on V.*

Proof. The result follows immediately from the fact that $[x, x] = 0$. $\qquad\square$

As a consequence of Lemma 2.6.30 we can immediately deduce the following:

Corollary 2.6.31. *A G-module $V \in \mathcal{O}$ is integrable if all positive norm simple root vectors and even negative norm and finite type root vectors act in a locally nilpotent way on V.*

The integrability condition for a highest weight module can therefore be expressed in algebraic terms.

Lemma 2.6.32. *A highest weight G-module $V = V(\Lambda)$ is integrable if and only if $\frac{2(\Lambda, \alpha_i)}{(\alpha_i, \alpha_i)}$ (resp. $\frac{(\Lambda, \alpha_i)}{(\alpha_i, \alpha_i)}$) is a non-negative integer when $i \in I \setminus S$ (resp. $i \in S$) such that $a_{ii} > 0$ and when there are roots of negative norm and finite type, $\frac{2(\Lambda, \alpha)}{(\alpha, \alpha)} \in \mathbf{Z}_+$, where α is the unique even positive root with negative norm.*

We want to find formulae giving the character and super-character of integrable modules in the category $\mathcal{O}$ as they are the ones generalizing finite dimensional modules. The formulae that we prove here hold for irreducible integrable highest weight modules $L(\Lambda)$ of highest weight $\Lambda \in H$ with the added technical condition that $(\Lambda, \alpha_i) \geq 0$ for all $i \in I$, namely $\Lambda \in P^+$.

When the BKM superalgebra G is infinite dimensional, Theorem 2.3.44 tells us that there are no roots of finite type and negative norm and hence the formulae we derive apply to a large class of integrable irreducible highest weight modules. When G is finite dimensional, a module is integrable if and only if it is finite dimensional.

When G is of $B(m,0)$ type, all roots have positive norm and hence the formulae apply to all integrable irreducible highest weight modules, in other words to all irreducible finite dimensional modules.

Suppose that G is finite dimensional with non-trivial odd part with no roots having negative norm but with roots having norm 0. Then the only finite dimensional representations the formulae apply to are those satisfying $(\Lambda, \alpha_i) \geq 0$ when $a_{ii} = 0$.

Suppose next that Δ contains a positive root α with negative norm and of finite type. We know from Theorem 2.3.33 that $\operatorname{supp}(\alpha) = I$, and so $(\Lambda, \alpha) > 0$ or $(\Lambda, \alpha_i) = 0$ for all $i \in I$. If the former holds then by Lemma 2.6.32, the module $L(\Lambda)$ is not integrable. So the only integrable irreducible module it applies to is the trivial one.

Therefore for finite dimensional BKM superalgebras with non-trivial odd part, the formulae we give mostly apply to infinite dimensional non-integrable irreducible highest weight modules. For the character and super-character formulae of finite dimensional modules of finite dimensional BKM Lie superalgebras with non-trivial odd part, see [Jeu], [JeuHKT], [Kac6,8], [KacWak1], [Ray5], [Serg2].

To state the main theorem, we first need a well known elementary result on the invariance of G-modules in the category $\mathcal{O}$ under the action of the Weyl group.

Lemma 2.6.33. *Let* $V \in \mathcal{O}$ *be an integrable G-module. Then, for any $w \in W$, $\dim V_{w\lambda} = \dim V_\lambda$ for all $\lambda \in H$.*

Proof. The result follows from integrability and induction on the length of the word w. $\qquad\square$

For $\mu = \sum_{i \in I} k_i \alpha_i$, write

$$\operatorname{ht}_0(\mu) = \sum_{i \in I \setminus S} k_i.$$

Considering the roots as elements of the formal root lattice Q, set

$$T_\Lambda = e(\Lambda + \rho) \sum \epsilon(\mu) e(-\mu) \quad \text{and} \quad T'_\Lambda = e(\Lambda + \rho) \sum \epsilon'(\mu) e(-\mu)$$

with

$$\epsilon(\mu) = (-1)^{\operatorname{ht}(\mu)} \quad \text{and} \quad \epsilon'(\mu) = (-1)^{\operatorname{ht}_0(\mu)}$$

if $\mu = \sum_i k_i \alpha_i$ and satisfies the following four conditions:

$$k_i \geq 0 \quad \forall\, i \in I$$
$$k_i \neq 0 \Longrightarrow a_{ii} \leq 0,\ (\alpha_i, \Lambda) = 0$$
$$k_j \neq 0 \text{ and } j \neq i \Longrightarrow a_{ij} = 0,$$
$$k_i = 1 \text{ unless } i \in S \text{ and } a_{ii} = 0.$$

Otherwise,

$$\epsilon(\mu) = 0 = \epsilon'(\mu).$$

When the roots are considered to be elements of H or H^*, we consider the image of T_Λ and T'_Λ under the functions f_Q and g_Q (see Proposition 2.3.9). For reasons of simplicity, we will also write the images T_Λ and T'_Λ.

Theorem 2.6.34. *For any weight* $\Lambda \in P^+$, *the character formula and super-character formula for the irreducible highest weight module* $L(\Lambda)$ *are respectively:*

$$\mathrm{ch} L(\Lambda) = e(-\rho) \sum_{w \in W} \det(w) w(T_\Lambda) R^{-1}$$

and

$$\mathrm{sch}\, L(\Lambda) = e(-\rho) \sum_{w \in W} \det(w) w(T'_\Lambda) R'^{-1},$$

where $\det(w) = (-1)^{l(w)}$ *and* $l(w)$ *is the length of the word* $w \in W$.

Before proving this Theorem, we need some technical results.

Lemma 2.6.35. *For any* $w \in W$,

$$w(e(\rho)R) = \epsilon(w)e(\rho)R \quad \text{and} \quad w(e(\rho)R') = \epsilon(w)e(\rho)R'.$$

Lemma 2.6.36. *Let* $\lambda = \Lambda - \sum_{i \in I} x_i \alpha_i$, $x_i \in \mathbf{Z}_+$ *be a weight of the irreducible highest weight* G-*module* $L(\Lambda)$. *Then, there exists* $i \in I$ *such that* $x_i \neq 0$ *and* $(\Lambda, \alpha_i) \neq 0$.

Proof. Let $\lambda = \Lambda - \sum_{i \in I} x_i \alpha_i \in P(L(\Lambda))$, $I_\lambda = \{i \in I : x_i \neq 0\}$ and N_λ be the Lie sub-superalgebra of G generated by the vectors f_i, $i \in I_\lambda$. Then, the weight space $L(\Lambda)_\lambda$ is contained in $U(N_-)N_\lambda v$, where v is a highest weight vector of the module $L(\Lambda)$. Hence as $L(\Lambda)_\lambda \neq 0$, $N_\lambda v \neq 0$, and so for some $i \in I_\lambda$, $f_i v \neq 0$. Since $e_j v = 0$ for all $j \in I$, $e_j f_i v = 0$ for all $j \neq i$. The G-module $L(\Lambda)$ being irreducible, it follows that

$$0 \neq e_i f_i v = [e_i, f_i]v = h_i v = (\Lambda, h_i)v.$$

$\square$

We are now ready to prove Theorem 2.6.34.

Proof of Theorem 2.6.34. For reasons of simplicity, in the proof we take the roots to be elements of the lattice Q. Note that in this case, a root is either odd or even.

Let $\Lambda \in P^+$. We first prove the character formula. From Corollary 2.6.28 and Lemma 2.6.19, we get:

$$e(\rho)\mathrm{ch}L(\Lambda)R = \sum_{\substack{\lambda \leq \Lambda \\ |\lambda+\rho|^2 = |\Lambda+\rho|^2}} c_\lambda e(\lambda + \rho). \tag{1}$$

Lemmas 2.6.33 and 2.6.35 imply that $c_\lambda = \epsilon(w)c_\mu$ if $w(\lambda + \rho) = \mu + \rho$ for some $w \in W$. Let λ be such that $c_\lambda \neq 0$. Hence,

$$\forall w \in W, \quad c_{w(\lambda+\rho)-\rho} \neq 0 \tag{2}$$

and so

$$w(\lambda + \rho) \leq \Lambda + \rho.$$

Let $\mu \in \{w(\lambda + \rho) - \rho | w \in W\}$ be such that $\mathrm{ht}\,(\Lambda - \mu)$ is minimal. Then $(\mu + \rho, \alpha_i) \geq 0$ for all $i \in I$ such that $a_{ii} > 0$. Let T be the sum of all terms in the right hand side of (1) for which $(\lambda + \rho, \alpha_i) \geq 0$ for all $i \in I$ with $a_{ii} > 0$. Hence (2) allows equation (1) to be re-written as

$$e(\rho)\mathrm{ch}L(\Lambda) \prod_{\alpha \in \Delta_0^+} (1 - e(-\alpha))^{\mathrm{mult}\alpha} = \sum_{w \in W} w(T) \prod_{\alpha \in \Delta_1^+} (1 + e(-\alpha))^{\mathrm{mult}\alpha}. \tag{3}$$

To prove the Character formula we need to compute T. Let $c_\lambda e(\lambda + \rho)$ be a non-trivial term in T. Then,

$$(\lambda + \rho, \alpha_i) \geq 0 \quad \forall i \in I \quad \text{such that} \quad a_{ii} > 0 \tag{4}$$

$$|\lambda + \rho|^2 = |\Lambda + \rho|^2. \tag{5}$$

Setting

$$\lambda = \Lambda - \sum_{i \in I} x_i \alpha_i, \quad \text{where} \quad x_i \in \mathbf{Z}_+, \tag{6}$$

equation (5) gives

$$\sum_{i \in I} x_i(\Lambda + \lambda + 2\rho, \alpha_i) = 0. \tag{7}$$

From Condition (4) and the definition of ρ and Λ,

$$a_{ii} > 0 \implies (\Lambda + \lambda + 2\rho, \alpha_i) > 0.$$

Next, consider $i \in I$ such that $a_{ii} \leq 0$. Suppose that $x_i \neq 0$. Then from (6), we get $x_i > 0$ and

$$(\lambda + 2\rho, \alpha_i) = (\lambda + \alpha_i, \alpha_i) = (\Lambda, \alpha_i) - \sum_{j \neq i} x_j(\alpha_j, \alpha_i) - (x_i - 1)(\alpha_i, \alpha_i) \geq 0. \tag{8}$$

Thus, it follows from equation (7) and the assumption that $\Lambda \in P^+$ that

$$(\Lambda + \lambda + 2\rho, \alpha_i) \geq 0.$$

This forces
$$x_i \neq 0 \implies a_{ii} \leq 0 \quad \text{and} \quad (\lambda + 2\rho + \Lambda, \alpha_i) = 0.$$
Finally (8) and the fact that $\Lambda \in P^+$ give
$$(\Lambda, \alpha_i) = 0, \quad (\alpha_i, \alpha_j) = 0 \quad \text{if} \quad x_j \neq 0 \quad \text{and} \quad j \neq i;$$
$$\text{and} \quad (\alpha_i, \alpha_i) = 0 \quad \text{if} \quad x_i > 1. \tag{9}$$
Set
$$\lambda = \Lambda - \alpha - \beta, \quad \text{where} \quad \alpha = \sum_{i \in I \setminus S} x_i \alpha_i, \quad \beta = \sum_{i \in S} x_i \alpha_i.$$

Claim: If $e(w(\lambda + \rho) - \gamma)$ is a term in T for $0 \neq \gamma = \sum_{i \in S} y_i \alpha_i$, $y_i \in \mathbf{Z}_+$ and $w \in W$, then $w(\lambda + \rho) = \lambda + \rho$.

From the above, $(\lambda + \rho, \alpha_i) \geq 0$ for all $i \in I$. Therefore
$$w(\lambda + \rho) = \lambda + \rho - \sum_{\substack{i \in I \\ a_{ii} > 0}} z_i \alpha_i,$$

where $z_i \in \mathbf{Z}_+$. Suppose that $e(w(\lambda + \rho) - \gamma)$ is a term in T. Then, applying Conditions (4) to the weight $\lambda - \sum_{\substack{i \in I \\ a_{ii} > 0}} z_i \alpha_i - \sum_{i \in S} y_i \alpha_i$, it follows that $z_i = 0$ for all $i \in I$ and so $w(\lambda + \rho) = \lambda + \rho$.

We finally show by induction on $\text{ht}(\beta)$ that $c_\lambda = \epsilon(\alpha)(-1)^{\text{ht}(\beta)}$, where $\epsilon(\alpha) = (-1)^n$ if α is the sum of n distinct pairwise perpendicular simple even roots of non-positive norm, α_i perpendicular to Λ, and $\epsilon(\alpha) = 0$ otherwise.

Suppose that $\text{ht}(\beta) = 0$. Then, there must be a term on the left hand side of (3) equal to $c_\lambda e(\lambda + \rho)$. So, either λ is a weight of the module $L(\Lambda)$ or
$$\lambda = \mu - (-1)^n \sum_{i=1}^{n} \mu_i,$$

where μ is a weight of the module $L(\Lambda)$, and for each i, μ_i is an even positive root. Now $\mu = \Lambda - \sum_i k_i \alpha_i$, where $k_i \in \mathbf{Z}_+$. Lemma 2.6.36 and (9) show that
$$\mu = \Lambda.$$

Furthermore the support of a root is connected (see Proposition 2.3.8), and so (9) forces the roots μ_i to be simple even, orthogonal to Λ and mutually orthogonal. We thus get the desired answer for c_λ.

Next assume that $\text{ht}(\beta) > 0$, and that the result holds for all weights λ with β of smaller height. Then, no term on the left hand side of (3) equals $c_\lambda e(\lambda + \rho)$. Since β is the sum of mutually orthogonal simple roots and if any appear more than once, it has norm 0, the only sub-sum of β that are roots are equal to simple roots. Hence the above claim gives

$$c_\lambda e(\lambda) + \sum_{s=1}^{r} c_{\lambda + \alpha_{i_1} + \dots \alpha_{i_s}} e(\lambda + \alpha_{i_1} + \dots \alpha_{i_s}) e(-\alpha_{i_1}) \dots e(-\alpha_{i_s}) = 0,$$

where r is the number of distinct simple roots α_i with $x_i \neq 0$ since in the product $\prod_{\alpha \in \Delta_1^+}(1 + e(-\alpha))^{\mathrm{mult}\alpha}$, the simple roots appear exactly once as they have multiplicity 1. It follows by induction that

$$c_\lambda + \epsilon(\alpha)(-1)^{ht\beta} \sum_{s=1}^r \binom{r}{s}(-1)^{-s} = 0,$$

giving the desired answer for c_λ.

We next prove the super-character formula. The above arguments lead to:

$$e(\rho)\mathrm{sch}\, L(\Lambda) \prod_{\alpha \in \Delta_0^+} (1 - e(-\alpha))^{\mathrm{mult}\alpha} = \sum_{w \in W} w(T') \prod_{\alpha \in \Delta_1^+} (1 - e(-\alpha))^{\mathrm{mult}\alpha},$$

where the terms in T' satisfy conditions (9). Keeping the above notation, we calculate T'. We show by induction on $\mathrm{ht}\,(\beta)$ that $c'_\lambda = \epsilon(\alpha)$, where $\epsilon(\alpha) = (-1)^n$ if α is the sum of n distinct pairwise perpendicular simple even roots α_i perpendicular to Λ, and $\epsilon(\alpha) = 0$ otherwise. When $\mathrm{ht}\,(\beta) = 0$, this follows as above. So assume $\mathrm{ht}\,(\beta) > 0$. Then, for the same reason as before,

$$c'_\lambda e(\lambda) + \sum_{s=1}^r (-1)^s c'_{\lambda + \alpha_{i_1} + \dots \alpha_{i_s}} e(\lambda + \alpha_{i_1} + \dots \alpha_{i_s}) e(-\alpha_{i_1}) \dots e(-\alpha_{i_s}) = 0,$$

where r is the number of distinct simple roots α_i with $x_i \neq 0$ since in the product $\prod_{\alpha \in \Delta_1^+}(1 + e(-\alpha))^{\mathrm{mult}\alpha}$, the simple roots appear exactly once as they have multiplicity 1. It follows by induction that

$$c'_\lambda + \epsilon(\alpha) \sum_{s=1}^r \binom{r}{s}(-1)^s = 0,$$

giving the desired answer for c'_λ. $\square$

Remark 2.6.36. Note that the assumption of irreducibility is only used in the proof of Theorem 2.6.34 through the application of Lemma 2.6.36. However, if G is a Kac-Moody Lie superalgebra, then Lemma 2.6.36 is not needed: as there are no simple roots of non-positive norm in this case, comparing the left hand and the right hand side of equality (3) in the proof, it follows that $T = e(\Lambda + \rho)$. Therefore when G is a Kac-Moody Lie superalgebra, since the character and super-character formulae hold for all highest weight modules $V(\Lambda)$, for each highest weight $\Lambda \in P^+$, there must be precisely one (up to isomorphism). In other words, for Kac-Moody Lie superalgebras, integrable highest weight modules are necessarily irreducible. However, as the next example shows this is not generally the case for BKM superalgebras.

Example 2.6.37. Let $S = \emptyset$ and $A = \begin{pmatrix} 2 & -1 \\ -1 & 0 \end{pmatrix}$. Define $e_i v = 0$, $h_1 v = v$, and $f_1^2 v = 0 = h_2 v$. Then the vector v generates a highest weight integrable module $V(\Lambda)$ with highest weight $\Lambda \in P^+$ satisfying $(\Lambda, \alpha_1) = 1$ and $(\Lambda, \alpha_2) = 0$.

However it is not irreducible if $f_2 v \neq 0$ since in this case $f_2 v$ generates a proper non-trivial G-submodule. Note that in this case, Lemma 2.6.36 does not hold. Indeed, the weight $\lambda = \Lambda - \alpha_2 \in P(V(\Lambda))$ but $(\lambda, \alpha_i) = 0$ for $i = 1, 2$. If $f_2 v = 0$, then the module $V(\Lambda)$ is irreducible.

The character and super-character formulae lead to the most important formulae associated to a BKM superalgebra:

Theorem 2.6.38. *For any BKM superalgebra G,*

$$\frac{\prod_{\alpha \in \Delta_0^+} (1 - e(-\alpha))^{m_0(\alpha)}}{\prod_{\alpha \in \Delta_1^+} (1 + e(-\alpha))^{m_1(\alpha)}} = e(-\rho) \sum_{w \in W} \det(w) w(T), \quad and$$

$$\frac{\prod_{\alpha \in \Delta_0^+} (1 - e(-\alpha))^{m_0(\alpha)}}{\prod_{\alpha \in \Delta_1^+} (1 - e(-\alpha))^{m_1(\alpha)}} = e(-\rho) \sum_{w \in W} \det(w) w(T').$$

These are respectively called the denominator formula and the super-denominator formula.

Proof. Since $L(0)$ is the irreducible trivial G-module, it has dimension 1, and so

$$\mathrm{ch} L(0) = 1 = \mathrm{sch}\, L(0).$$

On the other hand, the character and super-character formulae applied to $L(0)$ give an expression for $\mathrm{ch} L(0)$ and $\mathrm{sch}\, L(0)$. Equating the two give the denominator and super-denominator formulae. $\quad\square$

From the Cartan decomposition (Corollary 2.3.7), it follows that a complete information of the structure of the BKM superalgebra G is equivalent to knowing the dimension of its generalized Cartan subalgebra H, the set of roots Δ together with their multiplicity and parity, and which among them form a set of "generating" roots Π, in other words are the simple roots. Therefore, the character and super-denominator formulae essentially characterize the superalgebra G (up to the dimension of its generalized Cartan subalgebra): an expression for the product side is equivalent to a description of the roots and their multiplicities, and an expression for the sum side is equivalent to a description of the simple roots since the Weyl group is generated by simple roots of positive norm reflections and T is a sum of simple roots of non-positive norm in which each appears at least once. Together the character and super-character formulae allow us to differentiate the even from the odd roots.

The main characteristic of these two formulae is that one side is a sum and the other a product. This is why they give in many cases interesting identities.

We conclude this section by considering two examples of BKM algebras that are not Kac-Moody Lie algebras. For the affine Lie algebras, the denominator formula is equivalent to the Macdonald identities for powers of the Dedekind η function [Dys], [Kac14, §12.1], [Mac], [Mo5].

Example 2.6.39. The Monster Lie algebra M

We saw in Example 2.3.11.1 that the roots of M are elements in the Lorentzian lattice $II_{1,1}$ with bilinear form given by the matrix $\begin{pmatrix} 0 & -1 \\ -1 & 0 \end{pmatrix}$ with respect to the standard basis $(1,0)$, $(0,1)$. We also saw that the simple roots are the elements $(1, n)$ with multiplicity $c(n)$, where $J(q) = \sum_n c(n)q^n$ is the normalized modular invariant. We will show in Chapter 4 that (m, n) is a root with multiplicity $c(mn)$.

Let $p = e(-1, 0)$ and $q = e(0, -1)$ be formal exponentials. Then the product side of the denominator formula is:

$$\prod_{\substack{m>0 \\ n\in\mathbf{Z}}} (1 - p^m q^n)^{c(mn)}.$$

To compute the sum side, we first compute T. We find which simple roots have non-positive norm: Since $(m, n).(m, n) = -2mn$, the only roots with positive norm are $\pm(1, -1)$. Also $(1, n_1).(1, n_2) = -n_2 - n_1 = 0$ implies that $n_1 n_2 \leq 0$. Hence, no two simple roots of non-positive norm are orthogonal and neither have norm 0. From Example 2.3.44, we know that $(1, 0)$ is a Weyl vector. Hence,

$$T = p^{-1} - \sum_{n>0} c(n)q^n = p^{-1} + q^{-1} - J(q)$$

The Weyl group is $W = <r>$, where r is the reflection generated by the unique simple root $(1, -1)$ of positive norm. Since

$$r(1,0) = (1,0) - ((1,-1).(1,0))(1,-1) = (0,1),$$

$$r(p) = q \quad \text{and} \quad r(q) = p.$$

Therefore,

$$r(T) = q^{-1} + p^{-1} - J(p).$$

As a result, the sum side of the denominator formula is

$$p^{-1}(T - r(T)) = p^{-1}(j(p) - j(q))$$

and so the denominator formula is

$$\prod_{\substack{m>0 \\ n\in\mathbf{Z}}} (1 - p^m q^n)^{c(mn)} = p^{-1}(j(p) - j(q)).$$

Example 2.6.40. The Fake Monster Lie algebra F

The root lattice of the fake monster lie algebra F is the even unimodular Lorentzian lattice $II_{25,1}$ of rank 26 (see section 3.1 for details on lattices). Let Λ be the Leech lattice, i.e. the unique positive definite lattice of rank 24 with no vectors of norm 2. Then uniqueness of even unimodular Lorentzian lattices (see Corollary 3.1.6) implies that

$$II_{25,1} = \Lambda \oplus II_{1,1}.$$

We write vectors in $II_{25,1}$ as (λ, m, n), with $\lambda \in \Lambda$ and $(m, n) \in II_{1,1}$.

The roots are the non-zero vectors α of $II_{25,1}$ with multiplicity $p_{24}(1 - \alpha^2/2)$, where $p_{24}(n)$ is the number of partitions of n into parts of 24 colours, so that

$$\sum_n p_{24}(1 + n) = q^{-1} \Pi_{n>0}(1 - q^n)^{24} = \Delta(q)^{-1} = q^{-1} + 24 + 324q + \ldots$$

The simple roots of finite type (i.e. positive norm) are in bijective correspondence with points in the Leech lattice:

$$(\lambda, 1, \frac{\lambda^2}{2} - 1), \quad \lambda \in \Lambda.$$

The Weyl group W is thus isomorphic to the reflection group of the Leech lattice and so its the full automorphism group of the Lorentzian lattice $II_{1,1}$ [Con2]. The simple roots of infinite type are:

$$(0, 0, n), \quad n \in \mathbf{N}$$

with multiplicity $p_{24}(1) = 24$. Their restricted simple root multiplicity is equal to their multiplicity. Writing $(.,.)$ both for the bilinear form in $II_{1,1}$ and the one in Λ, the Weyl vector (see Definition 2.3.47) $\rho = (\mu, a, b)$ satisfies

$$0 = (\rho, (0, 0, n)) = an$$

for all $n \in \mathbf{N}$ since the root $(0, 0, n)$ has norm 0, and

$$1 = (\rho, (\lambda, 1, \frac{\lambda^2}{2} - 1)) = (\mu, \lambda) - b - a(\frac{\lambda^2}{2} - 1)$$

for all $\lambda \in \Lambda$ since the root $(\lambda, 1, \frac{\lambda^2}{2} - 1)$ has norm 2. Hence,

$$\rho = (0, 0, -1).$$

We next describe the set of positive roots. As positive roots are sums of simple roots, if $\alpha = (\lambda, m, n)$ is a positive root, then either $m > 0$ or $\alpha = (0, 0, n)$. Hence, as $\alpha, \rho = m$,

$$\alpha \in \Delta^+ = \{\alpha \in II_{25,1} : \alpha, \rho > 0 \quad \text{or} \quad \alpha = (0, 0, n)\}.$$

We will see in section 5.4 that Δ^+ is in fact the set of positive roots. The product side of the denominator formula is thus

$$\prod_{\alpha \in \Delta^+} (1 - e(-\alpha))^{p_{24}(1 - \alpha^2/2)}.$$

All the simple roots of infinite type have norm 0 and are mutually orthogonal. Hence,

$$
\begin{aligned}
T &= \sum_{n_1, n_2, \cdots} (-1)^{n_1+n_2+\cdots} e(n_1 \rho) \binom{24}{n_1} e(n_2 2\rho) \binom{24}{n_2} \cdots \\
&= (1 - e(\rho))^{24} (1 - e(2\rho))^{24} \cdots \\
&= \Pi_{n>0} (1 - e(n\rho))^{24} \\
&= \sum_{n \in \mathbf{Z}} \tau(n) e(n\rho)
\end{aligned}
$$

by definition of the Ramanujan tau function [Hard]. Hence the denominator formula is

$$
e(\rho) \prod_{\alpha \in \Delta^+} (1 - e(-\alpha))^{p_{24}(1-\alpha^2/2)} = \sum_{\substack{w \in W \\ n \in \mathbf{Z}}} \det(w) \tau(n) e(w(n\rho)).
$$

The Kac-Moody Lie subalgebra generated by the roots of finite type is the Feingold-Frenkel Lie algebra [FeinF], [Fre].

Remark 2.6.41. In the case of non-symmetrizable Kac-Moody Lie algebras, the character formula has been proved by Kumar [Kum1,2]. There are as yet no results on non-highest weight representations of BKM superalgebras. However, there is an obvious natural example, namely the adjoint representation for infinite dimensional BKM superalgebras.

Exercises 2.6.

1. Prove the character formula for Kac-Moody Lie algebras.

2. Find a different construction of the Verma module as an induced module.

3. Let H_1 be a generalized Cartan subalgebra of the Kac-Moody Lie algebra G. Set

$$
V = \{ v \in L(\Lambda) : v \neq 0, (\Lambda, \Lambda) v \otimes v = \sum_{\alpha \in \Delta \cup \{0\}} e_{\alpha,i}(v) \otimes e_{-\alpha,i}(v) \}.
$$

(i) Show that H_1 acts semisimply on the irreducible module $L(\Lambda)$ for every weight $\Lambda \in P^+$. Let v_Λ be a highest weight vector of the module $L(\Lambda)$ (with respect to H). For $\lambda \in W(\Lambda)$, set

$$
\mathcal{V}(\lambda)_+ = \{ v \in \mathcal{V} : \mathrm{supp}(v) \geq \lambda \quad \text{and} \quad \lambda \in \mathrm{supp}(v) \}
$$

and

$$
\mathcal{V}(\lambda)_- = \{ v \in \mathcal{V} : \mathrm{supp}(v) \leq \lambda \quad \text{and} \quad \lambda \in \mathrm{supp}(v) \}.
$$

(ii) Show that the $U(H_1)$-submodule V generated by the vector v_Λ is finite dimensional and that $V \cap \mathcal{V} \neq \emptyset$.

(iii) Deduce that there exists an eigenvector v of H_1 in $V \cap \mathcal{V}$.

(iv) Show that $\mathcal{V} = \mathcal{V}(\lambda)_+ \cup \mathcal{V}(\lambda)_-$.

(v) Deduce that there exists $g \in \mathcal{G}$ such that $g(v)$ is a multiple of v_Λ.

(vi) Deduce that $g(H_1) = H$.

For a solution, see [KacP6].

4. Let R_+ (resp. R^-) be the ideal of $\tilde{G}$ (see section 2.1, paragraph preceding Proposition 2.1.13) generated by the elements $(\operatorname{ad} e_i)^{\frac{-2a_{ij}}{a_{ii}}+1} e_j$, (resp. $(\operatorname{ad} f_i)^{\frac{-2a_{ij}}{a_{ii}}+1} f_j$), for $a_{ii} > 0$, $i \in I \setminus S$; $(\operatorname{ad} e_i)^{\frac{-a_{ij}}{a_{ii}}+1} e_j$, (resp. $(\operatorname{ad} f_i)^{\frac{-a_{ij}}{a_{ii}}+1} f_j$, for $a_{ii} > 0$, $i \in S$; and $[e_i, e_j]$ (resp. $[f_i, f_j]$) if $a_{ij} = 0$.

(i) Prove that the ideal R_+ is generated as an ideal by the subspaces $R_\alpha = \{x \in R_+ : [h, x] = (f_Q(\alpha), h)x\}$ for the elements $\alpha \in Q \setminus \Pi$ such that $\alpha > 0$ (resp. $\alpha < 0$ and $2(\rho, \alpha) = (\alpha, \alpha)$ (see paragraph preceding Proposition 2.3.9 for the definition of the map f_Q).

Hint: Define a Verma module $\tilde{M}(\Lambda)$ ($\Lambda \in H$) for the Lie superalgebra $\tilde{G}$ and show that the unique maximal submodule $\tilde{N}(0)$ of $\tilde{M}(0)$ is isomorphic to $\oplus_{i \in I} \tilde{M}(-\alpha_i)$. Deduce that there is an G-module homomorphism $R_-/[R_-, R_-] \to \oplus_{i \in I} M(-\alpha_i)$. Using the PBW Theorem, show this map is injective. Conclude.

(ii) Deduce that the ideal $R = R_+ \oplus R_-$ is the maximal ideal in $\tilde{G}$ such that $R \cap H = 0$.

For a solution, see [Kac14, §9.11] and [Ray3].

5. Find the denominator formula for the (untwisted) extended affine Lie algebras and show that for the extended affine Lie algebra $\hat{G}$ corresponding to the finite dimensional simple Lie algebra G it is equivalent to the Macdonald identity giving the Fourier series for $\eta^{\dim G}$.

For a solution, see [Kac14, §12.1].

6. Suppose that $S = \emptyset$. Let ϕ be the projection of the vector space G on N_-. Define the operator Ω_0' on the Lie subalgebra N_- to be

$$\Omega_0'(x) = \sum_{\alpha \in \Delta^+} \sum_{i=1}^{\operatorname{mult}(\alpha)} [f_{\alpha,i}, \phi([e_{\alpha,i}, x])].$$

(i) Show that for any $\alpha \in \Delta^+$ and $x \in G_{-\alpha}$, $\Omega_0'(x) = (2(\rho, \alpha) - (\alpha, \alpha))$.

Hint: Find $(\Omega_0'(x))(v)$, where v is a highest weight vector of the Verma module $M(0)$. Use Lemma 2.6.26.

(ii) Show that the contravariant bilinear form $(x, y)_0 = -(\omega_0(x), y)$ (see section 2.2) is positive definite on root spaces. When the Lie algebra G is finite dimensional show that it is positive definite on G.

For a solution, see [Kac14, Theorem 11.7].

Chapter 3

Singular Theta Transforms of Vector Valued Modular Forms

In this chapter we introduce vector valued modular forms and show how to derive their theta transforms. As we concentrate only on the properties needed for our classification purpose given in sections 5.2 and 5.3, for an in depth study of these objects, the reader can consult the following books: [Borc11], [EicZ], [Frei], [Mi], [Shi], [Serr2].

3.1 Lattices

We start with some classical properties of lattices.

Definition 3.1.1. *A (integral) lattice M is a free abelian group of finite rank with a symmetric $\mathbf{Z}$-valued bilinear form $(.,.)$. It is said to be even if for all $v \in M$, $(v,v) \equiv 0 \pmod 2$. Otherwise it is said to be odd. The rank or dimension $\dim M$ (resp. signature $\text{sign}(M)$) of the lattice M is the dimension (resp. signature) of the real vector space $M \otimes_{\mathbf{Z}} \mathbf{R}$ with bilinear form induced from that of M. The lattice M is Lorentzian if it has signature $(m,1)$ or $(1,n)$.*

Unless the lattice under consideration is not integral, we will sometimes omit to add this precision. We now briefly discuss dual lattices.

Definition 3.1.2. *The dual lattice M^* of the lattice M is the lattice of all vectors v^* in $M \otimes_{\mathbf{Z}} \mathbf{R}$ for which $(v^*, v) \in \mathbf{Z}$ for all $v \in M$. The lattice M is unimodular if $M = M^*$.*

The following result is an immediate consequence of the definition.

Lemma 3.1.3. *An integral lattice M is always contained in its dual M^* and the quotient group M^*/M is an abelian finite group.*

However one should be careful in case the lattice M is degenerate. Indeed if $z \in M$ is such that $(z, M) = 0$, then there are no vectors in the vector space $M \otimes_{\mathbf{Z}} \mathbf{R}$ whose inner product with z is non-trivial.

Lemma 3.1.4. *The dual lattice M^* of M is isomorphic to $Hom(M, \mathbf{Z})$ if and only if the lattice M is non-degenerate.*

We next give a well known property of the signature of unimodular lattices as the root lattices of some of the most interesting BKM algebras are indeed unimodular. This result is a consequence of Milgram's formula, which we first state. There are elementary proofs of this formula. Here we derive it as a consequence of results on the Siegel theta function for the lattice M in section 3.3.

Lemma 3.1.5. *For any nonsingular lattice M,*

$$\frac{1}{\sqrt{|M^*/M|}} \sum_{\lambda \in M^*/M} \mathbf{e}^{\lambda^2/2} = \mathbf{e}^{sign(M)/8},$$

where $\mathbf{e} = \exp(2\pi i)$.

Proof. Since the map ρ_M defined in Theorem 3.3.4 is a representation of the metaplectic group (see Definition 3.3.2) on the group ring $\mathbf{C}[M^*/M]$,

$$\rho_M(ST)e_\gamma = \mathbf{e}((\gamma, \gamma)/2) \frac{\sqrt{i}^{\,b^- - b^+}}{\sqrt{|M^*/M|}} \sum_{\delta \in M^*/M} \mathbf{e}(-(\gamma, \delta))e_\delta$$

for any $\gamma \in M^*/M$, where $(b^+, b^-) = sign(M)$. Hence,

$$
\begin{aligned}
\rho_M(ST)^2 e_\gamma &= \mathbf{e}((\gamma, \gamma)/2) \frac{i^{\,b^- - b^+}}{|M^*/M|} \sum_{\delta, \mu \in M^*/M} \mathbf{e}((\delta, \delta)/2)\mathbf{e}(-(\gamma, \delta))\mathbf{e}(-(\delta, \mu))e_\mu \\
&= \frac{i^{\,b^- - b^+}}{|M^*/M|} \sum_{\delta, \mu \in M^*/M} \mathbf{e}((\delta - \gamma)^2/2)\mathbf{e}(-\mu, \delta)e_\mu \\
&= \frac{i^{\,b^- - b^+}}{|M^*/M|} \sum_{\delta, \mu \in M^*/M} \mathbf{e}(\delta^2/2)\mathbf{e}(-\mu, \delta + \gamma)e_\mu.
\end{aligned}
$$

And so,

$$
\begin{aligned}
\rho_M(ST)^3 e_\gamma &= \frac{\sqrt{i}^{\,3(b^- - b^+)}}{\sqrt{|M^*/M|}^3} \\
&\quad \times \sum_{\mu, \delta, \lambda \in M^*/M} \mathbf{e}(\delta^2/2)\mathbf{e}((\mu, \mu)/2)\mathbf{e}(-(\mu, \delta + \gamma))\mathbf{e}(-\lambda, \mu)e_\lambda \\
&= \frac{\sqrt{i}^{\,3(b^- - b^+)}}{\sqrt{|M^*/M|}^3} \sum_{\mu, \delta, \lambda \in M^*/M} \mathbf{e}((\delta - \mu)^2/2)\mathbf{e}(-(\mu, \gamma + \lambda))e_\lambda
\end{aligned}
$$

$$= e_{-\gamma} \frac{\sqrt{i}^{-3(b^- - b^+)}}{\sqrt{|M^*/M|}^{-3}} \left(\sum_{\delta \in M^*/M} \mathbf{e}(\delta^2/2) \right) \sum_{\mu, \lambda \in M^*/M} \mathbf{e}(-(\mu, \lambda)) e_\lambda$$

$$= \frac{(i\sqrt{i})^{b^- - b^+}}{\sqrt{|M^*/M|}} \left(\sum_{\delta \in M^*/M} \mathbf{e}(\delta^2/2) \right) e_{-\gamma}.$$

By Lemma 3.3.3, $S^2 = (ST)^3$. Hence, from the action of S^2 on the group ring given in Theorem 3.3.4, we can deduce that

$$\frac{i^{b^-/2 - b^+/2}}{\sqrt{|M^*/M|}} \left(\sum_{\delta \in M^*/M} \mathbf{e}(\delta^2/2) \right) = 1.$$

The result follows since $i^{b^+/2 - b^-/2} = \exp(2\pi i (b^+ - b^-)/8)$. $\qquad\square$

Corollary 3.1.6. *If an even lattice M is unimodular, then $sign(M)$ is divisible by 8. Furthermore, for every $m, n \in \mathbf{Z}_+$ such that 8 divides $m - n$, there is a unique (up to isomorphism) lattice of rank $m - n$ and dimension $m + n$. It is written $II_{m,n}$.*

Proof. Applying Lemma 3.1.5 to the case when $M = M^*$, the first part follows. Uniqueness is a consequence of unimodularity. The lattice of rank 2 with bilinear form given by the matrix $\begin{pmatrix} 0 & -1 \\ -1 & 0 \end{pmatrix}$ is even and unimodular and hence must be $II_{1,1}$. The lattice $(E_8)^r \oplus II_{1,1}^n$ is clearly even and unimodular with rank $8r$ and dimension $8r + 2n$, proving existence. $\qquad\square$

In the particular case of a unimodular Lorentzian lattice, the existence of norm 0 vectors simplifies the proofs in the next sections and in chapter 5.

Definition 3.1.7. *A vector v in the lattice M (resp. M^*) is primitive if for all non-zero integers $n \in \mathbf{Z}$, $\frac{1}{n} v \notin M$ (resp M^*).*

Lemma 3.1.8. *If M is a unimodular even Lorentzian lattice, then it contains a primitive norm 0 vector.*

Proof. Let $s = sign(M)$. By the uniqueness part of Corollary 3.1.6 it follows that there is a positive (or negative) definite lattice K with signature s and that $M = K \oplus II_{1,1}$ is the unique Lorentzian lattice with signature s. The result is then an immediate consequence of the structure of the sublattice $II_{1,1}$. $\qquad\square$

3.2 Ordinary Modular Functions

Before developing the concept of vector valued modular forms, we remind the reader of some properties of ordinary modular functions. For $\tau \in \mathbf{C}$, we will write $x = \mathcal{R}(\tau)$ and $y = \mathcal{I}(\tau)$, i.e. $\tau = x + iy$.

The upper half plane

$$\mathcal{H} = \{\tau \in \mathbf{C} : Im(\tau) > 0\}$$

is a Riemannian manifold (see Lemma B.2.26) with isometry group the Lie group $SL_2(\mathbf{R})$ acting as follows:

$$A\tau = \frac{a\tau + b}{c\tau + d}$$

for $A = \begin{pmatrix} a & b \\ c & d \end{pmatrix} \in SL_2(\mathbf{R})$ and $\tau \in \mathcal{H}$. In particular the action is transitive.

Lemma 3.2.1.

(i) The stabilizer of $i \in \mathcal{H}$ is $SO_2(\mathbf{R}) = \{A \in SL_2(\mathbf{R}) : A^t A = 1\}$.

(ii) The spaces $SL_2(\mathbf{R})/SO_2(\mathbf{R})$ and $\mathcal{H}$ are homeomorphic.

Proof. (i): Let $A = \begin{pmatrix} a & b \\ c & d \end{pmatrix} \in SL_2(\mathbf{R})$ be such that $Ai = i$. Then $\frac{ai+b}{ci+d} = i$. Hence, $ai + b = di - c$. Therefore $a = d$ and $b = -c$. Since $ad - bc = 1$, $a^2 + b^2 = 1$. Hence, $A^t = \begin{pmatrix} a & -c \\ b & d \end{pmatrix}$ and $A^t A = 1$. Conversely, if $\det A = 1$ and $A^t A = 1$, then $ad - bc = 1$, $a^2 + c^2 = 1 = b^2 + d^2$, and $ab + cd = 0$. It follows that $a = d$ and $b = -c$ and so A stabilizes i.

(ii): The homeomorphism is given by the map

$$SL_2(\mathbf{R})/SO_2(\mathbf{R}) \to \mathcal{H}$$
$$A(SO_2(\mathbf{R})) \mapsto A(i).$$

$\square$

Definition 3.2.2. *The discrete subgroups*

$$\Gamma_0(N) = \{ \begin{pmatrix} a & b \\ c & d \end{pmatrix} \in SL_2(\mathbf{Z}) : c \equiv 0 \pmod{N} \}$$

for $N \in \mathbf{N}$ *are congruence subgroups of* $SL_2(\mathbf{R})$. *In particular,* $\Gamma_0(1) = SL_2(\mathbf{Z})$ *is the modular group.*

Lemma 3.2.3. *The modular group* $SL_2(\mathbf{Z})$ *is generated by the matrices* $S = \begin{pmatrix} 0 & -1 \\ 1 & 0 \end{pmatrix}$ *and* $T = \begin{pmatrix} 1 & 1 \\ 0 & 1 \end{pmatrix}$.

Definition 3.2.4. *A matrix* $A \in SL_2(\mathbf{R})$ *is parabolic if it has a unique fixed point in* $\mathbf{C}$.

If $A \in SL_2(\mathbf{R})$ is a parabolic element, then its unique fixed point is in $\mathbf{Q} \cup \{\infty\}$ since if a quadratic with integral coefficients has a unique root, then the latter must be in this set.

Definition 3.2.5. *A point in* $r \in \mathbf{R} \cup \{\infty\}$ *is a cusp point of a discrete subgroup* Υ *of the modular group* $SL_2(\mathbf{Z})$ *if it is fixed by a parabolic element of* Υ.

Note that ∞ is a cusp point of the congruence subgroup $\Gamma_0(N)$ for all $N \in \mathbf{N}$ since its the fixed point of the parabolic element $\begin{pmatrix} 1 & 1 \\ 0 & 1 \end{pmatrix}$. The next result follows from Exercise 3.2.3.

Lemma 3.2.6. *The set of cusp points of the modular* $SL_2(\mathbf{Z})$ *is* $\mathbf{Q} \cup \{\infty\}$ *and it acts transitively on it.*

Set $\mathcal{H}^* = \mathcal{H} \cup \mathbf{Q} \cup \{\infty\}$. Sometimes ∞ is written $i\infty$, i.e. it is considered to be the limit point on the imaginary axis rather than the real one.

Definition 3.2.7. *A fundamental domain for a discrete subgroup* Γ *of the Lie group* $SL_2(\mathbf{R})$ *is a connected open subset* D *of the upper half plane satisfying the following conditions:*

(i) *if* $x \neq y \in D$, *then for any* $A \in \Gamma$, $Ax \neq y$.
(ii) $\mathcal{H} = \bigcup_{A \in \Gamma} A\overline{D}$, *where* $\overline{D}$ *is the closure of* D *in* $\mathcal{H}$.

Before giving a fundamental domain for the modular group, we state an elementary result

Lemma 3.2.8. *For any* $\tau \in \mathcal{H}$, $A = \begin{pmatrix} a & b \\ c & d \end{pmatrix} \in SL_2(\mathbf{R})$, $\mathcal{I}(A\tau) = \dfrac{\mathcal{I}(\tau)}{|c\tau+d|^2}$.

Proof.

$$
\begin{aligned}
\mathcal{I}(A\tau) &= \mathcal{I}\left(\frac{(a\tau + b)(c\overline{\tau} + d)}{|c\tau + d|^2}\right) \\
&= \frac{\mathcal{I}(ad\tau + bc\overline{\tau})}{|c\tau + d|^2} \\
&= \frac{(ad - bc)\mathcal{I}(\tau)}{|c\tau + d|^2} \\
&= \frac{\mathcal{I}(\tau)}{|c\tau + d|^2}.
\end{aligned}
$$

$\square$

Lemma 3.2.9.
$$
D = \{\tau \in \mathcal{H} : |\tau| > 1, \mathcal{R}(z) < \frac{1}{2}\}
$$

is a fundamental domain for the modular group $SL_2(\mathbf{Z})$.

Proof. Since the group $SL_2(\mathbf{R})$ acts transitively on $\mathcal{H}$ and $2i \in D$, to prove this result we only need to show that

(i) if $A(2i) \in D$ for some $A = \begin{pmatrix} a & b \\ c & d \end{pmatrix} \in SL_2(\mathbf{Z})$, then $A(2i) = 2i$ and

(ii) For any $\tau \in \mathcal{H}$, there is some $A \in SL_2(\mathbf{R})$ such that $A(\tau) \in D$.

We first prove (i). $A(2i) = \frac{2ai+b}{2ci+d}$. Suppose that $A(2i) \in D$. So, by Lemma 3.2.8,

$$\frac{2}{4c^2 + d^2} = \frac{\mathcal{I}(2i)}{|2ci + d|^2} > 1.$$

Equivalently,

$$4c^2 + d^2 < 2.$$

Since $c, d \in \mathbf{Z}$, this forces $c = 0$ and $d = \pm 1$. As $\det A = 1$, we then get $a = d$. Hence $A(2i) = 2i + b$ or $2i - b$. Since $A(2i) \in D$, it follows that $b \in [-\frac{1}{2}, \frac{1}{2}] \cap \mathbf{Z}$, and so $b = 0$. As a result, $A(2i) = 2i$ as wanted.

(ii): Let $\tau \in \mathcal{H}$. For $A = \begin{pmatrix} a & b \\ c & d \end{pmatrix} \in SL_2(\mathbf{Z})$, by Lemma 3.2.8, $\mathcal{I}(A\tau) = \frac{\mathcal{I}(\tau)}{|c\tau + d|^2}$. For any $M \in \mathbf{R}$, there exists finitely many $(c, d) \in \mathbf{Z}^2$ such that $|c\tau + d| < M$. Hence, there is some matrix $A \in SL_2(\mathbf{Z})$ such that $\mathcal{I}(A\tau)$ is maximal. Since $TA\tau = A\tau + 1$, there is some $n \in \mathbf{Z}_+$ such that $\mathcal{R}(T^n Az) \in [\frac{-1}{2}, \frac{1}{2}]$. Then, $|T^n A\tau| \geq 1$ for otherwise

$$\mathcal{I}(-1/T^n A\tau) = \frac{\mathcal{I}(T^n A\tau)}{|T^n A\tau|^2} > \mathcal{I}(T^n A\tau).$$

However

$$\frac{\mathcal{I}(T^n A\tau)}{|T^n A\tau|^2} = \frac{\mathcal{I}(A\tau)}{|T^n A\tau|^2} \leq \mathcal{I}(A\tau)$$

by definition of A. Hence the subset D satisfies condition (ii). □

Definition 3.2.10. *A modular function $f(\tau)$ of weight k and with level N ($k \in \mathbf{Z}_+$ and $N \in \mathbf{N}$) is a function on the upper half plane $\mathcal{H}$, meromorphic on $\mathcal{H}$ and at the cusps such that for any matrix $\begin{pmatrix} a & b \\ c & d \end{pmatrix} \in \Gamma_0(N)$,*

$$f(\frac{a\tau + b}{c\tau + d}) = (c\tau + d)^k f(\tau).$$

If the function f is holomorphic on $\mathcal{H}$ and at infinity it is said to be a modular form of weight k with level N. If f is zero at the cusps, then it is a cusp form.

By Lemma 3.2.2, to check if a meromorphic function on the upper half plane is a modular function of level 1, we only need to consider how it transforms under the action of the matrices S and T.

Lemma 3.2.11. *At $i\infty$ the modular function f has Laurent expansion*

$$\sum_{n=-m}^{\infty} a_n e^{2\pi i \tau},$$

where

$$a_n = \int_0^{2\pi} \frac{f(x)}{e^{2\pi i x}} dx.$$

This series converges to $f(\tau)$ in the region $\tau > -\log R/2\pi$, for some real $0 < R < 1$. If f is a modular form, then $a(n) = 0$ for $n < 0$. If f is a cusp form, then $a(n) = 0$ for $n \leq 0$.

Proof. Since $T \in \Gamma_0(N)$ if f is a modular function with level N,

$$f(\tau + 1) = f(\tau) \tag{1}$$

for all $\tau \in \mathcal{H}$. Set $q = e^{2\pi i \tau}$. Because of equality (1), The function $\tilde{f}(q) = f(\tau)$ is a well defined function on the unit disc $D = \{q \in \mathbf{C} : 0 < |q| < 1\}$. Since f is meromorphic at infinity, the function $\tilde{f}$ is meromorphic at 0. Equivalently, $\tilde{f}$ has a pole of order $m \geq 0$ say at 0. Thus there is some $0 < R < 1$ such that for $|q| < R$

$$\tilde{f}(q) = \sum_{n=-m}^{\infty} a_n e^{2\pi i \tau},$$

where

$$a_n = \frac{1}{2\pi i} \oint_{C_r} \frac{\tilde{f}(q)}{q} dq,$$

where $0 < r < R$ (see Corollary C.2.7). The result follows. □

Lemma 3.2.12. *Let f be a modular function. Set $\tau = x + iy \in \mathcal{H}$*

(i) *There is a positive real $N > 0$ such that $\left|e^{-Ny} f(\tau)\right|$ is bounded as $\mathcal{I}(\tau) \to \infty$.*

(ii) *If the function f is invariant under the action of $SL_2(\mathbf{Z})$, then there is a positive real $N > 0$ such that $\left|e^{-N/y} f(\tau)\right|$ is bounded as $\mathcal{I}(\tau) \to 0$.*

Proof. Suppose first that f is holomorphic at infinity. Considering the Taylor series (see Theorem C.2.6) at $q = 0$, it follows that $|f(\tau)|$ is bounded at infinity. Dividing by q^m, where m is the lowest power appearing in the Laurent series of f at infinity, (i) then follows for arbitrary modular functions.

For any rational cusp p of f, there is an element $A \in SL_2(\mathbf{Z})$ such that $A(i\infty) = p$. Since $f(A\tau) = f(\tau)$ and $f(\tau)$ is meromorphic at infinity, (ii) follows.

□

An automorphic form on $\mathcal{H}$ is a generalization of the concept of a modular form. Exercise 3.2.1 says that the quotient space $SL_2(\mathbf{Z}) \backslash \mathcal{H}^*$ is the compact Riemann sphere. Hence the modular group is a Fuschian group.

Definition 3.2.13. *Let Υ be a discrete subgroup of the modular group and $\mathcal{H}_\Upsilon^* = \mathcal{H} \cup \{cusp\ points\ of\ \Upsilon\}$. If the Hausdorff topological space $\Upsilon \backslash \mathcal{H}_\Upsilon^*$ is compact, then Υ is a Fuschian group (of the first kind).*

Definition 3.2.14. *An automorphic form f on $\mathcal{H}$ of weight k ($k \in \mathbf{Z}_+$) for the Fuschian group Υ is a meromorphic function on $\mathcal{H}$, meromorphic at the cusps of Υ such that*

$$f(\frac{a\tau + b}{c\tau + d}) = (c\tau + d)^k f(\tau)$$

for all matrices $\begin{pmatrix} a & b \\ c & d \end{pmatrix} \in \Upsilon$.

Some of the most important examples of modular forms are theta functions of unimodular positive definite lattices. We state some of their properties.

Definition 3.2.15. *The theta function of a positive definite lattice K is the function on the upper half plane $\mathcal{H}$ defined by*

$$\theta_K(\tau) = \sum_{x \in K} q^{x^2/2}$$

for $\tau \in \mathcal{H}$, where $q = e^{2\pi i \tau}$.

Theorem 3.2.16. *Let K be a unimodular positive definite even integral lattice of rank n. Then θ_K is a modular form of weight $n/2$ and level 1 but it is not a cusp form.*

Proof.

$$\theta_K(\tau + 1) = \theta_K(\tau)$$

holds since the lattice K is even. We show that

$$\theta_K(-1/\tau) = (-i\tau)^{\frac{n}{2}} \theta_K(\tau).$$

For a fixed value of τ, set $g(x) = e^{\pi i x^2 \tau}$. By Lemma D.13, its Fourier transform is:

$$(-i\tau)^{-1/2} e^{\frac{-i\pi x^2}{\tau}}.$$

So the Poisson summation formula (see Corollary D.8) for $x = 0$ gives

$$
\begin{aligned}
\theta_K(\tau) &= \sum_{\lambda \in K^*} e^{\pi i \lambda^2 \tau} \\
&= (-i\tau)^{\frac{-n}{2}} \sum_{\lambda \in K} e^{\frac{-i\pi \lambda^2}{\tau}} \\
&= (-i\tau)^{\frac{-n}{2}} \theta_K(\frac{-1}{\tau}).
\end{aligned}
$$

Since the Lattice K is unimodular, n is divisible by 8 (see Lemma 3.1.5). Hence $(-i\tau)^{\frac{-n}{2}} = \tau^{\frac{-n}{2}}$.

To show that the function θ_K is holomorphic on $\mathcal{H}$, we need to show that the series $\sum_{x \in L} q^{x^2/2}$ converges uniformly on $\mathcal{H}$ (see Lemma C.2.5). Let $\tau \in \overline{D}$, where D is the fundamental domain of the modular group. Since $\mathcal{I}(\tau) > 0$ and $|\mathcal{R}(\tau)| \le 1/2$ and

$$|\tau|^2 = \mathcal{R}(\tau)^2 + \mathcal{I}(\tau)^2 \ge 1,$$

it follows that $\mathcal{I}(\tau) \ge \sqrt{3}/2$. Hence,

$$\left| q^{x^2/2} \right| = e^{-\pi \mathcal{I}(\tau) x^2} \le e^{-\pi \sqrt{3} x^2/2}.$$

The series

$$\sum_{x \in K} e^{-\pi \sqrt{3} x^2 / 2}$$

converges by the ratio test in the case when K has rank 1. The rank n case is an immediate consequence. Therefore the series $\sum_{x \in K} q^{x^2/2}$ converges uniformly in $\overline{D}$. Hence the function θ_K is analytic on $\overline{D}$. By Lemma 3.2.9, for any $\tau \in \mathcal{H}$, there exists $A = \begin{pmatrix} a & b \\ c & d \end{pmatrix} \in SL_2(\mathbf{Z})$ such that $A\tau \in \overline{D}$. Thus θ_K is analytic on $\mathcal{H}$ since $\theta_K(A\tau) = (c\tau + d)^{\frac{n}{2}} \theta_K(\tau)$. Since the convergence is uniform,

$$\lim_{\mathcal{I}(\tau) \to \infty} \theta_K(\tau) = \sum_{x \in K} \lim_{\mathcal{I}(\tau) \to \infty} q^{x^2/2}.$$

The lattice K being positive-definite,

$$\lim_{\mathcal{I}(\tau) \to \infty} q^{x^2/2} = \begin{cases} 0 & \text{if } x \neq 0, \\ 1 & \text{if } x = 0 \end{cases}.$$

As a result,

$$\lim_{\mathcal{I}(\tau) \to \infty} \theta_K(\tau) = 1.$$

In particular, the function θ_K is holomorphic at infinity. The function θ_K is not a cusp form since $\lim_{\mathcal{I}(\tau) \to \infty} \theta_K(\tau) \neq 0$. $\square$

We finish this section with the definition of Jacobi forms as their generalizations to higher dimensions will be needed to calculate the Fourier coefficients of the automorphic forms we construct in section 5.3. For details, see [EicZ] and [Borc9, §3].

3.2.17. *A Jacobi form of weight k and index m ($k, m \in \mathbf{Z}_+$) for the modular group $SL_2(\mathbf{Z})$ is a holomorphic function $\phi : \mathbf{C} \times \mathcal{H} \to \mathbf{C}$ satisfying*

(i)

$$\phi\left(\frac{z}{c\tau + \delta}, \frac{a\tau + b}{c\tau + d}\right) = (c\tau + d)^k e^m \left(\frac{c(z^2/2)}{c\tau + d}\right) \phi(z, \tau),$$

for all $\begin{pmatrix} a & b \\ c & d \end{pmatrix} \in SL_2(\mathbf{Z})$;

(ii)

$$\phi(z + \lambda\tau + \mu, \tau) = e^{-m}(z\lambda/2 + \tau\lambda^2/2)\phi(z, \tau),$$

for all $\lambda, \mu \in \mathbf{Z}$;
(iii) $\phi(z, \tau)$ *has a Fourier expansion of the form*

$$\sum_{l, n \in \mathbf{Q}} c(l, n) e^{2\pi i \tau l} e^{2\pi i z n}$$

with $c(l, n) = 0$ for $l \leq n^2/4m$. If $c(l, n) = 0$ for $l = n^2/4m$, then ϕ is said to be a cusp form.

Exercises 3.2

1. (i) Show that the modular group acts continuously on the quotient space $SL_2(\mathbf{Z})\backslash\mathcal{H}^*$ and that this space is isomorphic to $\overline{D} \cup \{\infty\}$.

 (ii) Deduce that the space $SL_2(\mathbf{Z})\backslash\mathcal{H}^*$ is compact and define a natural complex structure on it.

 (iii) Show that the space $SL_2(\mathbf{Z})\backslash\mathcal{H}^*$ has genus 0 and thus that it is isomorphic to the compact Riemann sphere $\mathbf{C} \cup \infty$.

2. Check that the definition of parabolic elements of $SL_2(\mathbf{Z})$ given in Definition 3.2.4 coincides with the usual definition of a parabolic matrix (namely of a matrix with precisely one eigenvalue).

3. Show that for any rational $p \in \mathbf{Q}$, there is a parabolic element $A \in SL_2(\mathbf{Z})$ such that $A(\infty) = p$. Deduce that the set $\mathbf{Q} \cup \{\infty\}$ is the set of cusps of the modular group.

3.3   Vector Valued Modular Functions

The root lattices of the most interesting and important known examples of BKM algebras that are not Kac-Moody Lie algebras are not only even Lorentzian but also unimodular. Their study does not require the theory of vector valued modular functions. However there are examples of BKM algebras with non unimodular Lorentzian lattices. So to have a complete classification of the BKM algebras to which one can associate a modular function, we need to generalize modular functions with values in $\mathbf{C}$ to vector valued modular forms. This is why we give an exposition of some of their properties that we will need in chapter 5. All the definitions and proofs can be simplified in the unimodular case. For a more general detailed study of section 2.3 and 2.4, see [Borc9,11] and [Bar].

In the previous section we considered theta functions for positive definite lattices. In order to motivate the definition of modular functions with values in a vector space, we consider a typical example which will play a fundamental role subsequently, namely the theta function of an arbitrary non-degenerate lattice, not necessarily positive definite nor necessarily unimodular. Let us fix some notation first.

Set M to be an even non-degenerate lattice with bilinear form $(.,.)$ and signature (b^+, b^-) and $G(M)$ to be the Grassmannian of maximal positive definite subspaces of the real vector space $M \otimes_{\mathbf{Z}} \mathbf{R}$. For $v^+ \in G(M)$, set v^- to be its orthogonal complement. So v^+ (resp. v^-) is a maximal positive (resp. negative) definite subspace. We will use the simplified notation

$$\mathbf{e} = \exp(2\pi i).$$

A theta function on M can be defined as a function with values in the group ring $\mathbf{C}[M^*/M]$. The quotient M^*/M being a finite abelian group (see Lemma 3.1.2), this is a finite dimensional vector space. For $\gamma \in M^*/M$, set $e_\gamma \in \mathbf{C}[M^*/M]$ to be the corresponding element. As γ runs through M^*/M, we get a basis $(e_\gamma)_\gamma$ such that for $\gamma, \mu \in [M^*/M]$, $e_\gamma e_\mu = e_{\gamma+\mu}$. There is an inner product on the space $\mathbf{C}[M^*/M]$ defined by

$$(e_\gamma, e_\delta) = \begin{cases} 1 & \text{if } \gamma + \delta = 0 \\ 0 & \text{otherwise.} \end{cases}$$

The complex conjugate of the basis vector e_γ is defined to be $e_{-\gamma}$.

Definition 3.3.1. *For $\tau \in \mathcal{H}$, $\alpha, \beta \in M \otimes \mathbf{R}$, $v^+ \in G(M)$, $\gamma \in M^*/M$, set*

$$\theta_{M+\gamma}(\tau, \alpha, \beta; v^+) = \sum_{\lambda \in M+\gamma} \exp(2\pi i(\tau(\lambda+\beta)_{v^+}^2/2 + \overline{\tau}(\lambda+\beta)_{v^-}^2/2 - (\lambda+\beta/2, \alpha))).$$

The generalized theta function for the lattice M is the $\mathbf{C}[M^*/M]$-valued function defined as follows:

$$\Theta_M(\tau, \alpha, \beta; v^+) = \sum_{\gamma \in M^*/M} \theta_{M+\gamma}(\tau, \alpha, \beta; v^+)e_\gamma.$$

When $\alpha = \beta = 0$, this is the Siegel theta function $\Theta_M(\tau; v^+)$ of the lattice M.

The generalized theta function is a generalization of the classical theta function defined on $\{(\tau, \alpha, \beta) : \alpha \in M \otimes \mathbf{R}, \tau, \beta \in \mathbf{C}, \mathcal{I}(\tau) > 0\}$ (see [Kac14, Chapter 13]).

We study how the modular group acts on the theta function. For $A = \begin{pmatrix} a & b \\ c & d \end{pmatrix} \in SL_2(\mathbf{Z})$, consider the expression

$$(c\tau + d)^{-b^+/2}(c\overline{\tau} + d)^{-b^-/2}\Theta_M(A\tau, a\alpha + b\beta, c\alpha + d\beta; v^+).$$

The rational numbers $b^+/2$ and $b^-/2$ need not be integers. As there are two possibilities for the square root of $c\tau + d$ and $c\overline{\tau} + d$, the above expression for a given matrix A is not in general uniquely defined. So we need to consider the double cover (see Definition B.3.1) of $SL_2(\mathbf{R})$. The next result is an immediate consequence of Theorems B.3.7 and B.3.8.

Corollary/Definition 3.3.2. *The group $SL_2(\mathbf{R})$ has a unique double cover, called the metaplectic group and written $Mp_2(\mathbf{R})$. Its elements are*

$$(\begin{pmatrix} a & b \\ c & d \end{pmatrix}, f),$$

where f is a holomorphic function on the upper half plane such that $f^2(\tau) = c\tau + d$ for all $\tau \in \mathcal{H}$. The group law is given by

$$(A, f(\tau))(B, g(\tau)) = (AB, f(B(\tau))g(\tau))$$

for (A, f), (B, g) in $Mp_2(\mathbf{R})$.

Since the group $SL_2(\mathbf{Z})$ is generated by the matrices $\begin{pmatrix} 1 & 1 \\ 0 & 1 \end{pmatrix}$ and $\begin{pmatrix} 0 & -1 \\ 1 & 0 \end{pmatrix}$, one can deduce the generators of its double cover. For reasons of simplicity we keep the same notation for the generators of $Mp_2(\mathbf{Z})$ and their images in $SL_2(\mathbf{Z})$. Which group they are elements of will be clear from the context.

Lemma 3.3.3.

$$Mp_2(\mathbf{R}) = \langle T, S : S^2 = (ST)^3, S^8 = 1 \rangle,$$

where

$$T = (\begin{pmatrix} 1 & 1 \\ 0 & 1 \end{pmatrix}, 1) \quad and \quad S = (\begin{pmatrix} 0 & -1 \\ 1 & 0 \end{pmatrix}, \sqrt{\tau}).$$

Proof.

$$S^2 = (\begin{pmatrix} -1 & 0 \\ 0 & -1 \end{pmatrix}, i)$$

and so

$$S^4 = (\begin{pmatrix} 1 & 0 \\ 0 & 1 \end{pmatrix}, -1) \quad and \quad S^8 = 1.$$

$$ST = (\begin{pmatrix} 0 & -1 \\ 1 & 1 \end{pmatrix}, \sqrt{\tau + 1}).$$

So

$$(ST)^2 = (\begin{pmatrix} -1 & -1 \\ 1 & 0 \end{pmatrix}, \sqrt{\tau}) \quad and \quad (ST)^3 = (\begin{pmatrix} -1 & 0 \\ 0 & -1 \end{pmatrix}, i).$$

□

We can now not only show how the theta function transforms under the action of the modular group, but also describe an action of the metaplectic group on the group ring.

Theorem 3.3.4. *For* $(A, f) \in Mp_2(\mathbf{Z})$, *where* $A = \begin{pmatrix} a & b \\ c & d \end{pmatrix} \in SL_2(\mathbf{Z})$, *set*

$$\rho_M((A, f))\Theta_M(\tau, \alpha, \beta; v^+) = (c\tau + d)^{-b^+/2}(c\bar{\tau} + d)^{-b^-/2}$$

$$\times \Theta_M(A\tau, a\alpha + b\beta, c\alpha + d\beta; v^+).$$

Then, ρ_M *is a representation of the group* $Mp_2(\mathbf{Z})$ *on the vector space* $\mathbf{C}[M^*/M]$ *given by:*

$$\rho_M(T)(e_\gamma) = \mathbf{e}((\gamma, \gamma)/2)e_\gamma,$$

$$\rho_M(S)(e_\gamma) = \frac{\sqrt{i}^{b^- - b^+}}{\sqrt{|M^*/M|}} \sum_{\delta \in M^*/M} \mathbf{e}(-(\gamma, \delta))e_\delta$$

$\forall \gamma \in M^*/M$. *In particular,* $\rho_M((S^2, i))e_\gamma = i^{b^- - b^+}e_{-\gamma}$.

Proof. The definition of ρ_M implies that it is a group homomorphism and that the identity element acts trivially. Hence ρ_M is a representation of the group

$Mp_2(\mathbf{Z})$. We now find the action of $\rho_M(T)$ and $\rho_M(S)$ on the basis vectors e_γ, $\gamma \in M^*/M$. Rewriting the definition of $\rho_M((A, f))$ for any $(A, f) \in Mp_2(\mathbf{Z})$ in terms of the coordinates of the theta function, we get

$$\sum_{\gamma \in M^*/M} \theta_{M+\gamma}(\tau, \alpha, \beta; v^+)\rho_M((A, f))e_\gamma$$

$$= \sum_{\gamma \in M^*/M} \theta_{M+\gamma}(A\tau, a\alpha + b\beta, c\alpha + d\beta; v^+)e_\gamma \qquad (1)$$

We first consider the element $T \in Mp_2(\mathbf{Z})$. For fixed $\gamma \in M^*/M$,

$$\theta_{M+\gamma}(\tau + 1, \alpha + \beta, \beta; v^+)$$
$$= \sum_{\lambda \in M+\gamma} \mathbf{e}((\tau+1)(\lambda+\beta)^2_{v+}/2 + (\overline{\tau}+1)(\lambda+\beta)^2_{v-}/2 - (\lambda+\beta/2, \alpha+\beta))$$
$$= \sum_{\lambda \in M+\gamma} \mathbf{e}(\tau(\lambda+\beta)^2_{v+}/2 + \lambda^2_{v+}/2 + \beta^2_{v+}/2 + (\lambda_{v+}, \beta_{v+}) \qquad (2)$$
$$+ \overline{\tau}(\lambda+\beta)^2_{v-}/2 + \lambda^2_{v-}/2 + \beta^2_{v-}/2 + (\lambda_{v-}, \beta_{v-}) - (\lambda+\beta/2, \alpha+\beta))$$
$$= \mathbf{e}((\gamma, \gamma)/2) \sum_{\lambda \in M+\gamma} \mathbf{e}(\tau(\lambda+\beta)^2_{v+}/2 + \overline{\tau}(\lambda+\beta)^2_{v-}/2 - (\lambda+\beta/2, \alpha))$$
$$- \mathbf{e}((\gamma, \gamma)/2)\theta_{M+\gamma}(\tau, \alpha, \beta; v^+)$$

since the spaces v^+ and v^- being orthogonal complements,

$$(\lambda_{v-}, \beta_{v-}) + (\lambda_{v+}, \beta_{v+}) = (\lambda, \beta)$$

and

$$\lambda^2_{v-}/2 + \lambda^2_{v+}/2 = \lambda^2/2,$$

and since the lattice M being even, $(\beta/2, \beta) \in \mathbf{Z}$.

Substituting the expression for $\theta_{M+\gamma}(\tau + 1, \alpha + \beta, \beta; v^+)$ given by equality (2) into equality (1) applied to T, we can deduce that

$$\sum_{\gamma \in M^*/M} \theta_{M+\gamma}(\tau, \alpha, \beta; v^+)\rho_M(T)e_\gamma = \sum_{\gamma \in M^*/M} \mathbf{e}((\gamma, \gamma)/2)\theta_{M+\gamma}(\tau, \alpha, \beta; v^+)e_\gamma.$$
$$(3)$$

Since the vectors e_γ are linearly independent and (3) holds for all $\tau \in \mathcal{H}$ and all vectors $\alpha, \beta \in M \otimes_{\mathbf{Z}} \mathbf{R}$, substituting the defining expression for the theta function in both sides of (3) gives the desired answer for $\rho_M(T)$.

We next compute the action of the element S on the basis vectors. For fixed $\gamma \in M^*/M$, by definition of the theta function,

$$\theta_{M+\gamma}(-1/\tau, -\beta, \alpha; v^+)$$
$$= \sum_{\lambda \in M} \mathbf{e}((-1/\tau)(\lambda + \alpha + \gamma)^2_{v+}/2 + (-1/\overline{\tau})(\lambda + \alpha + \gamma)^2_{v-}/2$$
$$- (\lambda + \gamma + \alpha/2, -\beta)).$$

As the coordinate variables of the vector λ_{v+} and λ_{v-} are complementary ones, applying Lemmas D.13 and D.14, we can deduce that the Fourier transform of a summand in the previous sum is:

$$(i/\tau)^{-b^+/2}(-i/\overline{\tau})^{-b^-/2}\mathbf{e}(\tau(\lambda+\beta)_{v+}/2+\overline{\tau}(\lambda+\beta)_{v-}/2-(\lambda+\beta/2,\alpha)-(\lambda,\gamma)).$$

Hence using the Poisson summation formula given in Corollary D.8, we get

$$
\begin{aligned}
&\theta_{M+\gamma}(-1/\tau,-\beta,\alpha;v^+)\\
&=\frac{\sqrt{i}^{\,b^- -b^+}\tau^{\frac{b^-}{2}}}{\sqrt{|M^*/M|}}\overline{\tau}^{\frac{b^+}{2}}\sum_{\lambda\in M^*}\mathbf{e}(\tau(\lambda+\beta)_{v+}^2/2+\overline{\tau}(\lambda+\beta)_{v-}^2/2\\
&\quad-(\lambda+\alpha/2,\beta)-(\lambda,\gamma))\\
&=\frac{\sqrt{i}^{\,b^- -b^+}\tau^{\frac{b^-}{2}}\overline{\tau}^{\frac{b^+}{2}}}{\sqrt{|M^*/M|}}\sum_{\delta\in M^*/M}\sum_{\lambda\in M+\delta}\mathbf{e}(\tau(\lambda+\beta)_{v-}^2/2+\overline{\tau}(\lambda+\beta)_{v+}^2/2\\
&\quad-(\lambda+\alpha/2,\beta)-(\lambda,\gamma))\\
&=\frac{\sqrt{i}^{\,b^- -b^+}\tau^{\frac{b^-}{2}}\overline{\tau}^{\frac{b^+}{2}}}{\sqrt{|M^*/M|}}\sum_{\delta\in M^*/M}\mathbf{e}(-(\delta,\gamma))\\
&\quad\times\sum_{\lambda\in M+\delta}\mathbf{e}(\tau(\lambda+\beta)_{v-}^2/2+\overline{\tau}(\lambda+\beta)_{v+}^2/2-(\lambda+\alpha/2,\beta)).
\end{aligned}
\tag{4}
$$

This last equality follows from the fact that for $\lambda\in M+\delta$, $(\lambda,\gamma)=(\lambda-\delta,\gamma)+(\delta,\gamma)$ and since $(\lambda-\delta,\gamma)\in\mathbf{Z}$ as $\lambda-\delta\in M$, we have $\mathbf{e}(\lambda-\delta,\gamma)=1$. By definition, the right hand side of equality (4) is equal to $\theta_\delta(\tau,\alpha,\beta;v^+)$, which for the same reasons as for T shows that $\rho_M(S)$ is as expected.

It only remains to check that $\rho_M(S^2)$ is as given.

$$
\begin{aligned}
\rho_M(S)^2 e_\gamma &= \frac{i^{b^- -b^+}}{|M^*/M|}\sum_{\delta,\mu\in M^*/M}\mathbf{e}(-(\gamma,\delta))\mathbf{e}(-(\delta,\mu))e_\mu\\
&= \frac{i^{b^- -b^+}}{|M^*/M|}e_{-\gamma}\sum_{\delta,\mu\in M^*/M}\mathbf{e}(-(\gamma+\mu,\delta))e_{\mu+\gamma}\\
&= \frac{i^{b^- -b^+}}{|M^*/M|}e_{-\gamma}\sum_{\mu\in M^*/M}\Big(\sum_{\delta\in M^*/M}\mathbf{e}(-\mu,\delta)\Big)e_\mu.
\end{aligned}
$$

$\square$

When the lattice M is unimodular, ρ_M is just the trivial representation on the one dimensional space $\mathbf{C}$. We now have an idea how to generalize ordinary modular functions to vector valued ones.

Definition 3.3.5. *Let ρ be a representation of the group $Mp_2(\mathbf{R})$, V the corresponding representation space, and $m^+, m^- \in \frac{1}{2}\mathbf{Z}$. A vector valued modular* *form of weight* (m^+, m^-) *and type* ρ *is a real analytic function* F *on the upper half plane* $\mathcal{H}$ *with values in* V *such that*

$$F(\frac{a\tau + b}{c\tau + d}) = (c\tau + d)^{m^+}(c\bar{\tau} + d)^{m^-} \rho((\begin{pmatrix} a & b \\ c & d \end{pmatrix}, f))F(\tau),$$

where $((\begin{pmatrix} a & b \\ c & d \end{pmatrix}, f) \in Mp_2(\mathbf{Z})$.

Note that in this generalized version, a modular form is allowed to be meromorphic (see Definition 3.2.10) at the cusps.

According to this definition, by Lemma 3.2.8, the function

$$\tau \mapsto y$$

defined on the upper half plane is a modular form of weight $(-1, -1)$ and type ρ, where ρ is the trivial representation on the one dimensional space $\mathbf{R}$.

Theorem 3.3.4 says that the Siegel theta function is a modular form of weight $(b^+/2, b^-/2)$ and type ρ_M. Indeed, its definition shows that it is holomorphic on $\mathcal{H}$. As this example shows, in general, a modular form f does not have a Laurent series at infinity (see Lemma 3.2.11) since the action of the element T gives

$$\theta_{M+\gamma}(\tau + 1) = \mathbf{e}((\gamma, \gamma)/2)\theta_{M+\gamma}(\tau)$$

for $\gamma \in M^*/M$ and $(\gamma, \gamma)/2$ may not be an integer. However, rewriting the theta function in a different form, we get

$$\theta_{M+\gamma}(\tau) = \sum_{\lambda \in M+\gamma} \mathbf{e}(\lambda^2/2\tau) \exp(\pi y(\lambda_{v+}^2 - \lambda_{v-}^2))$$

$$= \sum_{m \in \mathbf{Q}} \sum_{k \in \mathbf{Z}} c(m, k)q^m y^{-k}.$$

In the last line, we have replaced $\exp(\pi y(\lambda_{v+}^2 - \lambda_{v-}^2))$ by its infinite series expansion. Note that $c(m, k) = 0$ for $k > 0$ and that $m = \lambda^2/2$ is a rational but not necessarily an integer when $\gamma \neq 0$. This example motivates the next definition:

Definition 3.3.6. *A modular form* F *of weight* (m^+, m^-) *is almost holomorphic if at infinity, its components have Fourier expansions of type*

$$\sum_{m \in \mathbf{Q}} \sum_{k \in \mathbf{Z}} c(m, k)q^m y^{-k},$$

where $c(m, k) = 0$ *either for* $m << 0$ *or for* $k < 0$ *or for* $k >> 0$. *It is said to be holomorphic if* $m^- = 0$ *and* $c(m, k) = 0$ *for all* $k \neq 0$ *and* $m < 0$.

We give a partial justification for this definition. The modular forms we will encounter will be almost holomorphic (except in section 5.4).

Lemma 3.3.7. *If* F *is a modular form of type* ρ_M *then at infinity, its component* F_γ, $\gamma \in M^*/M$ *has Fourier expansion of type*

$$\sum_{m \in \mathbf{Z}} c_{m,\gamma}(y)q^{m+\gamma^2/2},$$

where the coefficients $c_{m,\gamma}(y)$ are functions of y. If F is meromorphic at the cusps, then these coefficients are constants and $c_{m,\gamma}(y) = 0$ for $m \ll 0$.

Proof. Let

$$F(\tau) = \sum_{\gamma \in M^*/M} f_\gamma e_\gamma.$$

Since for any $\gamma \in M^*/M$ $f_\gamma(\tau + 1) = \mathbf{e}((\gamma, \gamma)/2))f_\gamma(\tau)$, setting $g_\gamma(\tau) = \mathbf{e}(-(\gamma, \gamma)/2))$, we get

$$g_\gamma(\tau + 1) = g_\gamma(\tau).$$

Since as a function of x, f_γ is bounded but not necessarily as a function of y, $g_\gamma(\tau) = \sum_{n \in \mathbf{Z}} d_{n,\gamma}(y)e^{2\pi i n x}$, where the coefficients $d_n(y)$ are functions of y. Set $c_{n,\gamma}(y) = d_{n,\gamma}e^{2\pi n y}$. If f_γ is meromorphic at $i\infty$, then so is g. So in this case the coefficients c_n are constants as the Fourier series is the Laurent series at $q = 0$. Hence,

$$f_\gamma(\tau) = \sum_{n \in \mathbf{Z}} c_{n,\gamma}(y)q^{n + \gamma^2/2},$$

proving the result. □

In particular, Lemma 3.3.7 shows that the sum over m in definition 3.3.6 cannot in general be over $\mathbf{Z}$ but has to be over $\mathbf{Q}$ unless the lattice M is unimodular. This is the case even when the modular form F is holomorphic on $\mathcal{H}$ and meromorphic at $i\infty$.

If F and G are two modular functions of type ρ_M, then we will take the product of F and G to be the inner product of F and G.

$$F(\tau)G(\tau) = (F(\tau), G(\tau)),$$

where $(.,.)$ is the inner product on $\mathbf{C}[M^*/M]$ defined earlier in this section.

From now on, we restrict ourselves to the case of an even non-degenerate lattice M of the following type:

$$M = L \oplus L^\perp, \tag{I}$$

where L is a sublattice of rank $(b^+ - 1, b^- - 1)$ and the orthogonal complement of L in M, $L^\perp$ is isomorphic to $II_{1,1}$. Indeed, the most interesting BKM algebras have a root lattice of this type (see Examples 2.3.11.1 and 2.6.40). The lattice M being non-degenerate, so is L. Since the lattice $II_{1,1}$ is self-dual, it follows from (I) that

$$M^* = L^* \oplus L^\perp \tag{II}$$

and so

$$M^*/M = L^*/L. \tag{II'}$$

Set $z \in L^\perp$ to be a primitive vector of norm 0 and $z' \in L^\perp$ to be the vector satisfying $(z, z') = 1$. Note that

$$z'^2 = 0.$$

We want to find an expression for the Siegel theta function of M in terms of the Siegel theta function of the lattice L. For $v^+ \in G(M)$, set $z_{v\pm}$ to be the components of z in $v^\pm$, i.e.

$$z = z_{v^+} + z_{v^-},$$

and $w_1^\pm$ to be the orthogonal complement of the vector $z_{v\pm}$ in $v^\pm$.

Lemma 3.3.8. *The subspace $w_1{}^+$ (resp. $w_1{}^-$) is a maximal positive (negative) definite subspace of $((L \otimes_{\mathbf{Z}} \mathbf{R}) \oplus \mathbf{R}z)/\mathbf{R}z$.*

Proof. By definition of the subspace $w_1^\pm$, $(w_1^\pm, z) = 0$ since $z = z_{v^+} + z_{v^-}$. Hence the result follows by maximality of $v^\pm$. $\qquad\qquad\square$

Set $w^\pm \subset L \otimes \mathbf{R}$ to be the subspace of vectors $\lambda \in L \otimes_{\mathbf{Z}} \mathbf{R}$ such that $\lambda + az \in w_1^\pm$ for some $a \in \mathbf{R}$. By Lemma 3.3.8, w^+ (resp. w^-) is a maximal positive (negative) definite subspace of $L \otimes_{\mathbf{Z}} \mathbf{R}$. For $\lambda \in M^* \otimes \mathbf{R}$, let $\lambda_{w\pm} \in w^\pm$ be the orthogonal projection of $\lambda_{v\pm}$ in $w^\pm$.

Lemma 3.3.9. *For all $\lambda \in M \otimes \mathbf{R}$,*

$$\lambda_{v\pm} \equiv \lambda_{w\pm} + \frac{(\lambda, z_{v\pm})}{z_{v\pm}^2} z_{v\pm} \quad (\mathrm{mod}\ \mathbf{R}z)$$

and

$$\lambda_{v\pm}^2 = \lambda_{w\pm}^2 + \frac{(\lambda, z_{v\pm})^2}{z_{v\pm}^2}.$$

Proof.

$$\lambda_{v\pm} = \lambda_{w_1^\pm} + az_{v\pm}$$

by definition of the subspaces $w_1^\pm$. Hence

$$(\lambda_{v\pm}, z_{v\pm}) = az_{v\pm}^2$$

since

$$(\lambda_{w_1^\pm}, z_{v\pm}) = 0. \tag{1}$$

Moreover, by orthogonality of the vector space v^+ and v^-, $(\lambda_{v\pm}, z_{v\pm}) = (\lambda, z_{v\pm})$. Since by definition of the subspaces $w^\pm$, $\lambda_{w_1}^{\pm}{}^2 = \lambda_w^{\pm 2}$, the second equality then follows from (1). $\qquad\qquad\square$

The Siegel theta function $\Theta_M(\tau, v^+)$ is expressed in terms of vectors in M. The first step is to find an expression for it in terms of vectors modulo z. This in turn leads to the expression we want in terms of vectors in the lattice L. Set

$$J = L \oplus \mathbf{Z}z'.$$

We have to be careful since the bilinear form is degenerate on J.

By definition of Siegel theta functions, we need to characterize maximal positive and negative definite sublattices of L in terms of those of M. So we start by doing this.

Lemma 3.3.10. *For* $\tau \in \mathcal{H}$, $v^+ \in G(M)$, $\gamma \in M^*/M$,

$$\theta_{M+\gamma}(\tau, v^+) = (2yz_{v+}^2)^{\frac{-1}{2}} \sum_{\lambda \in J+\gamma} \sum_{n \in \mathbf{Z}} \mathbf{e}(\tau\lambda_{w+}^2/2 + \overline{\tau}\lambda_{w-}^2/2$$

$$-n(\lambda, (z_{v+} - z_{v-})/2z_{v+}^2) - \frac{|(\lambda, z)\tau + n|^2}{4iyz_{v+}^2}).$$

Proof. For $\tau \in \mathcal{H}$, $v^+ \in G(M)$, $\gamma \in M^*/M$, applying the definition of the Seigel theta function, we get

$$\theta_{M+\gamma}(\tau, v^+) = \sum_{\lambda \in J+\gamma} \sum_{n \in \mathbf{Z}} \mathbf{e}(\tau(\lambda + nz)_{v+}^2/2 + \overline{\tau}(\lambda + nz)_{v-}^2/2).$$

Set

$$g(\lambda, n) = \mathbf{e}(\tau(\lambda + nz)_{v+}^2/2 + \overline{\tau}(\lambda + nz)_{v-}^2/2)$$

for $\lambda \in (M/z) + \gamma$, $n \in \mathbf{Z}$. By the Poisson summation formula (Corollary D.8),

$$\theta_{M+\gamma}(\tau, v^+) = \sum_{\lambda \in J+\gamma} \sum_{n \in \mathbf{Z}} \hat{g}(\lambda, n), \tag{1}$$

where $\hat{g}(\lambda, n)$ is the Fourier transform of $g(\lambda, n)$ with respect to the variable n. Rewriting $g(\lambda, n)$ makes the calculation of $\hat{g}(\lambda, n)$ easier: since $\lambda = \lambda_{v-} + \lambda_{v+}$ and $0 = z_{v-}^2 + z_{v+}^2$,

$$g(\lambda, n) = \mathbf{e}((\tau - \overline{\tau})z_{v+}^2 n^2/2 + (\tau(\lambda, z_{v+}) + \overline{\tau}(\lambda, z_{v-}))n/2 + \tau\lambda_{v+}^2/2 + \overline{\tau}\lambda_{v-}^2/2).$$

Applying lemmas D.13 and D.14, we get

$$\hat{g}(\lambda, n) = (2yz_{v+}^2)^{\frac{-1}{2}}\mathbf{e}(-\frac{n^2}{2(\tau - \overline{\tau})z_{v+}^2} - n(\tau(\lambda, z_{v+}) + \frac{\overline{\tau}(\lambda, z_{v-})}{(\tau - \overline{\tau})z_{v+}^2}$$

$$+\tau\lambda_{v+}^2/2 + \overline{\tau}\lambda_{v-}^2/2 - \frac{(\tau(\lambda, z_{v+}) + \overline{\tau}(\lambda, z_{v+}))^2}{2(\tau - \overline{\tau})z_{v+}^2} \tag{2}$$

since

$$\tau - \overline{\tau} = 2iy \tag{3}$$

Let us compute the right hand side of equality (2).

$$\frac{(\tau(\lambda, z_{v+}) + \overline{\tau}(\lambda, z_{v-}))^2}{2(\tau - \overline{\tau})z_{v+}^2} = \frac{\tau^2(\lambda, z_{v+})^2 + \overline{\tau}^2(\lambda, z_{v-})^2 + 2|\tau|^2(\lambda, z_{v+})(\lambda, z_{v-})}{2(\tau - \overline{\tau})z_{v+}^2}$$

From Lemma 3.3.9 we get

$$\tau\lambda_{v+}^2/2 + \overline{\tau}\lambda_{v-}^2/2 = \tau\lambda_{w+}^2/2 + \overline{\tau}\lambda_{w-}^2/2 + \frac{\tau(\lambda, z_{v+})^2}{2z_{v+}^2} + \frac{\overline{\tau}(\lambda, z_{v-})^2}{2z_{v-}^2}$$

$$= \tau\lambda_{w+}^2/2 + \overline{\tau}\lambda_{w-}^2/2 + \frac{\tau(\lambda, z_{v+})^2 - \overline{\tau}(\lambda, z_{v-})^2}{2z_{v+}^2}$$

$$= \tau\lambda_{w+}^2/2 + \overline{\tau}\lambda_{w-}^2/2 + \frac{\tau(\tau - \overline{\tau})(\lambda, z_{v+})^2 - \overline{\tau}(\tau - \overline{\tau})(\lambda, z_{v-})^2}{2(\tau - \overline{\tau})z_{v+}^2}.$$

Hence

$$
\begin{aligned}
\hat{g}(\lambda, n) &= (2yz_{v+}^2)^{\frac{-1}{2}}\mathbf{e}\Big(-\frac{n^2}{2(\tau-\overline{\tau})z_{v+}^2} - \frac{n(\tau(\lambda, z_{v+}) + \overline{\tau}(\lambda, z_{v-}))}{(\tau-\overline{\tau})z_{v+}^2} \\
&\quad + \frac{\tau\lambda_{w+}^2}{2} + \frac{\overline{\tau}\lambda_{w-}^2}{2} - \frac{|\tau|^2((\lambda, z_{v+})^2 + (\lambda, z_{v-})^2 + 2(\lambda, z_{v+})(\lambda, z_{v-}))}{2(\tau-\overline{\tau})z_{v+}^2}\Big) \\
&= (2yz_{v+}^2)^{\frac{-1}{2}}\mathbf{e}\Big(\frac{\tau\lambda_{w+}^2}{2} + \frac{\overline{\tau}\lambda_{w-}^2}{2} \\
&\quad - \frac{n^2 + 2n(\tau(\lambda, z_{v+}) + \overline{\tau}(\lambda, z_{v-})) + |\tau|^2(\lambda, z)^2}{2(\tau-\overline{\tau})z_{v+}^2}\Big).
\end{aligned}
$$

The result then follows from equalities (3) and (1).   □

Lemma 3.3.10 now allows us to write the Siegel theta function of the lattice M in terms of a generalized theta function of the lattice L.

Theorem 3.3.11. *Let* $\mu = -z' + \frac{(z_{v-}, z')}{z_{v+}^2}z_{v+} + \frac{(z_{v+}, z')}{z_{v-}^2}z_{v-}$. *Then* $\mu \in L \otimes_{\mathbf{Z}} \mathbf{R}$ *and for all* $\tau \in \mathcal{H}$, $v^+ \in G(M)$, $\gamma \in M^*/M$,

$$
\theta_{M+\gamma}(\tau; v^+) = (2yz_{v+}^2)^{\frac{-1}{2}} \sum_{c,d \in \mathbf{Z}} \mathbf{e}\Big(-\frac{|c\tau+d|^2}{4iyz_{v+}^2}\Big)\theta_{L+\gamma}(\tau, d\mu, -c\mu; w^+).
$$

Proof. To show that $\mu \in L \otimes_{\mathbf{Z}} \mathbf{R}$, it suffices to check that the vector μ is orthogonal to both the vectors z and z':

$$
\begin{aligned}
(\mu, z) &= -1 + ((z_{v+}, z)(z_{v-}, z')/z_{v+}^2 + (z_{v-}, z)(z_{v+}, z')/z_{v-}^2 \\
&= -1 + (z_{v+}, z_{v+})(z_{v-}, z')/z_{v+}^2 + (z_{v-}, z_{v-})(z_{v+}, z')/z_{v-}^2 \\
&= -1 + (z, z') \\
&= 0
\end{aligned}
$$

and

$$
(\mu, z') = ((z_{v+}, z')(z_{v-}, z')(1/z_{v+}^2 + 1/z_{v-}^2) = 0
$$

since $z_{v+}^2 + z_{v-}^2 = z^2 = 0$.

For any $\lambda \in J + \gamma$, there is a unique vector $\lambda_L \in L + \gamma$ and a unique integer $c \in \mathbf{Z}$,

$$
\lambda = \lambda_L + cz'. \tag{1}
$$

From Lemma 3.3.10, we therefore get

$$
\theta_{M+\gamma}(\tau, v^+) = (2yz_{v+}^2)^{\frac{-1}{2}} \sum_{c,d \in \mathbf{Z}} \sum_{\lambda_L \in L+\gamma} \mathbf{e}\Big(-\frac{|c\tau+d|^2}{4iyz_{v+}^2}\Big) \tag{2}
$$

$$
\times \mathbf{e}(\tau(\lambda_L + cz')_{w+}^2/2 + \overline{\tau}(\lambda_L + cz')_{w-}^2/2 - d(\lambda_L + cz', (z_{v+} - z_{v-})/2z_{v+}^2)).
$$

$(z', z) = 1$.

On the other hand (remember that $L^*/L = M^*/M$ by (II')),

$$\theta_{L+\gamma}(\tau, d\mu, -c\mu; w^+)$$

$$= \sum_{\lambda_L \in L+\gamma} \mathbf{e}(\tau(\lambda_L - c\mu)^2_{w^+}/2 + \overline{\tau}(\lambda_L - c\mu)^2_{w^-}/2 - (\lambda_L - c\mu/2, d\mu)). \quad (3)$$

Since $-\mu + z'$ is a linear combination of the vectors $z_{v\pm}$, by definition of $w_1^{\pm}$,

$$(\lambda_L - c\mu)_{w_1^{\pm}} = (\lambda_L + cz')_{w_1^{\pm}}.$$

Hence, by Lemma 3.3.8 (ii),

$$(\lambda_L - c\mu)^1_{w^{\pm}} = (\lambda_L + cz')^2_{w^{\pm}}.$$

Since $(\lambda_L, z') = 0 = (\lambda_L, z)$ and $z_{v-} = z - z_{v+}$,

$$(\lambda_L, \mu) = (\lambda_L, (z_{v+} - z_{v-})/2z_{v+}^2). \quad (4)$$

We have already shown that $(\mu, z') = 0$. Hence, as $(z', z') = 0$ and $(z_{v+}, z_{v-}) = 0$, $z_{v+}^2 + z_{v-}^2 = z^2 = 0$,

$$(\mu, \mu) = (\frac{(z_{v-}, z')}{z_{v+}^2}z_{v+} + \frac{(z_{v+}, z')}{z_{v-}^2}z_{v-}, \frac{(z_{v-}, z')}{z_{v+}^2}z_{v+} + \frac{(z_{v+}, z')}{z_{v-}^2}z_{v-})$$

$$= \frac{(z_{v-}, z')^2}{z_{v+}^2} + \frac{(z_{v+}, z')^2}{z_{v-}^2}$$

$$= \frac{(z_{v-}, z')^2 - (z', z_{v+})^2}{z_{v+}^2}$$

$$= \frac{(z_{v-} - z_{v+}, z')(z', z_{v+} + z_{v-})}{z_{v+}^2}$$

$$= \frac{(z_{v-} - z_{v+}, z')}{z_{v+}^2}.$$

This together with (2), (3) and (4) prove the Theorem.   □

Exercises 3.3

1. Suppose that M is an arbitrary lattice with signature (b^+, b^-) (i.e. not necessarily satisfying condition (I)) containing a primitive vector z of norm 0. Let $N \in \mathbf{N}$ to be minimal such that $(z, \lambda) = N$ for some $\lambda \in M$. Set $z' \in M^*$ to be a vector satisfying $(z, z') = 1$ and write $L = (M \cap z^{\perp})/z$. Let $w^{\pm}$ be the orthogonal complement of $z_{v\pm}$ in $v^{\pm}$. Set $\mu = -z' + z_{v+}/2z_{v+}^2 + z_{v-}/2z_{v-}^2$.

(i) Show that $|M^*/M| = N^2|L^*/L|$.

(ii) Check that $\mu \in L \otimes_{\mathbf{Z}} \mathbf{R}$.

(iii) Show that for all $\tau \in \mathcal{H}$, $v^+ \in G(M)$, $\gamma \in M^*/M$,

$$\theta_{M+\gamma}(\tau; v^+, p) =$$

$$(2yz_{v^+}^2)^{\frac{-1}{2}} \sum_{\substack{c \equiv (\gamma, z) \pmod{N} \\ d \in \mathbf{Z}}} \mathbf{e}\left(-\frac{|c\tau + d|^2}{4iyz_{v^+}^2} - (\gamma, z')d + \frac{(z', z')cd}{2} \right)$$

$$\times \theta_{L+(\gamma-cz')}(\tau, \mu d, -c\mu; w^+),$$

where $L + (\gamma - cz')$ is the coset of L in L^* given by $L^* \cap (M + (\gamma - cz'))/\mathbf{Z}z$.
For a solution, see [Borc11, Theorem 5.2].

2. We suppose that the assumptions of Exercise 1 hold. Let

$$F_M = \sum_{\gamma \in M^*/M} = f_{M+\gamma} e_\gamma$$

be a modular form of type ρ_M and weight $(-b^-/2, -b^+/2)$ and $F_L = \sum_{\gamma \in L^*/L} = f_\gamma e_\gamma$, where

$$f_\gamma(\tau, \alpha, \beta) = \sum_{\substack{\delta \in M^*/M \\ \delta|L=\gamma}} \mathbf{e}(-(\delta, \alpha z') - \alpha\beta(z', z')/2) f_{M+\gamma+\beta z'}(\tau)$$

and $\delta|L$ is the restriction of the homomorphism $\delta : M \to \mathbf{Z}$ to L.

(i) Show that $\{\delta \in M^* : \delta|L = 0\} = \{mz/N : m \in \mathbf{Z}\}$.
(ii) Let $\gamma \in L^*/L$ and $\delta \in M^*/M$ such that $\delta|L = \gamma$. Deduce that $(\delta^2 - \gamma^2)/2 \in \mathbf{Z}$ and $\lambda \in M^*/M$ such that $\delta|L = \gamma$ if and only if $\lambda = \delta + mz/N$ for some $m \in \mathbf{Z}/N\mathbf{Z}$.
(iii) Show that for all $\left(\begin{pmatrix} a & b \\ c & d \end{pmatrix}, \sqrt{c\tau + d} \right) \in M_{p_2}(\mathbf{Z})$,

$$F_L\left(\frac{a\tau + b}{c\tau + d}, a\alpha + b\beta, c\alpha + d\beta \right)$$

$$= (c\tau + d)^{-b^-/2}(c\bar{\tau} + d)^{-b^+/2}\rho_L((a \quad b)\, cd\, \sqrt{c\tau + d})F_L(\tau, \alpha, \beta).$$

For a solution, see [Borc11, Theorem 5.3].

3.4 The Singular Theta Correspondence

In this section we explain a generalized version of the theory of theta correspondence for lattices that are not necessarily positive definite. In other words, we define a correspondence between modular forms of weight $(\frac{b^+ - b^-}{2}, 0)$ and type ρ_M and a class of meromorphic functions on the Grassmannian $G(M)$, which has the structure of a Hermitian symmetric space (see Definition B.2.22 and Lemma 3.5.1). In this generalization, the functions obtained from the modular forms have singularities. The method explained here is a variant of the Rankin-Selberg one (frequently used in the theory of modular forms) and was given by Borcherds in [Borc11]. It is based on the work of Harvey and Moore [HarvM, Appendix A].

Let F be an almost holomorphic modular function on the upper half plane (i.e. holomorphic on $\mathcal{H}$ and meromorphic at the cusps) of weight $(\frac{b^+ - b^-}{2}, 0)$ and of type ρ_M. Then,

$$F(\tau) = \sum_{\gamma \in M^*/M} f_\gamma(\tau) e_\gamma$$

for some complex valued functions f_γ on $\mathcal{H}$. By Lemma 3.3.7, for each $\gamma \in M^*/M$,

$$f_\gamma = \sum_{m \in \mathbf{Q}} c_\gamma(m) q^n,$$

where $c_\gamma(m) \in \mathbf{R}$ for all $\gamma \in M^*/M$ and $c_\gamma(m) = 0$ for sufficiently small values of m. Set

$$F_M(\tau) = y^{b^+/2} F(\tau).$$

Then, Lemma 3.2.8 implies that F_M is a vector valued modular function on $\mathcal{H}$ of weight $(-b^-/2, -b^+/2)$ and type ρ_M which is also almost holomorphic.

$$F_M(\tau) = \sum_{\gamma \in M^*/M} f_{M+\gamma}(\tau) e_\gamma,$$

for some complex valued functions $f_{M+\gamma}$ on $\mathcal{H}$. Comparing with the coordinates of F, we get

$$
\begin{aligned}
f_{M+\gamma}(\tau) &= f_\gamma(\tau) y^{b^+/2} \\
&= \sum_{m \in \mathbf{Z}} c_\gamma(m) q^m y^{b^+/2} \\
&= \sum_{m \in \mathbf{Q}} c_{\gamma,m}(y) \mathbf{e}(mx),
\end{aligned}
$$

where

$$c_{\gamma,m}(y) = c_\gamma(m) y^{b^+/2} \exp(-2\pi y).$$

We can easily deduce the following relation between the components of the modular function F. We will not need it in this chapter but in chapter 5.

Lemma 3.4.1. *For any* $\gamma \in M^*/M$, $m \in \mathbf{Q}$, $c_\gamma(m) = c_{-\gamma}(m)$.

Proof. Apply the element $(S^2, i) \in Mp_2(\mathbf{Z})$ to the modular form F of weight $(b^+ - b^-)/2$. We get

$$
\begin{aligned}
f_{-\gamma}(\tau) &= (-1)^{(b^+ - b^-)/2} i^{b^- - b^+} f_\gamma(\tau) \\
&= (-1) i^{-b^-} i^{b^-} i^{-2} f_\gamma(\tau) \\
&= (-1)^2 f_\gamma(\tau) \\
&= f_\gamma(\tau).
\end{aligned}
$$

The result then follows. $\square$

For a reminder about (Lebesgue) integrals on the upper half plane, see Definition C.1.9 and Corollary C.1.12.

Definition 3.4.2. *The theta transform of the function F is the following integral:*

$$\Phi_M(v^+, F) = \int\int_{SL_2(\mathbf{Z})\backslash\mathcal{H}} F_M(\tau)\overline{\Theta_M}(\tau, v^+) \, dxdy/y^2$$

For a definition of the product $F_M\overline{\Theta}_M$, see §3.3 (paragraph following Corollary 3.3.7). From the definition we see that, the modular function F being fixed, Φ_M is a function on the Grassmannian $G(M)$. Foremost, we need to make sure that this integral makes sense. First we show that it is independent of the choice of a fundamental domain (see Definition 3.2.7) of $\Gamma = SL_2(\mathbf{Z})$. Then, we study its convergence.

Lemma 3.4.3. *The differential* $F_M(\tau)\overline{\Theta_M}(\tau, v^+) \, dxdy/y^2$ *is invariant under the action of* $SL_2(\mathbf{R})$.

Proof. Let $A = \begin{pmatrix} a & b \\ c & d \end{pmatrix}$ and $(A, f) \in Mp_2(\mathbf{Z})$. Then,

$$F_M(A\tau) = (c\tau + d)^{-b^-/2}(c\bar{\tau} + d)^{-b^+/2}\rho_M((A, f))F_M(\tau)$$

and applying Theorem 3.3.4,

$$\overline{\Theta_M}(A\tau; v^+) = (c\tau + d)^{b^-/2}(c\bar{\tau} + d)^{b^+/2}\overline{\rho_M((A, f))\Theta_M}(\tau; v^+).$$

From Lemma B.2.25 we can deduce that

$$(A, f)^*(dxdy/y^2) = dxdy/y^2.$$

Hence,

$$F_M(A\tau)\overline{\Theta_M}(A\tau; v^+) = \rho_M((A, f))F_M(\tau)\overline{\rho_M((A, f))\Theta_M}(\tau; v^+).$$

So to finish the proof, by Lemma 3.3.3, it suffices to show that the right hand side of this equality is invariant when $A = T$ and $A = S$. Note that

$$\overline{\Theta_M}(\tau; v^+) = \sum_{\gamma \in M^*/M} \overline{\theta_{M+\gamma}}e_{-\gamma}.$$

Suppose first that $A = T$. Then,

$$\rho_M((T, f))F_M(\tau)\overline{\Theta_M}(\tau; v^+)$$

$$= \sum_{\gamma \in M^*/M} f_{M+\gamma}(\tau)\overline{\theta_{M+\gamma}}(\tau; v^+)e((\gamma, \gamma)/2)e(-(\gamma, \gamma)/2)$$

$$= \sum_{\gamma \in M^*/M} f_{M+\gamma}(\tau)\overline{\theta_{M+\gamma}}(\tau; v^+)$$

$$= F_M(\tau)\overline{\Theta_M}(\tau; v^+).$$

Suppose now that $A = S$. Then,

$$\rho_M((S,f))F_M(\tau)\overline{\rho_M((S,f))\Theta_M}(\tau;v^+)$$

$$= \frac{1}{|M^*/M|} \sum_{\gamma,\delta \in M^*/M} f_{M+\gamma}(\tau)\mathbf{e}(-\gamma,\delta)e_\delta \sum_{\gamma,\delta \in M^*/M} \overline{\theta_{M+\gamma}}(\tau)\mathbf{e}(\gamma,\delta)e_{-\delta}$$

$$= \frac{1}{|M^*/M|} \sum_{\gamma \in M^*/M} |M^*/M| f_{M+\gamma}(\tau)\overline{\theta_{M+\gamma}}(\tau)$$

$$= F_M(\tau)\overline{\Theta_M}(\tau;v^+).$$

$\square$

Corollary 3.4.4. *The Theta transform of the modular function F is independent of the choice of the fundamental domain of the modular group.*

Both the modular function F_M and the Seigel theta function are holomorphic and hence continuous on the upper half plane and are possibly monomorphic only at cusp points. Hence, if we exclude from the closure of the fundamental domain $\overline{D}$ a neighbourhood of $i\infty$, the latter being the only cusp point in $\overline{D}$, then the resulting integral is convergent. So we only need to study what happens near $i\infty$, or equivalently as $y \to \infty$. Set

$$D_l = \{\tau \in D : Im(\tau) \le l\}.$$

Lemma 3.4.5. *The integral*

$$G_M(l,s;v^+) = \int\int_{D_l} F_M(\tau)\overline{\Theta}_M(\tau;v^+)y^{-s}\,dxdy/y^2.$$

exists for all $l > 0$, $s \in \mathbf{C}$ and $v^+ \in G(M)$.

Proof. Both the functions F_M and Θ_M are analytic on $\mathcal{H}$. Hence, the existence follows from the fact that D_l is a compact closed subset of $\overline{D}$ on which the integrand is analytic. $\square$

Definition 3.4.6. *Let f be a function on the upper half plane. If the limit*

$$g(s) = \lim_{l\to\infty} \int\int_{D_l} f(\tau)y^{-s}\,dxdy/y^2$$

exists for $s \in \mathbf{C}$ such that $\mathcal{R}(s) >> 0$ and can be continued to a meromorphic function on $\mathbf{C}$, then

$$\int\int_{SL_2(\mathbf{Z})\backslash\mathcal{H}} f(\tau)y^{-s}\,dxdy/y^2$$

is defined to be the constant term of the Laurent expansion of the function $g(s)$ at $s = 0$.

We next show that for any $v^+ \in G(M)$, the limit function of s

$$G_M(s;v^+) = \lim_{l\to\infty} G_M(l,s;v^+)$$

exists and that, in particular, it satisfies the conditions of Definition 3.4.6. We know what a singularity of a complex function (see Definition C.2.1) or more generally of a function on a Hermitian symmetric space is. We need a further definition in order to deal with the problems that arise from singular points.

Definition 3.4.7. *Let f be a function on a Hermitian symmetric space V. A point $v \in V$ is a singularity point of type g for the function f, where g is a function on V if the function $f - g$ can be defined on a subset of V of co-dimension 1 in such a way that it is analytic at the point v.*

We next study the singularities of a real function given by an improper integral which we will encounter frequently.

Lemma 3.4.8. *For $s \in \mathbf{C}$, $r \in \mathbf{R}$, $r \geq 0$, set*

$$f(r) = \int_1^\infty e^{-ry} y^{s-1} dy.$$

The function f has a singularity at $r = 0$ of type

$$\begin{cases} -(-r)^{-s} \log(r)/(-s)! & \text{if } s \in \mathbf{Z}_- \\ r^{-s} \Gamma(s), & \text{otherwise} \end{cases}$$

where Γ is the Gamma function (see §C.4 (i)). In particular, the function f has no singularities at $r > 0$ when $\mathcal{R}(s) > 0$.

Proof. To study the singularity at $r = 0$, we consider three cases.
Case 1: $\mathcal{R}(s) > 0$.
   The integral

$$\int_0^1 \exp(-ry) y^{s-1} dy$$

exists at all $r \geq 0$. This follows by replacing $\exp(-ry)$ by it Taylor series and integrating termwise since

$$\sum_{n=0}^\infty \frac{(-r)^n}{n!} \int_0^1 y^{n+s-1} dy = -\sum_{n=0}^\infty \frac{(-r)^n}{n!(n+s)}$$

and the latter series converges absolutely when $\mathcal{R}(s) > 0$.
   Hence the singularity of f at any point $r \geq 0$ is of type

$$\int_0^\infty \exp(-ry) y^{s-1} dy = r^{-s} \int_0^\infty e^{-y} y^{s-1} dy$$

$$= r^{-s} \Gamma(s)$$

by Definition C.4.1.
Case 2: $s = 0$.
   Integrating by parts, we get

$$f(r) = \int_1^\infty \exp(-ry) y^{-1} dy = r \int_1^\infty \exp(-ry) \log(y) dy.$$

As above, the singularity of $f(r)$ at $r = 0$ is of type

$$r \int_0^\infty \exp(-ry) \log(y) dy = \int_0^\infty \exp(-y) \log(y/r) dy$$

$$= \int_0^\infty \exp(-y) \log(y) dy - \int_0^\infty \exp(-y) \log(r) dy.$$

Since

$$\int_0^\infty \exp(-y) \log(y) dy \leq \int_0^\infty \exp(-y) y dy = \Gamma(2) = 2,$$

the integral $\int_0^\infty \exp(-y) \log(y) dy$ exists. As it is independent of r, $f(r)$ has a singularity of type

$$-\log(r) \int_0^\infty \exp(-y) dy = -\log(r)$$

at $r = 0$.

Case 3: $s < 0$.

Integrating by parts, we get

$$f(r) = \frac{1}{s} + \frac{r}{s} \int_1^\infty \exp(-ry) y^s dy.$$

As $s \neq 0$, the function f has a singularity of type $r \int_1^\infty \exp(-ry) y^s dy$ at $r = 0$. Hence continuing m times where m is the smallest positive integer such that $s + m \geq 0$ and applying the second case when s is an integer and the first one otherwise, the result follows.   □

We continue with two technical calculations. We state them as separate Lemmas as they will be used in several subsequent proofs.

Lemma 3.4.9.

$$\sum_{m \in \mathbb{Q}} c_\gamma(m) \int_{-1/2}^{1/2} \exp(2\pi i x(m - \lambda^2/2)) dx = \begin{cases} \sum_{m \in \mathbb{Z}} c_\gamma(\lambda^2/2) & \text{if } m = \lambda^2/2 \\ 0 & \text{otherwise.} \end{cases}$$

Proof. As the modular forms $\overline{\theta}_{M+\gamma}$ and f_γ have weight (b^-, b^+) and $(-b^-, -b^+)$ respectively, $f_\gamma \overline{\theta}_{M+\gamma}$ is a scalar of weight 0. In other words, it is invariant under the action of $SL_2(\mathbb{Z})$. Hence,

$$\int_{y \geq 1} \int_{-1/2}^{1/2} f_\gamma(\tau) \overline{\theta}_{M+\gamma}(\tau; v^+) y^{b^+/2} \, dx dy/y^2$$

$$= \int_{y \geq 1} \int_{-1/2}^{1/2} f_\gamma(\tau + 1) \overline{\theta}_{M+\gamma}(\tau + 1; v^+) y^{b^+/2} \, dx dy/y^2.$$

This implies that

$$\sum_{m\in Q} c_\gamma(m) \int_{-1/2}^{1/2} \exp(2\pi i x(m - \lambda^2/2))dx$$

$$= \sum_{m\in Q} c_\gamma(m) \int_{-1/2}^{1/2} \exp(2\pi i(x+1)(m - \lambda^2/2))dx$$

$$= \sum_{m\in Q} c_\gamma(m)\exp(2\pi i(m - \lambda^2/2)) \int_{-1/2}^{1/2} \exp(2\pi i x(m - \lambda^2/2))dx.$$

The result follows. $\qquad\square$

Lemma 3.4.10.

(i) As a function of s,

$$\int\int_{D_1} F(\tau)\overline{\Theta}(\tau; v^+) y^{b^+/2-s}\, dxdy/y^2$$

is analytic on **C**.

(ii)

$$\int_{y\geq 1}\int_{|x|\leq 1/2} F(\tau)\overline{\Theta}(\tau; v^+) y^{b^+/2-s}\, dxdy/y^2$$

$$= \sum_{\lambda\in M^*} c_\lambda(\lambda^2/2) \int_1^\infty \exp(-2\pi y\lambda_{v+}^2) y^{-2+b^+/2-s} dy.$$

(iii) As a function of s, $\sum_{\substack{\lambda\in M^* \\ \lambda_{v+}\neq 0}} \int_1^\infty \exp(-2\pi y\lambda_{v+}^2) y^{-2+b^+/2-s} dy$ is analytic on **C**.

Proof. Since the region

$$D_1 = \{y\in D : y \leq 1\} = \{1/2 \leq y \leq 1, |x| \leq 1/2, x^2 + y^2 \geq 1\}$$

is closed and compact and the functions F and Θ are holomorphic on the upper half plane $\mathcal{H}$, part (i) follows.
We next prove (ii).

$$\lim_{l\to\infty} \int_{1\leq y\leq l}\int_{|x|\leq 1/2} F(\tau)\overline{\Theta}(\tau; v^+) y^{b^+/2-s}\, dxdy/y^2 \qquad (1)$$

$$= \sum_{\gamma\in M^*/M} \int_{y\geq 1}\int_{|x|\leq 1/2} f_\gamma(\tau)\overline{\theta}_{M+\gamma}(\tau; v^+) y^{b^+/2-s}\, dxdy/y^2$$

since the group M^*/M being finite, the summation and integration order can be interchanged. By definition,

$$\theta_{M+\gamma}(\tau; v^+) = \sum_{\lambda\in M+\gamma} \mathbf{e}(\tau\lambda_{v+}^2/2 + \overline{\tau}\lambda_{v-}^2/2)$$

and

$$f_\gamma(\tau) = \sum_{m \in \mathbb{Q}} c_\gamma(m) \mathbf{e}(m\tau).$$

Both these series are absolutely convergent on $\mathcal{H}$ since both the functions are holomorphic on the upper half plane. In (1), we replace $\theta_{M+\gamma}$ and f_γ with their respective series. Before doing this, note that

$$2\pi i(m\tau - \bar{\tau}\lambda_{v+}^2/2 + \tau\lambda_{v-}^2/2 = 2\pi i x(m - \lambda^2/2) + 2\pi y(-m - \lambda_{v+}^2/2 + \lambda_{v-}^2/2) \quad (2)$$

since $\lambda^2 = \lambda_{v+}^2 + \lambda_{v-}^2$. The above two series being absolutely convergent, so is their product

$$\sum_{\lambda \in M+\gamma} \sum_{m \in \mathbb{Q}} c_\gamma(m) \exp(-2\pi y\lambda_{v+}^2) \exp(2\pi i x(\lambda^2/2 + m)).$$

Therefore for any $l > 0$,

$$\int_{1 \le y \le l} \int_{-1/2 \le x \le 1/2} \sum_{\lambda \in M+\gamma} \sum_{m \in \mathbb{Q}} c_\gamma(m)$$

$$\times \exp(-2\pi y\lambda_{v+}^2) \exp(2\pi i x(\lambda^2/2 + m)) y^{b^+/2-s-2} dx dy$$

$$= \sum_{\lambda \in M+\gamma} \sum_{m \in \mathbb{Q}} c_\gamma(m) \left(\int_1^l \exp(-2\pi y\lambda_{v+}^2) y^{b^+/2-s-2} dy \right)$$

$$\times \int_{-1/2 \le x \le 1/2} \exp(2\pi i x(\lambda^2/2 + m)) dx$$

as the integrals being definite ones, we can integrate termwise and the series remains convergent. Hence by Lemma 3.4.9, we can deduce that the right hand side of (1) is equal to

$$\sum_{\lambda \in M^*} c_\lambda(\lambda^2/2) \lim_{l \to \infty} \int_1^l \exp(-2\pi y\lambda_{v+}^2) y^{-2+b^+/2-s} dy$$

$$= \sum_{\lambda \in M^*} c_\lambda(\lambda^2/2) \int_1^\infty \exp(-2\pi y\lambda_{v+}^2) y^{-2+b^+/2-s} dy.$$

Once again in the above expression absolute convergence allows the limit to be taken termwise. (ii) now follows.

Finally suppose that $\lambda_{v+} \ne 0$. For $y \ge 1$, $\exp(-2\pi y\lambda_{v+}^2)$ is a decreasing function of y and when $\mathcal{R}(s) \ge b^+/2 - 2$, $y^{-2+b^+/2-s}$, so is $y^{b^+/2-s-2}$. Hence,

$$G_1(s) = \int_1^\infty e^{-2\pi\lambda_{v+}^2 y} y^{b^+/2-s-2} dy$$

$$\le \int_1^\infty e^{-2\pi\lambda_{v+}^2 y}$$

$$= \frac{1}{2\pi\lambda_{v+}^2} e^{-2\pi\lambda_{v+}^2}.$$

The series

$$\sum_{\substack{\lambda \in M^* \\ \lambda_{v+} \neq 0}} c_\lambda(\lambda^2/2) \frac{1}{2\pi\lambda_{v+}^2} e^{-2\pi\lambda_{v+}^2}$$

is absolutely convergent. It follows that the series

$$\sum_{\substack{\lambda \in M^* \\ \lambda_{v+} \neq 0}} c_\lambda(\lambda^2/2) G_1(s)$$

is absolutely and uniformly convergent when $\mathcal{R}(s) \geq b^+/2 - 2$. Moreover,

$$\frac{\delta G_1(s)}{\delta s} = \int_1^\infty e^{-2\pi\lambda_{v+}^2 y} \frac{\delta y^{b^+/2-s-2}}{\delta s} dy = -\int_1^\infty e^{-2\pi\lambda_{v+}^2 y} \log(y) y^{b^+/2-s-1} dy$$

and $\log(y) \leq y$ for all $y \geq 1$. Hence the above implies that

$$\sum_{\substack{\lambda \in M^* \\ \lambda_{v+} \neq 0}} c_\lambda(\lambda^2/2) G_1(s)$$

is absolutely and uniformly convergent when $\mathcal{R}(s) \geq b^+/2 - 1$. Therefore,

$$\sum_{\substack{\lambda \in M^* \\ \lambda_{v+} \neq 0}} c_\lambda(\lambda^2/2) G_1(s)$$

is analytic for $\mathcal{R}(s) \geq b^+/2 - 1$, proving (iii) in this case. When $\mathcal{R}(s) < b^+/2 - 1$, (iii) follows from Lemma 3.4.8. $\qquad\square$

The next observation will ensure that various sums are finite and so not problematic.

Lemma 3.4.11. *For any $m \in \mathbf{Z}_+$, there are only finitely many vectors $\lambda \in M^*$ satisfying $-m \leq \lambda^2 \leq 0$.*

Proof. Let $\lambda_1, \cdots, \lambda_n$ be a basis for the lattice M and $\lambda_1^*, \cdots, \lambda_n^*$ its dual basis in M^*. Since $(\lambda_i^*, \lambda_i) = 1$, for each i, λ_i^* is a primitive vector of M^*. Hence, for any $\lambda \in M^*$, $\lambda = \sum_i a_i \lambda_i^*$ with $a_i \in \mathbf{Z}$. Also for each i, there is a positive integer b_i such that $\mu_i = \frac{1}{b_i}\lambda_i$ is a primitive vector of M^*. Hence, $\lambda = \sum_i c_i \mu_i$ with $c_i \in \mathbf{Z}$. Let $m \in \mathbf{Z}_+$. $\lambda^2 = \sum_i a_i c_i / b_i$. Suppose that $\lambda^2 \leq 0$. Then $-m \leq \lambda^2 \leq 0$ if and only if

$$-m \leq \sum_i a_i c_i / b_i \leq 0.$$

If this holds then, for each i, $|a_i c_i| \leq bm$, where $b = max\{b_1, \cdots, b_n\}$. Since any discrete bounded set is finite, the result follows. $\qquad\square$

Corollary 3.4.12. *For any $v^+ \in G(M)$, as a function of s, $G_M(s, v^+)$ is analytic for $\mathcal{R}(s) \gg 0$ and can be analytically continued to a meromorphic function on $\mathbf{C}$.*

Proof. By definition of the fundamental domain D (see Lemma 3.2.9),

$$G_M(s, v^+) = \lim_{l \to \infty} \int_{0 < y \le l} \int_{\substack{|x| \le 1/2 \\ x^2 + y^2 \ge 1}} F(\tau)\overline{\Theta}(\tau; v^+) y^{b^+/2 - s} \, dx \, dy / y^2.$$

By assumption, there are only finitely many coefficients $c_\gamma(m) \ne 0$ for $m < 0$. Hence by Lemmas 3.4.10 (i), (ii) and 3.4.11, to prove that as a function of s, $G_M(s, v^+)$ is analytic when $\mathcal{R}(s) > b^+/2 - 1$ it suffices to show that in this region, the integral

$$G_1(s) = \int_1^\infty \exp(-2\pi y \lambda_{v^+}^2) y^{-2 + b^+/2 - s} \, dy$$

converges when $\lambda_{v+} = 0$. In this case

$$G_1(s) = \int_1^\infty y^{-2 + b^+/2 - s} \, dy = [\frac{1}{b^+/2 - s - 1} y^{b^+/2 - s - 1}]_1^\infty = \frac{1}{b^+/2 - s - 1} \quad (1)$$

if $\mathcal{R}(s) > b^+/2 - 1$. Moreover, (1) and implies that if we set

$$G_1(s) = \frac{1}{b^+/2 - s - 1}$$

for all $s \in \mathbf{C}$ gives an analytic continuation (see §C.2) of G_1 to a meromorphic function on $\mathbf{C}$. Hence by Lemma 3.4.10 (ii) and (iii), $G(s, v^+)$ can also be analytically continued to a meromorphic function on $\mathbf{C}$. □

As a result, the theta transform $\Phi_M(v^+, F)$ of F on the Grassmannian $G(M)$ as defined in Definition 3.4.6 exists. Before computing the function $G_M(s, v^+)$, we study the singularities of $\Phi_M(v^+, F)$.

Theorem 3.4.13. *In a neighbourhood of the point $v_0^+ \in G(M)$, the point $v^+ \in G(M)$ is a singular point of the theta transform $\Phi_M(v^+, F)$ of F of type*

$$- \sum_{\substack{\lambda \in M^* \cap v_0^- \\ \lambda \ne 0}} c_\lambda(\lambda^2/2)(-2\pi \lambda_{v+}^2)^{1 - b^+/2} \log(\lambda_{v+}^2)/(1 - b^+/2)!$$

if $b^+ = 0$ or $b^+ = 2$, and of type

$$\sum_{\substack{\lambda \in M^* \cap v_0^- \\ \lambda \ne 0}} c_\lambda(\lambda^2/2)(2\pi \lambda_{v+}^2)^{1 - b^+/2} \Gamma(-1 + b^+/2)$$

otherwise.

Proof. The result follows from Lemma 3.4.10 applied to $s = 0$ and from Lemma 3.4.9. The term in the sum given in Lemma 3.4.10 corresponding to $\lambda = 0$ is independent of v^+ and so does not contribute to the singularity of $\Phi_M(v^+, F)$. Note that $b^{+2}/2 - 1 < 0$ if and only if $b^+ = 0$ or $b^+ = 1$. □

Note that by Lemma 3.4.11, the sums in Theorem 3.4.13 are finite since for any $\gamma \in M^*/M$, $c_\gamma(m) \neq 0$ when $m \leq 0$ is sufficiently small.

Now that we know the singularities of the function $\Phi_M(v^+, F)$, we can begin computing it.

Theorem 3.4.14. *Suppose that* Φ_M *is analytic at* $v^+ \in G(M)$. *For sufficiently small values of* $z_{v^+}^2$,

$$
G_M(s; v^+) = \frac{1}{\sqrt{2}|z_{v^+}|} G_L(s; w^+) + \frac{\sqrt{2}}{|z_{v^+}|} \sum_{\lambda \in L^*} \sum_{n>0} \mathbf{e}((n\lambda, \mu))
$$

$$
\times \int_{y>0} c_{\lambda, \lambda^2/2}(y) \exp(-\pi n^2/2 y z_{v^+}^2 - \pi y(\lambda_{w^+}^2 - \lambda_{w^-}^2)) y^{-s-5/2} dy.
$$

Proof. Using Theorem 3.3.11, we rewrite the theta function of $M + \gamma$ in terms of the generalized theta function of $L + \gamma$. This gives

$$
G_M(s; v^+) = \lim_{l \to \infty} \int \int_{D_l} \sum_{\gamma \in M^*/M} f_{M+\gamma}(\tau)(2y z_{v^+}^2)^{\frac{-1}{2}}
$$

$$
\times \sum_{c,d \in \mathbf{Z}} \mathbf{e}(-\frac{|c\tau + d|^2}{4iy z_{v^+}^2}) \overline{\theta}_{L+\gamma}(\tau, 0, -c\mu; w^+) y^{-s} dx dy/y^2 \tag{1}
$$

The summand in the above sum corresponding to $c = d = 0$ is

$$
(2z_{v^+}^2)^{\frac{-1}{2}} \sum_{\gamma \in M^*/M} \lim_{l \to \infty} \int \int_{D_l} f_{M+\gamma}(\tau) y^{-1/2} \overline{\theta}_{L+\gamma}(\tau; w^+) y^{-s} dx dy/y^2. \tag{2}
$$

We first show that this is equal to $(2z_{v^+}^2)^{\frac{-1}{2}} G_L(s, w^+)$.

Remember that $L^*/L = M^*/M$ (see equality (II) in §3.2). Similarly to the definition of F_M given at the start of §3.4 for the lattice M, for the lattice L, F_L is defined as follows:

$$
F_L(\tau) = \sum_{\gamma \in L^*/L} f_{L+\gamma} e_\gamma, \quad \text{where} \quad f_{L+\gamma} = f_{M+\gamma}(\tau) y^{-1/2} \tag{3}
$$

Indeed, by Lemma 3.2.8, the right hand side of equality (3) is a modular form of weight $(-(b^- - 1)/2, -(b^+ - 1)/2)$ and type ρ_L. Moreover, this is equal to

$$
\sum_{\gamma \in L^*/L} f_\gamma(\tau) y^{\frac{b^+ - 1}{2}}.
$$

Since the lattice L has signature $(b^+ - 1, b^- - 1)$, we can deduce equality (3). This takes care of the term corresponding to $(c, d) = (0, 0)$. We next consider the sum of the summands on the right hand side of (1) corresponding to $(c, d) \neq 0$.

It is equal to

$$
(2z_{v^+}^2)^{\frac{-1}{2}} \lim_{l\to\infty} \int\int_{D_l} \sum_{\gamma\in M^*/M} f_{M+\gamma}(\tau) \sum_{c,d\in\mathbf{Z},(c,d)\neq 0} \exp\left(-\pi\frac{|c\tau+d|^2}{2yz_{v^+}^2}\right)
$$
$$
\times \overline{\theta}_{L+\gamma}(\tau,\mu d,-c\mu;w^+)y^{-s-1/2}dxdy/y^2
$$

$$
= (2z_{v^+}^2)^{\frac{-1}{2}} \lim_{l\to\infty} \int\int_{D_l} \sum_{\gamma\in L^*/L} \sum_{\substack{c,d\in\mathbf{Z}\\(c,d)\neq 0}} \exp\left(-\pi\frac{|c\tau+d|^2}{2yz_{v^+}^2}\right)
$$
$$
\times f_{L+\gamma}(\tau)\overline{\theta}_{L+\gamma}(\tau,0,-c\mu;w^+)y^{-s}dxdy/y^2. \tag{4}
$$

Substituting $c,d\in\mathbf{Z}$ with $(c,d)\neq(0,0)$ by (nc,nd) with c,d coprime and $n\in\mathbf{N}$ in (4), we get

$$
(2z_{v^+}^2)^{\frac{-1}{2}} \lim_{l\to\infty} \int\int_{D_l} \sum_{\gamma\in L^*/L} \sum_{\substack{c,d\in\mathbf{Z}\\(c,d)=1}} \sum_{n\in\mathbf{N}} \exp\left(-\pi\frac{|c\tau+d|^2}{2yz_{v^+}^2}\right)
$$
$$
\times f_{L+\gamma}(\tau)\overline{\theta}_{L+\gamma}(\tau,0,-c\mu;w^+)y^{-s}dxdy/y^2. \tag{5}
$$

For $(c,d)=1$, there exist $a,b\in\mathbf{Z}$ such that $ad-bc=1$ and $a_1 d-b_1 c=1$ with $a_1,b_1\in\mathbf{Z}$ if and only if $a_1 = a+mc$ and $b_1 = b+md$ for some $m\in\mathbf{Z}$. As a result, (5) is equal to

$$
(2z_{v^+}^2)^{\frac{-1}{2}} \Big(\lim_{l\to\infty} \int\int_{D_l} \sum_{\gamma\in L^*/L} \sum_{\begin{pmatrix} a & b \\ c & d \end{pmatrix}\in SL_2(\mathbf{Z})/\mathbf{Z}}
$$
$$
\sum_{n\in\mathbf{N}} \exp(-n^2\pi\frac{|c\tau+d|^2}{2yz_{v^+}^2})F_L(\tau)\overline{\Theta}_L(\tau,0,-nc\mu;w^+)y^{-s}dxdy/y^2\Big). \tag{6}
$$

Now, $a\alpha+b\beta=n\mu$ and $c\alpha+d\beta=0$ implies that $\alpha=n\mu d$ and $\beta=0$. So, the lattice L having signature $(b^+ - 1, b^- - 1)$,

$$
\overline{\Theta}_L(\frac{a\tau+b}{c\tau+d},n\mu,0;w^+) = (c\overline{\tau}+d)^{\frac{b^+ -1}{2}}(c\tau+d)^{\frac{b^- -1}{2}}\overline{\Theta}_L(\tau,0,-n\mu c;w^+)
$$

by Theorem 3.3.4 and

$$
F_L(\frac{a\tau+b}{c\tau+d}) = (c\tau+d)^{\frac{-(b^- -1)}{2}}(c\overline{\tau}+d)^{\frac{-(b^+ -1)}{2}} F_L(\tau).
$$

We can then deduce that (6) is equivalent to

$$
\sum_{\gamma\in L^*/L}(2z_{v^+}^2)^{\frac{-1}{2}} \lim_{l\to\infty} \int\int_{D_l} \sum_{A\in SL_2(\mathbf{Z})/\mathbf{Z}} \sum_{n\in\mathbf{N}} \exp\left(\frac{-\pi n^2}{2\mathcal{I}(A\tau)z_{v^+}^2}\right)
$$
$$
\times F_L(A\tau)\overline{\Theta}_L(A\tau,n\mu,0;w^+)y^{-s}dxdy/y^2 \tag{7}
$$

since the group L^*/L being finite, the summation and integration order can be exchanged. In order to evaluate $G_M(s; v^+)$ we need to integrate term by term in the sum over $SL_2(\mathbf{Z})/\mathbf{Z}$. For this to be possible, the integral $G(s, v^+)$ needs to be convergent. However by Corollary 3.4.12, this is only the case for $\mathcal{R}(s) \gg 0$. So we need to justify this exchange some other way. Note that

$$\left|\overline{\Theta}_L(A\tau, n\mu, 0; w^+)\right| = \left|\overline{\Theta}_L(A\tau; w^+)\right| \tag{8}$$

and

$$F_L(A\tau)\overline{\Theta}_L(A\tau; w^+) = F_L(\tau)\overline{\Theta}_L(\tau; w^+) \tag{9}$$

for all matrices $A \in SL_2(\mathbf{Z})$.

Claim: In equality (8), we can exchange the sum over $SL_2(\mathbf{Z})/\mathbf{Z}$ with the integral.

By Lemma 3.4.10 (i), since D_1 is a close compact region,

$$\int\int_{D_1} F(\tau)\overline{\Theta}(\tau; v^+)y^{b^+/2-s}dxdy/y^2$$

$$= \sum_{A\in SL_2(\mathbf{Z})/\mathbf{Z}} \int\int_{D_1} \sum_{n\in\mathbf{N}} \exp\left(\frac{-\pi n^2}{2\mathcal{I}(A\tau)z_{v^+}^2}\right)$$
$$F_L(A\tau)\overline{\Theta}_L(A\tau, n\mu, 0; w^+)y^{-s}dxdy/y^2.$$

As

$$\left(\int\int_{D_1} + \int_{y\geq 1}\right) \int_{-1/2\leq x\leq 1/2} F(\tau)\overline{\Theta}(\tau; v^+)y^{b^+/2-s}dxdy/y^2$$

$$= \int\int_{D_1} F(\tau)\overline{\Theta}(\tau; v^+)y^{b^+/2-s}dxdy/y^2,$$

we only need to show that the double integration $\int_{y\geq 1}\int_{-1/2\leq x\leq 1/2}$ can be exchanged with the sum over $SL_2(\mathbf{Z})/\mathbf{Z}$. Let us consider

$$G_1(s, v^+, A)$$

$$= \int_{y\geq 1}\int_{-1/2\leq x\leq 1/2} \sum_{n\in\mathbf{N}} \exp\left(\frac{-\pi n^2}{2\mathcal{I}(A\tau)z_{v^+}^2}\right)$$
$$\left|F_L(A\tau)\right|\left|\overline{\Theta}_L(A\tau, n\mu, 0; w^+)\right|y^{-s}dxdy/y^2$$

$$= \int_{y\geq 1} \sum_{n\in\mathbf{N}} \exp\left(\frac{-\pi n^2}{2\mathcal{I}(A\tau)z_{v^+}^2}\right) \left|F_L(\tau)\right|\left|\overline{\Theta}_L(\tau; w^+)\right|y^{-s-2}dy.$$

This equality follows from (8) and (9) and using the fact that the variable x only appears in expressions of type $\exp(2\pi ipx)$ with $p \in \mathbf{R}$ having absolute value 1.

For any $A \in SL_2(\mathbf{Z})$, by Lemma 3.2.8, if $y = \infty$ then $A(y) = y$ or 0. Moreover, $A(D)$ is a fundamental domain and each fundamental domain contains precisely one cusp point and it is the only possible point of singularity of the modular form F. Hence if $A(i\infty) = i\infty$, then $A(D) = D$ and $A = \pm \begin{pmatrix} 1 & 0 \\ 0 & 1 \end{pmatrix}$ modulo $\mathbf{Z}$ (see above for the action of $\mathbf{Z}$ on $SL_2(\mathbf{Z})$). In this case,

$$G_1(s, v^+, A) = \int_1^\infty \sum_{n \in \mathbf{N}} \exp\left(\frac{-\pi n^2}{2yz_{v^+}^2}\right) |F_L(\tau)| |\overline{\Theta}_L(\tau; w^+)| y^{-s-2} dy$$

and from Corollary 3.4.12, $G_1(s, v^+, A)$ is analytic for $\mathcal{R}(s) \gg 0$. Moreover for any $s \in \mathbf{C}$, $G(s, v^+) - G_1(s, v^+, A)$ is analytic.

Next suppose that $A(i\infty) = 0$. Then,

$$G_1(s, v^+, A) = \int_p^0 \sum_{n \in \mathbf{N}} \exp\left(\frac{-\pi n^2}{2yz_{v^+}^2}\right) |F(\tau)| |\overline{\Theta}_L(\tau, ; w^+)| y^{b^2/2-s-5/2} dy$$

for some non-zero rational p. From Lemma 3.2.12 we know that there is a constant $m \geq 0$ such that $e^{-m/y} F(\tau)$ is bounded as $y \to 0$. Take $z_{v^+}^2 < 2/\pi(m+1)$. Since

$$|\overline{\theta}_{L+\gamma}(\tau, ; w^+)| = \sum_{\lambda \in L+\gamma} e^{-\pi y(\lambda_{v^+}^2 - \lambda_{v^-}^2)},$$

where $\lambda_{v^+}^2 - \lambda_{v^-}^2 \geq 0$, it follows that $G_1(s, v^+, A)$ converges for all $s \in \mathbf{C}$. This can be directly seen when $\mathcal{R}(s) \gg 0$. Otherwise it is a consequence of Lemma 3.4.8.

As a result, since $G(s, v^+)$ is analytic for $\mathcal{R}(s) \gg 0$, the integral and the sum can be exchanged in this region, in other words

$$G(s, v^+) = \sum_{A \in SL_2(\mathbf{Z})/\mathbf{Z}} G_1(s, v^+, A) \qquad (10)$$

for $\mathcal{R}(s) \gg 0$. Moreover, the above also implies that $G(s, v^+) - \sum_{A \in SL_2(\mathbf{Z})/\mathbf{Z}} G_1(s, v^+, A)$ is analytic on $\mathbf{C}$. Therefore, equality (10) holds for all $s \in \mathbf{C}$. This proves our claim.

Since D is a fundamental domain, by definition (see Definition 3.2.7), for any $\tau \in \mathcal{H}$, there is a unique $\tau_D \in \overline{D}$ such that $A\tau = \tau_D$ for some $A \in SL_2(\mathbf{Z})$. Setting $A_n = \begin{pmatrix} a+nc & b+nd \\ c & d \end{pmatrix}$, for all $n \in \mathbf{N}$,

$$\mathcal{I}(A_n\tau) = \mathcal{I}(A_0\tau)$$

by Lemma 3.2.8 and

$$\mathcal{R}(A_n\tau) = \mathcal{R}\left(\frac{((a+nc)\tau + (b+nd))(c\overline{\tau} + d)}{|c\tau + d|^2}\right)$$

$$= \mathcal{R}(A_0\tau) + n\frac{c^2(\tau\overline{\tau}) + d^2}{|c\tau + d|^2}$$

$$= \mathcal{R}(A_0\tau) + n.$$

Moreover, $A\tau = \tau$ for all $\tau \in \mathcal{H}$ if and only if $A = \begin{pmatrix} \pm 1 & 0 \\ 0 & \pm 1 \end{pmatrix}$. Hence, (8) becomes

$$(2z_{v+}^2)^{\frac{-1}{2}} 2 \int_{y>0} \int_{x \in \mathbf{R}/\mathbf{Z}} \sum_{n \in \mathbf{N}} \exp\left(\frac{-\pi n^2}{2yz_{v+}^2}\right) F_L(\tau)\overline{\Theta}_L(\tau, n\mu, 0; w^+) y^{-s-5/2} dx dy.$$

(11)

We replace Θ_L in the above integral by the defining expression for it. We can exchange the order of integration and summation since the following integral converges except for one term: As $\mathbf{R}/\mathbf{Z}$ is the unit circle,

$$\int_{y>0} \int_{x \in \mathbf{R}/\mathbf{Z}} \sum_{\lambda \in L+\gamma} |\exp(2\pi i(\tau\lambda_{w+}^2/2 + \overline{\tau}\lambda_{w-}^2/2))| dx dy$$

$$= \int_{y>0} \sum_{\lambda \in L} \exp(-\pi y(\lambda_{w+}^2 - \lambda_{w-}^2)) dy$$

$$= \sum_{\lambda \in L+\gamma} \frac{-1}{\pi(\lambda_{w+}^2 - \lambda_{w-}^2)}$$

and $\lambda_{w+}^2 - \lambda_{w-}^2 \neq 0$ unless $\lambda = 0$. Therefore, (11) is equal to

$$(2z_{v+}^2)^{\frac{-1}{2}} 2 \sum_{\gamma \in L^*/L} \sum_{\lambda \in L+\gamma} \int_{y>0} \int_{x \in \mathbf{R}/\mathbf{Z}} \sum_{n \in \mathbf{N}} \exp\left(\frac{-\pi n^2}{2yz_{v+}^2}\right)$$

$$\times f_{L+\gamma}(\tau) \mathbf{e}(\tau\lambda_{w+}^2/2 + \overline{\tau}\lambda_{w-}^2/2 - (\lambda, n\mu)) y^{-s-5/2} dx dy.$$

Replacing f_{L+g} with the expression on the right hand side of (3'), we get

$$(2z_{v+}^2)^{-\frac{1}{2}} + 2 \sum_{\gamma \in L^*/L} \sum_{\lambda \in L+\gamma}$$

$$\int_{y>0} \sum_{n \in \mathbf{N}} \exp\left(\frac{-\pi n^2}{2yz_{v+}^2}\right) \mathbf{e}((\lambda, n\mu)) \exp(-\pi y(\lambda_{w+}^2 - \lambda_{w-}^2))$$

$$\times \int_{x \in \mathbf{R}/\mathbf{Z}} \sum_{m \in \mathbf{Z}} c_{\lambda,m}(y) \mathbf{e}(-x\lambda^2/2 + mx) y^{-s-5/2} dx dy$$

$$= (2z_{v+}^2)^{-\frac{1}{2}} 2 \sum_{\gamma \in L^*/L} \sum_{\lambda \in L+\gamma} \int_{y>0} \sum_{n \in \mathbf{N}} \mathbf{e}((\lambda, n\mu)) \sum_{m \in \mathbf{Z}} c_{\lambda,m}(y) \exp\left(\frac{-\pi n^2}{2yz_{v+}^2}\right)$$

$$\times \exp(-\pi y(\lambda_{w+}^2 - \lambda_{w-}^2)) y^{-s-5/2} \int_{x \in \mathbf{R}/\mathbf{Z}} \mathbf{e}(-x\lambda^2/2 + mx) dx dy$$

since $|\mathbf{e}(mx)| = 1$ for all $m \in \mathbf{Z}$ implies that the order of the integral $\int_{x \in \mathbf{R}/\mathbf{Z}}$ and of the sum of terms of the Laurent series can be interchanged for the same reason as before. Since the interval $[-1/2, 1/2)$ represents $\mathbf{R}/\mathbf{Z}$, by Lemma 3.4.9,

$$G_M(s; v^+) = (2z_{v+}^2)^{\frac{-1}{2}} (G_L(s; w^+) + 2 \sum_{\lambda \in L^*} \int_{y>0} \sum_{n \in \mathbf{N}} e((\lambda, n\mu))$$

$$\times c_{\lambda, \lambda^2/2}(y) \exp\left(\frac{-\pi n^2}{2yz_{v+}^2}\right) \exp(-\pi y(\lambda_{w+}^2 - \lambda_{w-}^2)) y^{-s-5/2} dy$$

This last integral is equal to the one we are looking for. $\square$

In the previous Theorem the condition on z_{v+}^2 is needed because the modular form F may not be holomorphic at the rational cusps.

Note that in sections 3.3 and 3.4, calculations and formulae are simplified in the case when M is unimodular since then $M = M^*$. Before closing this section, we define the subgroup of the isometry group of M fixing the automorphic form.

Definition 3.4.15. *If σ is an isometry of the lattice M, then σ acts on the vector valued modular form F and its theta transform $\Phi_M(v^+, F)$ for $v^+ \in G(M)$ as follows:*

$$\sigma(F) = \sum_{\gamma \in M^*/M} f_\gamma e_{\sigma(\gamma)}$$

and

$$\sigma(\Phi_M(v^+, F)) = \Phi_M(\sigma v^+, tF).$$

The above makes sense since the isometry group of M acts on the Grassmannian $G(M)$. We will write $Aut(M, F)$ to be the stabilizer of the theta transform $\Phi_M(v^+, F)$ of F, i.e. for $\sigma \in Aut(M, F)$,

$$\Phi_M(\sigma v^+, \sigma F) = \Phi_M(v^+, F)$$

for any $v^+ \in G(M)$ that is not a singular point of the function Φ_M.

Exercises 3.4

1. Suppose that M is an arbitrary lattice with signature (b^+, b^-) (i.e. not necessarily satisfying condition (I)) containing a primitive vector z of norm 0. Let $N \in \mathbf{N}$ to be minimal such that $(z, \lambda) = N$ for some $\lambda \in M$. Set $z' \in M^*$ to be a vector satisfying $(z, z') = 1$ and write $L = (M \cap z^\perp)/z$. Let $w^\pm$ be the orthogonal complement of $z_{v\pm}$ in $v^\pm$. Set $\mu = -z' + z_{v+}/2z_{v+}^2 + z_{v-}/2z_{v-}^2$. Let F and F_M be the modular forms defined in section 3.4. Suppose that Φ_M is analytic at $v^+ \in G(M)$. Show that, for sufficiently small values of z_{v+}^2,

$$G_M(s; v^+) = \frac{1}{\sqrt{2}|z_{v+}|} G_L(s; w^+) + \frac{\sqrt{2}}{|z_{v+}|} \sum_{\lambda \in L^*} \sum_{n>0} e((n\lambda, \mu)) \times \sum_{\substack{\delta \in M^*/M \\ \delta|L = \lambda}} e((n\delta, z'))$$

$$\int_{y>0} c_{\delta, \lambda^2/2}(y) \exp(-\pi n^2/2yz_{v+}^2 - \pi y(\lambda_{w+}^2 - \lambda_{w-}^2)) y^{-s-5/2} dy.$$

Chapter 4

Γ-Graded Vertex Algebras

The aim of this chapter is to present the aspects of the theory necessary for the construction of BKM superalgebras. The purpose of this book is not the study of Γ-graded vertex algebras per se, so we only give the details needed for our purpose. Hence the exposition does not pretend to be a complete treatment of Γ-graded vertex algebras – a vast subject with a rich literature dedicated to it. For an in depth study, the interested reader can in particular consult [Borc2], [FrenB], [FrenLM2], [Kac15], [LepL] for vertex operator and vertex algebras, and [DonL], [KacB2], [Sch2] for the more general context of Γ-graded vertex algebras. We follow closely the methodology of the above mentioned books and papers written by the several authors who developed the subject and keep the approach and notation of [Kac15].

In this chapter Γ will denote an abelian group.

4.1 The Structure of Γ-graded Vertex Algebras

We start with some basic facts. We first need the notion of a generalized vertex operator but before we give an elementary result.

Lemma 4.1.1. *Let* Γ *be an abelian group of exponent* N *and* Δ *be a* **Z**-*bilinear symmetric map* $\Gamma \times \Gamma \to \mathbf{Q}/\mathbf{Z}$. *Then,* $\Delta(\Gamma \times \Gamma) \subset \frac{1}{N}\mathbf{Z}/\mathbf{Z}$.

Proof. Since the map Δ is **Z**-bilinear, $\mathbf{Z} = \Delta(0, h) = \Delta(Ng, h) = N\Delta(g, h)$ for all $g, h \in G$. □

For reasons of simplicity, when the context allows it without any possible confusion, we will write $\Delta(g, h)$ not just for a coset in $\mathbf{Q}/\mathbf{Z}$ itself but also for any rational representing the coset $\Delta(g, h)$.

Remember in Example 2.1.2 we saw that if a vector space V is graded by the abelian group $\mathbf{Z}_2$, then so is the associative algebra of endomorphisms $gl(V)$. More generally, this clearly holds for all abelian groups Γ: If V is Γ-graded, then so is $gl(V)$. In particular for all $g, h \in \Gamma$, $\phi \in gl(V)_g$, $v \in V_h$, $\phi(v) \in V_{g+h}$.

Since vertex operators are formal power series, we remind the reader of the notation for the different types of sets. For a set F,

$F[X]$ is the set of polynomials in X with coefficients in F, i.e. $\sum_{i=0}^{n} a_i X^i$, $a_i \in F$,

$F[[X]]$ is the set of power series $\sum_{i=0}^{\infty} a_i X^i$, $a_i \in F$, and

$F((X))$ is the set of Laurent series (see Corollary C.2.5) in X, i.e. $\sum_{i=-N}^{\infty} a_i X^i$, $a_i \in F$ and $N \geq 0$.

Definition 4.1.2. *Let Γ be an abelian group of exponent N and Δ be a **Z**-bilinear symmetric map $\Gamma \times \Gamma \to \mathbf{Q}/\mathbf{Z}$. A generalized vertex operator on the Γ-graded complex vector space*

$$V = \oplus_{\gamma \in \Gamma} V_g$$

is a formal power series in $gl(V)[[z^{\frac{1}{N}}, z^{-\frac{1}{N}}]]$, where N is a positive integer, written:

$$X(z) = \sum_{n \in \frac{1}{N}\mathbf{Z}} x_n z^{-n-1}$$

such that for any $v \in V$, $x_n(v) = 0$ for n large enough.

The vertex operator $X(z)$ is said to be of parity $g \in \Gamma$ if

(i) for all $n \in \mathbf{Z}$, $x_n \in gl(V)_g$ and

(ii) for any $v \in V_h$, $h \in \Gamma$, $x_n(v) = 0$ unless $n \in \mathbf{Z} + \Delta(g, h)$.

We then write $p(X) = g$. If $\Gamma \leq \mathbf{Z}_2$ and $\Delta = 0$, the operator $X(z)$ is said to be a vertex operator.

Thus generalized vertex operators are generating series for endomorphisms of V satisfying a convergence condition. Note that when we are considering vertex operators, the vector space V is $\mathbf{Z}_2$-graded and so $gl(V)$ is a Lie superalgebra. Moreover, in the case $\Gamma = \mathbf{Z}_2$, the exponent of Γ must be $N = 2$.

Before giving the definition of a Γ-graded vertex algebra, we define a concept generalizing the notion of commutativity for two generalized vertex operators. For this we need some basic facts about the Laurent expansion of the expression $(z - w)^n$.

Remark 4.1.3. When n is a rational or a negative integer, the formal expression $(z-w)^n$ can have several interpretations and so it needs to be precised. Consider it as a complex function. As a function of w, it is differentiable in the domain $|w| < |z|$ and this is a neighbourhood of $w = 0$. Hence for $|w| < |z|$, we can write its Taylor series (see Definition D.2.6) as a function of w:

$$(z - w)^n = \sum_{j=0}^{\infty} (-1)^j \binom{n}{j} z^{n-j} w^j.$$

Similarly for $|z| < |w|$, considering $(z - w)^n$ as a function of z, we get

$$(z - w)^n = \sum_{j=0}^{\infty} (-1)^{n-j} \binom{n}{j} w^{n-j} z^j = (-1)^n (w - z)^n$$

Here an extra precision is needed when n is not an integer: we have to define $(-1)^n$. In other words, if $n = \frac{p}{q}$, then $(-1)^n$ is a $2q$-th root of unity. So we have to choose and fix a $2q$-th root of unity. We take $e^{i\pi n}$.

Reverting back to formal power series, we define

$$i_{z,w}(z - w)^n = \sum_{j=0}^{\infty}(-1)^j \binom{n}{j} z^{n-j} w^j$$

and

$$i_{w,z}(z - w)^n = e^{\pi in} i_{w,z}(w - z)^n = e^{\pi in} \sum_{j=0}^{\infty}(-1)^j \binom{n}{j} w^{n-j} z^j.$$

Note that $i_{z,w}(z + w)^n = i_{z,-w}(z + w)^n$.

When n is a positive integer, the above reduces to

$$i_{z,w}(z - w)^n = (z - w)^n$$

and

$$i_{w,z}(z - w)^n = (w - z)^n e^{\pi in} = (z - w)^n.$$

Before giving the generalization of commutativity, we need some more tools as we are working in a general context of vector spaces graded by an abelian group Γ.

Definition 4.1.4. *A bi-multiplicative map $\nu : \Gamma \times \Gamma \to \mathbf{C}^*$ is a map satisfying*

$$\nu(g + h, k) = \nu(g, k)\nu(g, k) \quad and \quad \nu(k, g + h, k) = \nu(k, g)\nu(k, h)$$

for $g, h, k \in \Gamma$.

Next we give an elementary technical result.

Lemma 4.1.5. *Let Γ be an abelian group of exponent N and $\nu : \Gamma \times \Gamma \to \mathbf{C}^*$ a bi-multiplicative map. Then, for any $g, h \in \Gamma$, $\nu(g, h)^N = 1$.*

Proof. As the map ν is bi-multiplicative, $1 = \nu(0, h) = \nu(Ng, h) = \nu(g, h)^N.\square$

We can now define the notion generalizing commutativity .

Definition 4.1.6. *Let Γ be an abelian group of exponent N, Δ a $\mathbf{Z}$-bilinear symmetric map $\Gamma \times \Gamma \to \mathbf{Q}/\mathbf{Z}$ and $\nu : \Gamma \times \Gamma \to \mathbf{C}^*$ a bi-multiplicative map. Generalized vertex operators $X(z), Y(z)$ on a Γ-graded complex vector space V are said to be mutually local (with respect to ν and Δ) if for large enough positive n in $\mathbf{Z} + \Delta(p(X), p(Y))$,*

$$i_{z,w}(z - w)^n X(z)Y(w) = i_{w,z}(z - w)^n \nu(p(X), p(Y))Y(w)X(z) \qquad (*)$$

in $gl(V)[[z^{\frac{1}{N}}, z^{-\frac{1}{N}}, w^{\frac{1}{N}}, w^{-\frac{1}{N}}]]$.

Notation: When $f \in gl(V)_0$ (resp. X is a generalized vertex operator of parity 0), for any $g \in gl(V)$ (resp. any generalized vertex operator Y), we will write $[f, g]$ (resp. $[X, Y]$) for $fg - gf$ (resp. $XY - YX$).

When $\Gamma \leq \mathbf{Z}_2$, $gl(V)$ is a Lie superalgebra. So for $f, g \in gl(V)$ (resp. homogeneous vertex operators X, Y), we will write $[f, g]$ (resp. $[X, Y]$) for $fg - (-1)^{p(f)p(g)}gf$ (resp. $XY - (-1)^{p(X)p(Y)}YX$).

Remark 4.1.7. Suppose that X and Y are vertex operators. We then define

$$\nu(a, b) = (-1)^{p(a)p(b)} = \begin{cases} -1 & \text{if } a = b = \bar{1} \\ 1 & \text{otherwise} \end{cases}.$$

Hence X and Y are mutually local if $(z-w)^n[X(z), Y(w)] = 0$, where the bracket is taken coefficient wise and is the Lie super-bracket on the Lie superalgebra $gl(V)$.

Remark 4.1.8. We check that the exponent n in equality $(*)$ is indeed in $\Delta(p(X), p(Y))$.

For homogeneous generalized vertex operators X, Y and homogeneous vectors $a \in V$, from Condition (ii) of Definition 4.1.2, we see that the exponents of z (resp. w) in $X(z)Y(w)a$ belong to

$$-\Delta(p(X), p(Y) + p(a)) \quad (\text{resp.} \quad -\Delta(p(Y) + p(a)))$$

and those in $Y(w)X(z)a$ to

$$-\Delta(p(X) + p(a)) \quad (\text{resp.} \quad -\Delta(p(Y), p(x) + p(a))).$$

Hence, since the exponents of z (resp. w) on the left hand side of equality $(*)$ given in Definition 4.1.6 have to belong to the same coset as those on the right hand side, the exponent n of both $i_{z,w}(z - w)^n$ and $i_{w,z}(z - w)^n$ has to belong to the coset $\Delta(p(X), p(Y))$.

Remark 4.1.9 on locality. It is at first hard to see why commutativity is generalized in this manner. So we give some explanations.

Remark 4.1.3 shows that, considered as complex functions, the left (resp. right) hand side of equality $(*)$ in Definition 4.1.6 is the Taylor series of $(z - w)^n X(z)Y(w)$ in the domain $|w| < |z|$ (resp. $|z| < |w|$). Hence locality means that these complex functions can be analytically continued on the boundary $|w| = |z|$ with equal continuations (up to a factor of $\nu(p(X), p(Y))$.

Commutativity would be too strong a condition to impose: For any vector space U, let $U((z^{\frac{1}{N}}))$ be the set of Laurent series in z with coefficients in U, i.e. series $\sum_{m \geq l} u_m z^m$.

$$X(z)Y(w) \in gl(V)((z^{\frac{1}{N}}))((w^{\frac{1}{N}})) \quad \text{and} \quad Y(w)X(z) \in gl(V)((w^{\frac{1}{N}}))((z^{\frac{1}{N}})).$$

However these two sets are not the same. For example (see Remarks 4.1.3),

$$i_{z,w}(z - w)^{\frac{-1}{N}} \in gl(V)((z^{\frac{1}{N}}))((w^{\frac{1}{N}}))$$

but does not belong to $gl(V)((w^{\frac{1}{N}}))((z^{\frac{1}{N}}))$. If the operators X and Y commute, then $X(w)Y(z)$ and $Y(z)X(w)$ are both in

$$gl(V)((z^{\frac{1}{N}}))((w^{\frac{1}{N}})) \cap gl(V)((w^{\frac{1}{N}}))((z^{\frac{1}{N}})) = gl(V)[[z^{\frac{1}{N}}, w^{\frac{1}{N}}]][z^{\frac{-1}{N}}, w^{\frac{-1}{N}}],$$

i.e. they are series with finitely many terms in negative powers of w and z, which is very restrictive. However, this may well be the case for

$$i_{z,w}(z-w)^n X(z)Y(w) \quad \text{and} \quad i_{w,z}(z-w)^n \nu(p(X), p(Y))Y(w)X(z)$$

when $n \in \mathbf{Z} + \Delta(p(X), p(Y))$ is large enough.

As the next typical example shows, in general $i_{z,w}(z-w)^n X(z,w) = 0$ does not necessarily imply that $X(z,w) = 0$ – further evidence that locality is much weaker than commutativity.

Example 4.1.10. Considered as complex functions, $i_{z,w}(z-w)^{-1}$ is the Taylor series of $(z-w)^{-1}$ in the domain $|w| < |z|$. Thus as formal power series,

$$(z-w)i_{z,w}(z-w)^{-1} = 1 = (w-z)i_{w,z}(w-z)^{-1}.$$

In other words,

$$(z-w)(i_{z,w}(z-w)^{-1} + i_{w,z}(w-z)^{-1}) = 0$$

but

$$i_{z,w}(z-w)^{-1} + i_{w,z}(w-z)^{-1} = z^{-1} \sum_{n \in \mathbf{Z}} (\frac{z}{w})^n \neq 0.$$

We will have more to say about this series, written $\delta(z-w)$, later. It will prove of much use. The reason $\delta(z-w) = 0$ does not follow from $(z-w)\delta(z-w) = 0$ is that it is not a Laurent series in w. The latter behave in a nice way as we now verify.

Lemma 4.1.11. *Let Γ be an abelian group with exponent $N \in \mathbf{N}$ and V be a Γ-graded vector space. Suppose that $X(z,w) \in gl(V)[[w^{\frac{1}{N}}, w^{-\frac{1}{N}}, z^{\frac{1}{N}}, z^{-\frac{1}{N}}]]$ satisfy*

$$i_{z,w}(z-w)^n X(z,w) = 0$$

for some $n \in \mathbf{Z}_+ + \frac{1}{N}$ and that either X is a Laurent series in w. Then,

$$X(z,w) = 0.$$

Proof. Suppose that X is a Laurent series in w and that

$$i_{z,w}(z-w)^n X(z,w) = 0$$

for some $n \in \mathbf{Z}_+ + \frac{1}{N}$. Equivalently,

$$\sum_{j=0}^{\infty} (-1)^j \binom{n}{j} z^{n-j} w^j X(z,w) = 0.$$

Writing $X(z,w) = \sum_k x_k(z)w^k$, this becomes

$$\sum_{k \in \frac{1}{N}\mathbf{Z}} \sum_{j=0}^{\infty} (-1)^j \binom{n}{j} z^{n-j} x_{k-j} w^k = 0.$$

Since $X(z, w)$ is a Laurent series in w, there is some $K \in \frac{1}{N}\mathbf{Z}$ such that $x_k = 0$ for all $k \leq K$. Hence,

$$\sum_{j=0}^{\infty}(-1)^j \binom{n}{j} z^{n-j} x_{k-j} = 0$$

for all $k \in \frac{1}{N}\mathbf{Z}$ and so $X(z, w) = 0$ □

A consequence of locality is that imposes a condition on $\nu(g_2, g_1)$ in terms of $\nu(g_1, g_2)$.

Lemma 4.1.12. *Let Γ be an abelian group of exponent N, Δ a $\mathbf{Z}$-bilinear symmetric map $\Gamma \times \Gamma \to \mathbf{Q}/\mathbf{Z}$ and $\nu : \Gamma \times \Gamma \to \mathbf{C}^*$ a bi-multiplicative map. Suppose that the generalized vertex operators X, Y on a Γ-graded complex vector space V are mutually local with respect to ν and Δ and that $X(z)Y(w) \neq 0$. Then,*

$$\nu(p(X), p(Y))\nu(p(Y), p(X)) = e^{-2\pi i \Delta(p(X), p(Y))}.$$

Proof. Suppose that X and Y are mutually local generalized vertex operators. By definition of locality and Remark 4.1.3, for large enough positive rational $n \in \mathbf{Z} + \Delta(p(X), p(Y))$, for all $a \in V$,

$$
\begin{aligned}
i_{z,w}(z - w)^n X(z)Y(w)a &= i_{w,z}(z - w)^n e^{\pi i n} \nu(p(Y), p(X))Y(w)X(z)a \\
&= i_{w,z}(w - z)^n e^{\pi i n} \nu(p(X), p(Y))Y(w)X(z)a \\
&= e^{\pi i n} \nu(p(X), p(Y))\nu(p(Y), \\
&\quad P_w(X))i_{z,w}(w - z)^n X(z)Y(w)a \\
&= e^{2\pi i n} \nu(p(X), p(Y))\nu(p(Y), \\
&\quad p(X))i_{z,w}(z - w)^n X(z)Y(w)a.
\end{aligned}
$$

By definition of a generalized vertex operator, $X(z)Y(w)a$ is a Laurent series in w. Hence, by Lemma 4.1.11, for all $a \in V$,

$$e^{2\pi i n} \nu(p(X), p(Y))\nu(p(Y), p(X))X(z)Y(w)a = X(z)Y(w)a.$$

As this holds for all $a \in V$,

$$e^{2\pi i n} \nu(p(X), p(Y))\nu(p(Y), p(X))X(z)Y(w) = X(z)Y(w).$$

Since by assumption $X(z)Y(w) \neq 0$, it follows that

$$\nu(p(X), p(Y))\nu(p(Y), p(X)) = e^{-2\pi i n}.$$

Now, $n = n_Z + \Delta(p(X), p(Y))$ for some $n_Z \in \mathbf{Z}$ and so $e^{-2\pi i n} = e^{-2\pi i \Delta(p(X), p(Y))}$, which proves the result.   □

Note that when X and Y are vertex operators, Lemma 4.1.12 is satisfied for the map ν defined in Remark 4.1.7.

We are now ready to define Γ-graded vertex algebras.

Definition 4.1.13. *Let Γ be an abelian group of exponent N, Δ a $\mathbf{Z}$-bilinear symmetric map $\Gamma \times \Gamma \to \mathbf{Q}/\mathbf{Z}$ and $\nu : \Gamma \times \Gamma \to \mathbf{C}^*$ a bi-multiplicative map satisfying*

$$\nu(g, h)\nu(h, g) = e^{-2\pi i \Delta(g,h)} \quad \forall g, h \in \Gamma.$$

A Γ-graded vertex algebra is defined to be a Γ-graded vector space V with

(i) a vacuum vector $1 \in V_0$,
(ii) an endomorphism $d \in gl(V)$ called the translation operator and
(iii) For all $a \in V$, a generalized vertex operator

$$Y(a, z) = \sum_{n \in \frac{1}{N}\mathbf{Z}} a_n z^{-n-1} \in gl(V)[[z^{\frac{1}{N}}, z^{-\frac{1}{N}}]]$$

depending linearly on a and such that whenever $a \in V_g$, $Y(a, z)$ has parity g and satisfies the following axioms:

(1) (Translation) $[d, Y(a, z)] = \frac{d}{dz}Y(a, z)$ and $d(1) = 0$,
(2) (Vacuum) $Y(1, z) = I_V$, $Y(v, z).1 \in V[[z^{\frac{1}{N}}]]$ and $Y(a, z).1\big|_{z=0} = a$,
(3) (Locality) For all $a, b \in V$, the vertex operators $Y(a, z)$ and $Y(b, z)$ are mutually local with respect to ν and Δ.

If $\Gamma \leq \mathbf{Z}_2$, $\Delta = 0$ and $\nu(a, b) = \begin{cases} -1 & \text{if } a = b = \overline{1} \\ 0 & \text{otherwise} \end{cases}$, then V is a vertex (super)algebra.

Remark 4.1.14. The vacuum axiom makes sense since, by definition of a generalized vertex operator $Y(a, z)$, $a_n.1 = 0$ unless $n \in \mathbf{Z}$ since $\mathbf{Z} = \mathbf{Z} + \Delta(p(a), p(1))$ and $p(1) = 0$ by definition.

Our aim now is to find an expression for the "commutator"

$$a_n b_m - e^{\pi i n} \nu(p(a), p(b)) b_m a_n$$

of any two elements $a, b \in V$, called the vertex algebra identity (also known as the Jacobi identity or Borcherds identity). This fundamental identity in the context of Γ-graded vertex algebras is analogous to the Jacobi identity for Lie algebras. In fact in the original definition of a vertex algebra (see [Borc2]), it is given as a defining axiom – as stated in [FrenLM2], "it is implicit" in [Borc2] – and in that of a vertex operator algebra (see [FrenLM2]), it is also taken as one of the main defining axioms of a vertex operator algebra (see section 4.2 for the difference between these two structures). As stated earlier, our purpose for presenting some aspects of the theory of vertex algebras is the construction of BKM algebras from objects known as lattice vertex algebras. The Lie algebra structure on appropriate subquotients of lattice vertex algebras is a consequence of the vertex algebra identity (see section 4.3). However, rather than proving directly that it holds for lattice vertex algebras, it is easier to show that they are vertex algebras according to the above definition and that as a consequence the identity is valid.

We derive the vertex algebra identity in several steps. Our first goal is an associativity result. We start with some elementary properties of generalized vertex operators that we will need for this, and which also explain the above axioms.

Lemma 4.1.15. *Let Γ be an abelian group of exponent $N \in \mathbf{N}$, V be a Γ-graded vector space, d an endomorphism of V and 1 a vector in V_0.*

(i) *Let $X(z) = \sum_{n \in \frac{1}{N}\mathbf{Z}} x_n z^{-n-1} \in End(V)[[z^{\frac{1}{N}}, z^{-\frac{1}{N}}]]$ a vertex operator on V. Then, $[d, X(z)] = \frac{d}{dz}X(z)$ if and only if for all $n \in \mathbf{Z}$, $[d, x_n] = -n x_{n-1}$.*

(ii) *$Y(1, z) = I_V$ if and only if $1_n = \delta_{n,-1} I_V$*

(iii) *For $a \in V$, $Y(a, z).1 \in V[[z]]$ and $Y(a, z).1\big|_{z=0} = v$ if and only if $a_n.1 = 0$ for all $n \geq 0$ and $a_{-1}.1 = a$.*

(iv) *Let $X(z)$ and $Y(z)$ be mutually local vertex operators on V. Then, the vertex operators $\frac{d}{dz}X(z)$ and $Y(z)$ are also mutually local.*

Proof. (i) follows from the fact that $\frac{d}{dz}X(z) = \sum_{n \in \mathbf{Z}} -n x_{n-1} z^{-n-1}$. (ii) and (iii) are easily shown (see Remark 4.1.14 for (iii)).

We next prove (iv): For large enough positive n in $\mathbf{Z} + \Delta(p(X), p(Y))$,

$$(z - w)^n X(z)Y(w) = (w - z)^n e^{\pi i n} \nu(p(X), p(Y))Y(w)X(z). \tag{1}$$

We next apply $\frac{d}{dz}$ to both sides of this equality.

$$n(z - w)^{n-1} X(z)Y(w) + (z - w)^n \frac{d}{dz}X(z)Y(w)$$

$$= \nu(p(X), p(Y))(-n(w - z)^{n-1} e^{\pi i(n-1)} e^{\pi i} Y(w)X(z)$$

$$+ (w - z)^n e^{\pi i n} Y(w)\frac{d}{dz}X(z)).$$

As $e^{\pi i} = -1$, from (1), for large enough positive n in $\mathbf{Z} + \Delta(p(X), p(Y))$,

$$(z - w)^n \frac{d}{dz}X(z)Y(w) = (w - z)^n e^{\pi i n} \nu(p(X), p(Y))Y(w)\frac{d}{dz}X(z).$$

As $p(\frac{d}{dz}X(z)) = p(X(z))$, this proves (iv). $\square$

In the rest of section 4.1, we will assume that V is a Γ-graded vertex algebra and we will keep the notation of Definition 4.1.13.

Lemma 4.1.16. *The endomorphism d has parity 0 and for all $a \in V$, $d(a) = a_{-2}.1$.*

Proof. From Lemma 4.1.15 (i) and the vacuum axiom, for all $a \in V$,

$$d(a) = d a_{-1}.1 = a_{-2}.1.$$

Hence as $1 \in V_0$, the parity condition on the vertex operator $Y(a, z)$ implies that d has parity 0, proving the Lemma. $\square$

Lemma 4.1.17. *For all $a \in V$, $Y(a, z).1 = e^{zd}a$.*

Proof. Applying the translation operator d, $n - 1$ times to both sides of the equality in Lemma 4.1.16 leads to $\frac{d^n}{n!}(a) = a_{-n-1}.1$ for $n \in \mathbf{N}$, from which the result follows. □

We can now prove Goddard's uniqueness result. It is very useful as it says that any generalized vertex operator mutually local with each of the operators $Y(a, z)$ is fully determined by its action on the vacuum vector.

Lemma 4.1.18. *If for all $a \in V$, $X(z)$ and $Y(a, z)$ are mutually local vertex operators on V and $X(z).1 = Y(b, z).1$ for some $b \in V$, then $X(z) = Y(b, z)$.*

Proof. Let $X(z)$ be a vertex operator on V satisfying the conditions of the Lemma. Then for large enough positive $n \in \mathbf{Z} + \Delta(p(X), p(a))$, for any $a \in V$,

$$i_{z,w}(z - w)^n X(z)e^{wd}a$$

$$= i_{z,w}(z - w)^n X(z)Y(a, w).1 \quad \text{by Lemma 4.1.17}$$

$$= i_{w,z}(z - w)^n \nu(p(X), p(a))Y(a, w)Y(b, z).1 \quad \text{by assumption}$$

$$= e^{\pi in}i_{w,z}(w - z)^n \nu(p(X), p(a))Y(a, w)Y(b, z).1$$

$$= e^{\pi in}i_{z,w}(w - z)^n \nu(p(X), p(a))\nu(p(a), p(b))Y(b, z)Y(a, w).1$$

by locality

$$= e^{2\pi in}i_{z,w}(z - w)^n \nu(p(X), p(a))\nu(p(a), p(b))Y(b, z)e^{wd}a$$

by Lemma 4.1.17

$$= i_{z,w}(z - w)^n Y(b, z)e^{wd}a.$$

The last equality follows from the definition of the map ν and the fact that $p(X) = p(b)$ holds since $X(z).1 = e^{zd}b$ by assumption and Lemma 4.1.17 and since the translation operator has parity 0 by Lemma 4.1.16. As all powers of w in $X(z)e^{wd}u - Y(v, z)e^{wd}u$ are non-negative, by Lemma 4.1.11,

$$X(z)e^{wd}a = Y(v, z)e^{wd}a.$$

Again, as all powers of w in this equality are non-negative, we may set $w = 0$. So for any $a \in V$,

$$X(z)a = Y(v, z)a,$$

which proves the result. □

Lemma 4.1.19. *For any $a \in V$, $e^{wd}Y(a, z)e^{-wd} = Y(a, i_{z,w}(z - w))$.*

Proof.

$$e^{wd}Y(a, z)e^{-wd} = \sum_{n=0}^{\infty} \frac{1}{n!}(\text{ad } wd)^n Y(a, z).$$

By the translation axiom, this is

$$\sum_{n=0}^{\infty} \frac{w^n}{n!}(\frac{d}{dz})^n Y(a, z),$$

which is the Taylor series of $Y(a, z + w)$ in the domain $|w| < |z|$ since, as a complex function of w, $Y(a, z + w)$ is differentiable in this neighbourhood of $w = 0$. Therefore, the Lemma follows from the fact that the Taylor series in this neighbourhood of $(z + w)$, considered as a complex function of w, is $i_{z,w}(z + w)$. $\square$

We can now prove a semi-symmetry property.

Lemma 4.1.20. *For any* $a, b \in V$,

$$Y(a, z)b = \nu(p(a), p(b))e^{zd}Y(b, -z)a.$$

Equivalently

$$a_n b = (-1)^{n+1}\nu(p(a), p(b)) \sum_{j=0}^{\infty}(-1)^j \frac{d^j}{j!}(b_{j+n}a).$$

Proof. Lemmas 4.1.19 and 4.1.17 imply that

$$
\begin{aligned}
e^{(z+w)d}Y(b, -z)a &= Y(b, w)e^{(z+w)d}a \\
&= Y(b, w)Y(a, z + w)1.
\end{aligned}
$$

Hence, by the locality axiom and Lemma 4.1.17, for any large enough positive rational n in $\mathbf{Z} + \Delta(p(b), p(a))$,

$$i_{w,z+w}(-z)^n e^{(z+w)d}Y(b, -z)a = i_{z+w,w}(-z)^n e^{\pi i n}\nu(p(b), p(a))Y(u, z + w)e^{wd}b. \tag{1}$$

As all exponents of w in the left hand side of equality (1) are non-negative, this must also be the case on the right hand side. Hence, we can put $w = 0$. This gives

$$i_{w,z+w}(-z)^n e^{zd}Y(b, -z)a = i_{z+w,w}(-z)^n e^{\pi i n}\nu(p(b), p(a))Y(a, z)b. \tag{2}$$

Since $b_j a = 0 = a_j b$ for $j >> 0$, for large enough n, there are no negative powers of z. Hence we can divide both sides of equality (2) by z^n. However, we have to be careful with the value of $(-1)^n$ that appears on both sides of (2). By assumption (see Remark 4.1.3), $i_{w,z+w}(-z)^n = e^{\pi i n}i_{w,z+w}z^n = e^{\pi i n}z^n$ and $i_{z+w,w}(-z)^n = e^{\pi i n}z^n$. By definition of ν,

$$\nu(p(b), p(a))^{-1}e^{\pi i n} = \nu(p(b), p(a))e^{2\pi i \Delta(p(a),p(b))} = \nu(p(a), p(b)).$$

Therefore, the first equality now follows. The second equality is just the first equality written in terms of the coefficients of z^{-n-1}. $\square$

One of the fundamental properties of Γ-graded vertex algebras, namely "associativity", now follows easily. First note that

$$
\begin{aligned}
Y(Y(a, z)b, w)c &= \sum_{j \in \mathbf{Z} + \Delta(p(a+b), p(c))} (Y(a, z)b)_j cw^{-j-1} \\
&= \sum_{j \in \mathbf{Z} + \Delta(p(a+b), p(c))} \sum_{k \in \mathbf{Z} + \Delta(p(a)+p(b))} (a_k b)_j z^{-k-1}w^{-j-1} \\
&= \sum_{k \in \mathbf{Z} + \Delta(p(a)+p(b))} Y(a_k b, w)z^{-k-1}.
\end{aligned}
$$

Proposition 4.1.21. *For any $a, b, c \in V$ and $n \in \mathbf{Z} + \Delta(p(a), p(c))$ large enough,*

$$i_{z,w}(w-z)^n Y(a, i_{z,w}(z-w))Y(b, -w)c = i_{w,z}(w-z)^n Y(Y(a, z)b, -w)c.$$

Proof. Applying the locality axiom, for large enough positive $n \in \mathbf{Z} + \Delta(p(a), p(c))$,

$$i_{z,w}(z-w)^n Y(a, z)Y(c, w)b = i_{w,z}(z-w)^n Y(c, w)Y(a, z)b. \tag{1}$$

By Lemma 4.1.20,

$$Y(c, w)b = e^{wd}Y(b, -w)c \quad \text{and} \quad Y(c, w)Y(a, z)b = e^{wd}Y(Y(a, z)b, -w)c.$$

By Lemma 4.1.19,

$$Y(a, z)e^{wd} = e^{wd}Y(a, i_{z,w}(z-w)).$$

These equalities together with (1) give the result since

$$\nu(p(a), p(c))\nu(p(c), p(a)) = e^{-2\pi i n}, \quad i_{w,z}(z-w)^n = i_{w,z}(w-z)^n e^{\pi i n}$$

and $i_{z,w}(z-w)^n = i_{z,w}(w-z)^n e^{-\pi i n}$. $\qquad\qquad\square$

We next give some basic properties of $\delta(z-w) = z^{-1}\sum_{n\in\mathbf{Z}} z^{-n}w^n$ or more generally of

$$\delta_r(z-w) = \delta(z-w)\left(\frac{w}{z}\right)^r$$

for any rational $r \in \mathbf{Q}$.

Lemma 4.1.22.

(i) $\delta(z-w) = \delta(w-z)$

(ii) $\delta(z - i_{w,v}(w+v)) = \delta(v - i_{z,w}(z-w)) + \delta(-v, i_{w,z}(w-z))$.

Proof. The first equality is easy to check. Hence we only show the second one. By definition of $\delta(z, w)$,

$$\delta(z - i_{w,v}(w+v)) = \sum_{n\in\mathbf{Z}} z^{-n-1}i_{w,v}(w+v)^n$$

$$= \sum_{n\in\mathbf{Z}} z^{-n-1}\sum_{k=0}^{\infty}\binom{n}{k}w^{n-k}v^k \tag{1}$$

$$= \sum_{k=0}^{\infty}\frac{1}{k!}\frac{d}{dw}^k \delta(z, w)v^k.$$

Now,

$$i_{z,w}(z-w)^{-k-1} = \sum_{j=0}^{\infty} \binom{-k-1}{j} (-1)^j z^{-k-1-j} w^j$$

$$= \sum_{j=0}^{\infty} \binom{k+j}{j} z^{-k-1-j} w^j$$

$$= \sum_{n=k}^{\infty} \binom{n}{n-k} z^{-n-1} w^{n-k}$$

$$= \sum_{n=0}^{\infty} \binom{n}{k} z^{-n-1} w^{n-k}$$

since $\binom{n}{k} = \frac{n\cdots(n-k+1)}{k!} = 0$ for $0 \le n \le k-1$. Also,

$$i_{w,z}(w-z)^{-k-1} = \sum_{j=0}^{\infty} \binom{j+k}{j} w^{-k-1-j} z^j$$

$$= \sum_{n=-1}^{-\infty} \binom{-n-1+k}{-n-1} w^{n-k} z^{-n-1}$$

$$= \sum_{n=-1}^{-\infty} \binom{-n-1+k}{k} w^{n-k} z^{-n-1}$$

$$= (-1)^k \sum_{n=-1}^{-\infty} \binom{n}{k} w^{n-k} z^{-n-1}.$$

Hence, substituting into (1) and using part (i), we get

$$\delta(z - i_{w,v})(w+v)$$

$$= \sum_{k=0}^{\infty} v^k i_{z,w}(z-w)^{-k-1} + (-1)^k v^k i_{w,z}(w-z)^{-k-1}$$

$$= \delta(v - i_{z,w}(z-w)) - \sum_{k=-1}^{-\infty} v^k i_{z,w}(z-w)^{-k-1} + \sum_{k=0}^{\infty} (-v)^k i_{w,z}(w-z)^{-k-1}$$

$$= \delta(v - i_{z,w}(z-w)) - \sum_{k=-1}^{-\infty} v^k i_{w,z}(z-w)^{-k-1} + \sum_{k=0}^{\infty} (-v)^k i_{w,z}(w-z)^{-k-1}$$

$$= \delta(v - i_{z,w}(z-w)) + \sum_{k=-1}^{-\infty} (-v)^k i_{w,z}(w-z)^{-k-1} + \sum_{k=0}^{\infty} (-v)^k i_{w,z}(w-z)^{-k-1}$$

$$= \delta(v - i_{z,w}(z-w)) + \delta(-v - i_{w,z}(w-z)).$$

since whenever $-k-1$ is a non-negative integer, $i_{w,z}(w-z)^{-k-1} = i_{z,w}(w-z)^{-k-1}$. □

Definition 4.1.23. *If* $f(z,w) = \sum_{m,n\in\mathbf{Q}} f_{m,n} z^m w^n$ *is a formal series, then we will say that* $f(w,w)$ *is well defined if for any* $l \in \mathbf{Q}$, *only finitely many coefficients* $f_{m,l-m}$ *are non-zero.*

Note that for a well defined series $f(z,w) = \sum_{m,n\in\mathbf{Q}} f_{m,n} z^m w^n$, the series $f(w,w) = \sum_{l\in\mathbf{Q}} (\sum_{m\in\mathbf{Q}} f_{m,l-m}) w^l$ makes sense. Set

$$gl(V)[[z,z^{-1}]]\{w\} = \{f(z,w) = \sum_{r\in\mathbf{Q}} f_r(z)w^r : \quad f_r(z) \in gl(V)[[z,z^{-1}]]\}.$$

Lemma 4.1.24. *Let* $s \in \mathbf{Q}$ *and* $z^s f(z,w) \in gl(V)[[z,z^{-1}]]\{w\}$ *such that* $f(w,w)$ *is well defined. Then, for any* $r \in \mathbf{Q}$,

$$\delta_r(z-w)f(z,w) = \delta_{r+s}(z-w)f(w,w).$$

Proof. Let $z^s f(z,w) \in gl(V)[[z,z^{-1}]]\{w\}$. By definition of $\delta_r(z-w)$, we only need to prove that

$$\delta(z-w)f(z,w) = \delta_s(z-w)f(w,w). \tag{1}$$

Suppose that (1) holds for $s = 0$ and any $g \in gl(V)[[z,z^{-1}]]\{w\}$. Then,

$$\delta_s(z-w)f(w,w) = \frac{w}{z}^s \delta(z-w)f(w,w) = z^{-s}\delta(z-w)(w^s f(w,w))$$

$$= z^{-s}\delta(z-w)z^s f(z,w)$$

since $z^s f(z,w) \in gl(V)[[z,z^{-1}]]\{w\}$. And so the Lemma follows. Hence we only need to show that

$$\delta(z-w)f(z,w) = \delta(z-w)f(w,w)$$

for $f \in gl(V)[[z,z^{-1}]]\{w\}$. Any series f in this set can be written as follows:

$$f(z,w) = \sum_{n\in\mathbf{Q}} \sum_{m\in\mathbf{Z}} f_{m,n} z^m w^n.$$

As $f(z,w)$ is well defined,

$$f(z,w) = \sum_{l\in\mathbf{Q}} (\sum_{m\in\mathbf{Z}} f_{m,l-m}(\frac{z}{w})^m) w^l.$$

Therefore,

$$\delta(z-w)f(z,w) = \sum_{l\in\mathbf{Q}} (\sum_{m,n\in\mathbf{Z}} f_{m,l-m}(\frac{z}{w})^{m+n}) w^l z^{-1}$$

$$= (\sum_{m\in\mathbf{Z}} f_{m,l-m}) \sum_{j\in\mathbf{Z}} (\frac{z}{w})^j w^l z^{-1}$$

$$= \delta(z-w)f(w,w).$$

$\square$

We are now ready to prove that associativity together with locality imply the vertex algebra identity.

Theorem 4.1.25. *For any $a, b, c \in V$, any $n \in \mathbf{Z} + \Delta(p(a), p(b))$, $m \in \mathbf{Z} + \Delta(p(a), p(c))$, $k \in \mathbf{Z} + \Delta(p(b), p(c))$,*

$$\sum_{j=0}^{\infty} \binom{m}{j} (a_{n+j}b)_{m+k-j}c$$

$$= \sum_{j=0}^{\infty} (-1)^j \binom{n}{j} \left(a_{n+m-j}b_{k+j} - e^{i\pi n}\nu(p(a), p(b))b_{k+n-j}a_{m+j} \right)c.$$

Proof. Let $a, b \in V$ and for $r \in \mathbf{Z} + \Delta(p(a), p(b))$, set

$$f(z, i_{w,v}(w + v)) = (w + v)^r v^{-r} Y(a, z)Y(b, w).$$

(As a series in $i_{w,v}(w + v)$, f only contains a constant term). Applying Lemma 4.1.22,

$$\delta(z - i_{w,v}(w + v))f(z, i_{w,v}(w + v)) = \delta(v - i_{z,w}(z - w))f(z, i_{w,v}(w + v))$$
$$+ \delta(-v - i_{w,z}(w - z))f(z, i_{w,v}(w + v)). \tag{1}$$

We evaluate the three summands in this equality.

Let $c \in V$. By definition of generalized vertex operators, the coefficients on z in $f(z, i_{w,v}(w + v))c$ are in

$$\mathbf{Z} + \Delta(p(a), p(b)) - \Delta(p(a), p(b) + p(c)) = \mathbf{Z} - \Delta(p(a), p(c)).$$

Hence for $s \in \mathbf{Z} + \Delta(p(a), p(c))$, since the series f is clearly well defined, Lemma 4.1.24 implies that

$$\delta(z - i_{w,v}(w+v))f(z, i_{w,v}(w+v))c = \delta_s(z - i_{w,v}(w+v))f(i_{w,v}(w+v), i_{w,v}(w+v))c.$$

For $r \gg 0$ locality gives

$$f(z, i_{w,v}(w + v)) = i_{w,z}(z - w)^r \nu(p(a), p(b))v^{-r}Y(b, w)Y(a, z). \tag{2}$$

Hence f is a Laurent series in z. So for $r \gg 0$ and $s \gg 0$, all the exponents of $i_{w,v}(w + v)$ in $f(i_{w,v}(w + v), i_{w,v}(w + v))c$ are non-negative integers. Hence,

$$i_{w,v}(w + v) = i_{v,w}(w + v) = (w + v)$$

in this series. Therefore,

$$f(i_{w,v}(w + v), i_{w,v}(w + v))c = Y(a, i_{v,w}(w + v))Y(b, w)c, \tag{3}$$

$$f(i_{w,v}(w + v), i_{w,v}(w + v))c = i_{w,v}(w + v)^s Y(a, i_{w,v}(w + v))Y(b, w)c \tag{4}$$

and by associativity (Proposition 4.1.21),

$$f(i_{w,v}(w+v), i_{w,v}(w+v))c = i_{v,w}(w+v)^s Y(Y(a,v)b, w)c. \qquad (5)$$

Now $Y(Y(a,v)b, w)c$ and $Y(a, i_{w,v}(w+v))Y(b, w)c$ are Laurent series in v. So from Lemma 4.1.11 and equalities (4) and (5) we get

$$Y(Y(a,v)b, w)c = Y(a, i_{w,v}(w+v))Y(b, w)c.$$

As a result, for $r \gg 0$ and $s \gg 0$,

$$\delta(z - i_{w,v}(w+v))f(z, i_{w,v}(w+v))c = \delta_s(z - i_{w,v}(w+v))Y(Y(a,v)b, w)c. \qquad (6)$$

Consider the series f as a series $f(v, i_{z,w}(z-w))$ in v and $i_{z,w}(z-w)$. It is well defined and

$$v^r f(v, i_{z,w}(z-w)) \in gl(V)[[v, v^{-1}]]i_{z,w}(z-w).$$

So by Lemma 4.1.24,

$$\delta(v - i_{z,w}(z-w))f(v, i_{z,w}(z-w)) = \delta_r(v - i_{z,w}(z-w))Y(a, z)Y(b, w). \qquad (7)$$

Next consider the series f as a series $f(-v, i_{w,z}(w-z))$ in $-v$ and $i_{w,z}(w-z)$. Then for $r \gg 0$, by (2)

$$f(-v, i_{w,z}(w-z)) = i_{w,z}(w-z)^r e^{\pi i r} \nu(p(a), p(b))v^{-r}Y(b, w)Y(a, z).$$

So the previous arguments imply that for $r \gg 0$,

$$\delta(-v - i_{w,z}(w-z))f(-v, i_{w,z}(w-z))$$
$$= \delta_r(-v - i_{w,z}(w-z))e^{2\pi i r}\nu(p(a), p(b))Y(b, w)Y(a, z) \qquad (8)$$

since $i_{w,z}(w-z)^r = e^{\pi i n}i_{w,z}(z-w)^r$. Therefore equalities (1), (6), (7) and (8) lead to

$$\delta_s(z - i_{w,v}(w+v))Y(Y(a,v)b, w)c = \delta_r(v - i_{z,w}(z-w))Y(a, z)Y(b, w)c$$
$$+ \delta_r(-v - i_{w,z}(w-z))e^{2\pi i r}\nu(p(a), p(b))Y(b, w)Y(a, z)c \qquad (9)$$

for $r \gg 0$ and $s \gg 0$. To conclude the proof, we only need to check that this is equivalent to the vertex algebra identity. The right hand side of (9) is equal to

$$\sum_{j \in \mathbf{Z}} v^{-j-1-r}i_{z,w}(z-w)^{j+r}Y(a, z)Y(b, w)c$$

$$+ \sum_{j \in \mathbf{Z}} (-v)^{-j-1-r}i_{w,z}(w-z)^{j+r}\nu(p(a), p(b))e^{2\pi i r}Y(b, w)Y(a, z)c$$

$$= \sum_{n \in \mathbf{Z}+\Delta(p(a),p(b))} v^{-1-n}i_{z,w}(z-w)^n Y(a, z)Y(b, w)c$$

$$- \sum_{n \in \mathbf{Z}+\Delta(p(a),p(b))} v^{-1-n}i_{w,z}(w-z)^n \nu(p(a), p(b))e^{\pi i r}Y(b, w)Y(a, z)c$$

since for $r, n \in \mathbf{Z} + \Delta(p(a), p(b))$, $e^{2\pi i r} = e^{2\pi i n}$ and by assumption $(-1)^n = e^{-\pi i n}$. So the coefficient of v^{-n-1} in the right hand side of (9) is

$$i_{z,w}(z-w)^n Y(a,z)Y(b,w)c - i_{w,z}(w-z)^n \nu(p(a),p(b))e^{\pi i n}Y(b,w)Y(a,z)c$$

$$= \sum_{j=0}^{\infty}(-1)^j \binom{n}{j} \sum_{l \in \mathbf{Z}+\Delta(p(a),p(b+c))} a_l z^{-l-1+n-j} w^j Y(b,w)c$$

$$- \sum_{j=0}^{\infty}(-1)^j \binom{n}{j} \nu(p(a),p(b))e^{\pi i n}Y(b,w)w^{n-j} \sum_{l \in \mathbf{Z}+\Delta(p(a),p(c))} a_l z^{-l-1+j}c$$

And thus the coefficient of $v^{-n-1}z^{-m-1}$ in the right hand side of (9) is

$$\sum_{j=0}^{\infty}(-1)^j \binom{n}{j}(a_{n+m-j}w^j Y(b,w)c - \nu(p(a),p(b))e^{\pi i n}Y(b,w)w^{n-j}a_{m+j}c. \quad (10)$$

The coefficient of z^{-m-1} in the left hand side of (9) is equal to

$$i_{w,v}(w+v)^m Y(Y(a,v)b,w)c = \sum_{j=0}^{\infty}\binom{m}{j}w^{m-j}\sum_{l \in \mathbf{Z}+\Delta(p(a),p(b))}Y(a_l b,w)v^{j-l-1}c.$$

Hence the coefficient of $v^{-n-1}z^{-m-1}$ in the left hand side of (9) is

$$\sum_{j=0}^{\infty}\binom{m}{j}w^{m-j}Y(a_{n+j}b,w)c. \quad (11)$$

Equating the coefficients of w^{-k-1} in (10) and (11) we can deduce the vertex algebra identity. $\qquad\square$

We have shown that locality and associativity imply the vertex algebra identity. This is in fact an equivalence [DonL], [LepL], [FrenHL], [Kac15]. Indeed, the definition of a Γ-graded vertex algebra can be re-stated so as to have this identity as a defining axiom (see Exercise 4.1.2).

As we will be using the vertex algebra identity in various particular cases, we list them in the following result.

Corollary 4.1.26.

(i) For any $b, c \in V$, $a \in V_0$, $m \in \mathbf{Z}$, $k \in \mathbf{Z} + \Delta(p(b), p(c))$,

$$[a_m, b_k]c = \sum_{j=0}^{\infty}\binom{m}{j}(a_j b)_{m+k-j}c$$

or equivalently

$$[a_m, Y(b,z)] = \sum_{j=0}^{\infty}\binom{m}{j}Y(a_j b, z)z^{m-j}.$$

(ii) *Suppose that V is a vertex algebra. Then, for any $a, b \in V$, $m, k \in \mathbf{Z}$*

$$a_m b_k - (-1)^{p(a)p(b)} b_k a_m = \sum_{j=0}^{\infty} \binom{m}{j} (a_j b)_{m+k-j}.$$

Proof. We keep the notation of Theorem 4.1.25. If $a \in V_0$ or $\Delta = 0$, then $n \in \mathbf{Z}$. So (i) and (ii) follow by setting $n = 0$ in the vertex algebra identity. $\square$

We next discuss briefly the basic notion of a product of two generalized vertex operators. This requires some care. If $X(z)$ and $Y(z)$ are vertex operators on a vector space U, then the formal series $X(z)Y(z)$ is not usually a generalized vertex operator as it does not necessarily satisfy the convergence condition: For $a \in U$,

$$X(z)Y(z)a = \sum_k \left(\sum_m x_{k-m-1} y_m \right) z^{-k-1} a.$$

If $k - m - 1 < 0$, convergence of the operator Y implies that $\sum_m x_{k-m-1} y_m a = 0$ for $k \gg 0$. The problem arises from the coefficients x_{k-m-1} for $k - m - 1 \geq 0$. Since for any $a \in V$, $x_n a = 0$ for $n \gg 0$, we would want to put all endomorphisms x_n with $n \geq 0$ on the right hand side. To do this, we define the normal order of products of vertex operators as follows.

Definition 4.1.27. *Let $X(z) = \sum_{n \in \mathbf{Z}} x_n z^{-n-1}$ and $Y(z)$ be generalized vertex operators on V. The normally ordered product of X and Y is defined to be as follows:*

$$: X(z)Y(z) := \sum_{n<0} x_n z^{-n-1} Y(z) + \nu(p(X), p(Y)) Y(z) \sum_{n \geq 0} x_n z^{-n-1}.$$

More generally, the normally ordered product of the generalized vertex operators $X_i(z)$, $i = 1, \cdots, r$, where $X_i(z)$, $i = 1, \ldots r - 1$ is defined to be

$$: X_1(z)X_2(z) \cdots X_r(z) :=: X_1(z)(: X_2(z)...X_r(z) :) :$$

Remark 4.1.28. The above discussion shows that the formal series $: X(z)Y(z) :$ is a generalized vertex operator, i.e. satisfies the convergence condition. It clearly has the parity $p(X) + p(Y)$. Using usual conventions, we write

$$X(z)_+ = \sum_{n<0} x_n z^{-n-1}, \quad X(z)_- = \sum_{n \geq 0} x_n z^{-n-1}.$$

Although the next result is easy to check, we state it. It is needed to show that the generalized vertex operators we will construct in section 4.2 for each element of a Γ-graded vector space satisfy the axioms of a Γ-graded vertex algebra.

Lemma 4.1.29. *Let U be a Γ-graded vector space with a derivation $d \in EndU$ of parity 0. Let Y_i, $i = 1, 2$, be mutually local generalized vertex operators on U satisfying*

$$[d, Y_i(z)] = \frac{d}{dz}Y_i(z), \quad Y_i(z).1 \in EndV[[z]], \quad i = 1, 2.$$

Then,

$$\frac{d^2}{dz}Y_i(z) = [d, \frac{d}{dz}Y_i(z)], \quad \frac{d}{dz}Y_i(z).1 \in EndV[[z]].$$

If moreover Y_1 has parity 0, then the generalized vertex operators $\frac{d}{dz}Y_1(z)$ and $Y_2(z)$ are mutually local, and

$$: [d, Y_1(z)]Y_2(z) : + : Y_1(z)[d, Y_2(z)] :$$

$$= \frac{d}{dz} : Y_1(z)Y_2(z) :, \quad : Y_1(z)Y_2(z) : .1 \in EndV[[z]].$$

Proof. The first two statements are obvious.

$$i_{z,w}(z-w)^n \frac{d}{dz}Y_1(z)Y_2(w) = \frac{d}{dz}(i_{z,w}(z-w)^n \frac{d}{dz}Y_1(z))Y_2(w)$$
$$- ni_{z,w}(z-w)^{n-1}Y_1(z)Y_2(w)$$
$$= \frac{d}{dz}(i_{w,z}(z-w)^n \frac{d}{dz}Y_2(w)Y_1(z))$$
$$- ni_{w,z}(z-w)^{n-1}Y_2(w)Y_1(z)$$
$$= i_{w,z}(z-w)^n \frac{d}{dz}Y_2(w)\frac{d}{dz}Y_1(z)$$

for large enough $n \in \mathbf{Z} + \Delta(p(X), p(Y))$.

We next check the third equality.

$$\frac{d}{dz} : Y(v_1, z)Y(v_2, z) :$$
$$= -\sum_{n<0} n(v_1)_{n-1}z^{-n-1}Y(v_2, z) - (-1)^{\nu(p(v_1), p(v_2))}Y(v_2, z)\sum_{n\geq 0} n(v_1)_{n-1}z^{-n-1}$$
$$+ : Y(v_1, z)[d, Y(v_2, z)] :$$
$$=: [d, Y(v_1, z)]Y(v_2, z) : + : Y(v_1, z)[d, Y(v_2, z)] :$$

$\square$

We next prove Dong's Lemma [Li2]. It is crucial as it concerns the locality of normally ordered products of generalized vertex operators.

Lemma 4.1.30. *For any $a, b, c \in V$, the generalized vertex operators $Y(c, z)$ and $: Y(a, z)Y(b, z) :$ are mutually local.*

Proof. Let $a, b, c \in V$. Set

$$X(z, w) = i_{z,w}(z - w)^{-1+\Delta(p(a),p(b))} Y(a, z)Y(b, w) - i_{w,z}(z - w)^{-1+\Delta(p(a),p(b))}$$

$$\times \nu(p(a), p(b))Y(b, w)Y(a, z).$$

We show that for large enough $r \in \mathbf{Z} + \Delta(p(c), p(a) + p(b))$,

$$i_{v,w}(v - w)^r Y(c, v)X(z, w) = i_{w,v}(v - w)^r \nu(p(c), p(b))X(z, w)Y(c, v). \quad (1)$$

Since the parity of $Y(a, z)$ is 0, by locality, there exists $M > 0$ such that for any integer $n \geq M$

$$i_{z,w}(z - w)^{n+\Delta(p(a),p(b))} Y(a, z)Y(b, w)$$
$$= i_{w,z}(z - w)^{n+\Delta(p(a),p(b))} \nu(p(a), p(b))Y(b, w)Y(a, z), \quad (2)$$

$$i_{v,z}(v - z)^{n+\Delta(p(a),p(c))} Y(c, v)Y(a, z)$$
$$= i_{z,v}(v - z)^{n+\Delta(p(a),p(c))} \nu(p(c), p(a))Y(a, z)Y(c, v), \quad (3)$$

and

$$i_{v,w}(v - w)^{n+\Delta(p(b),p(c))} Y(c, v)Y(b, w)$$
$$= i_{w,v}(v - w)^{n+\Delta(p(b),p(c))} \nu(p(c), p(b))Y(b, w)Y(c, v). \quad (4)$$

Now, for any integer $n \geq N$,

$$(v - w)^{2n} = \sum_{k=0}^{\infty} \binom{2n}{k} (v - z)^{2n-k}(z - w)^k.$$

If $k < M$, then $2n - k > M$. Therefore, taking $r > 3n$ and $n \geq N$, (1) follows from equalities $(2) - (4)$.

Taking Res_z of both sides of equality (1) shows that the generalized vertex operators $Y(c, v)$ and $: Y(a, w)Y(b, w) :$ are mutually local. $\qquad\square$

We next give an elementary but useful consequence of the translation axiom.

Lemma 4.1.31. *For any $a \in V$, $Y(d(a), z) = \frac{d}{dz}Y(a, z)$.*

Proof. By Lemma 4.1.27, for any $a \in V$,

$$\frac{d}{dz}Y(a, z)1 = \frac{d}{dz}(e^{zd}a) = e^{zd}d(a) = Y(d(a), z)1.$$

Hence the result follows from Lemmas 4.1.28 and 4.1.29. $\qquad\square$

This implies a property that will be necessary in the next section where we construct generalized vertex operators.

Corollary 4.1.32. *For any elements $a \in V_0$, $b \in V$, $n \in \mathbf{Z}$,*

$$Y(a_{-n-1}b, z) =: \frac{1}{n!}(\frac{d^n}{dz^n}Y(a, z))Y(b, z) :$$

Proof. In Theorem 4.1.25, since $a \in V_0$, n and m are integers. So for any $c \in V$, setting, $n = -1$ and $m = 0$ in the vertex algebra identity, we get for all $k \in \mathbf{Z} + \Delta(p(c), p(b))$,

$$(a_{-1}b)_k c = \sum_{j=0}^{\infty} (a_{-1-j}b_{k+j} + b_{k-1-j}a_j)c$$

$$= \sum_{j<0}(a_j b_{k-j-1} + \sum_{j\geq 0} b_{k-1-j}a_j)c.$$

Equivalently,

$$Y(a_{-1}b, z) =: Y(a, z)Y(b, z) : \tag{1}$$

Now, by Lemma 4.1.31,

$$Y(d^n a, z) = (\frac{d}{dz})^n Y(a, z) \quad \forall n \in \mathbf{Z}_+ \tag{2}$$

So by Lemma 4.1.15 (i), $\frac{1}{n!}(d^n a)_{-1} = a_{-n-1}$. Since d is an endomorphism of parity 0 (see Lemma 4.1.16), $d^n(a) \in V_0$. Therefore, applying (1) to $d^n(a)$ instead of a and using (2), we get the result. $\qquad\square$

As this is all we need to know about arbitrary Γ-graded vertex algebras for our purpose of constructing BKM superalgebras, we can now apply this theory in order to construct Γ-graded lattice vertex algebras.

Exercises 4.1

1. Let R be a commutative ring with a derivations d (i.e. $d(ab) = d(a)b + ad(b)$ for $a, b \in R$).

(i) Show that

$$d^{(i)}d^{(j)} = \binom{i+j}{i} d^{(i+j)},$$

where $d^{(i)}$ is the i-th derivative of d, and

$$d^{(n)}(ab) = \sum_{i=0}^{n} d^{(i)}(a)d^{(n-i)}(b) \quad \forall a, b \in R.$$

(ii) Deduce that the ring R can be given the structure of a vertex algebra.

Hint: For any $a, b \in V$, define $a_n(b) = (d^{(-n-1)}(a))b$.

2. Suppose that conditions $(i) - (iii)$ of Definition 4.1.13 hold for the Γ-graded vector space V. Show that if V satisfies the vertex algebra identity given in Theorem 4.1.25, $Y(1, z) = 1$ and $a_{-1}.1 = a$ for all $a \in V$, then V is a Γ-graded vertex algebra.

For a solution, see [Kac15, §4.8].

3. Let Γ be a finite abelian group of exponent N. Show that the tensor product $U \otimes V$ of two Γ-graded vertex algebras U and V is a Γ-graded vertex algebra with vacuum vector $1 \otimes 1$ and translation operator $d_U \otimes 1 + 1 \otimes d_V$, where d_U (resp. d_V) is the translation operator of U (resp. V).

4.2 Γ-Graded Lattice Vertex Algebras

From now on we only consider particular examples of Γ-graded vertex algebras, namely Γ-graded lattice vertex algebras. The starting point is a lattice L. When L is even, we can take $\Gamma = 0$. Lattice vertex algebras have a nice description and as we will see in section 5.5, when the even lattice is Lorentzian of rank at most 26, we can derive BKM algebras from them. The aim is however not just to find all the BKM algebras but the entire class of BKM superalgebras that can be explicitly constructed. When the odd part is non-trivial, this is possible in rank 10 and has been done in [Sch2]. We do not give this construction in this book but the background necessary for the interested reader to understand it. In order to construct this Lie superalgebra with a non-zero odd part, we have to start with an odd Lorentzian lattice. It may seem at a first glance that this construction could be based on lattice vertex superalgebras (i.e. corresponding to the simpler case $\Gamma = \mathbf{Z}_2$). However though they reflect to a certain extent the existence of odd norm vectors in the root lattice, they only allow the derivation of the even part of the Lie superalgebra. For the odd part, we need the more complex structure of a Γ-graded lattice vertex algebra, where Γ is a larger abelian group. Thus Dong and Lepowsky's axiomatic development of the theory of Γ-graded-vertex algebras is fundamental for the construction of BKM superalgebras with non-trivial odd part.

Let L be a finitely generated rational lattice with non-degenerate bilinear form $(.,.)$ and L_0 the maximal integral even sublattice of L, i.e.

$$L_0 = \{v \in L : (v, v) \equiv 0 \pmod 2\}.$$

Assume that $L \subseteq L_0^*$, where $L_0^* = \{x \in L \otimes_{\mathbf{Q}} \mathbf{R} : (x, L_0) \subseteq \mathbf{Z}\}$ is the dual of L. Set

$$\Gamma = L/L_0.$$

Since the dual lattice L_0^* has the same rank as L_0 and its elements are rational linear combinations of the generators of the lattice L_0, for any $\alpha \in L_0^*$, there is some $n \in \mathbf{N}$ such that $n\alpha \in L_0$. Hence, considered as free abelian groups, the index $[L_0^* : L_0] < \infty$. In other words, the abelian group Γ is finite since $L \subseteq L_0^*$. Let N be the exponent of Γ. Set

$$\Gamma = \{g_0, \cdots, g_n\}, g_0 = 0 \quad g_i = L_0 + \gamma_i, \quad \gamma_i \in L, \quad \forall 0 \le i \le n.$$

Hence, setting $L_i = \gamma_i + L_0$, $L = L_0 \cup \cdots \cup L_n$.

We are going to construct a Γ-graded vertex algebra. So we next need maps Δ and ν with the properties defined in the previous section. Let

$$\Delta : \ \Gamma \times \Gamma \to \mathbf{Q}/\mathbf{Z}$$
$$(g_i, g_j) \mapsto \mathbf{Z} - (\gamma_i, \gamma_j).$$

The map Δ is clearly a $\mathbf{Z}$-bilinear symmetric map.

Define the parity of $\alpha \in L_i$ to be $p(\alpha) = g_i$.

Fix a bi-multiplicative map

$$\nu : \Gamma \times \Gamma \to \mathbf{C}^*$$

satisfying

$$\nu(g_i, g_j)\nu(g_j, g_i) = e^{2\pi i(\gamma_i, \gamma_j)} \tag{1}$$

and

$$\nu(g_i, g_i) = e^{\pi i(\gamma_i, \gamma_i)}. \tag{2}$$

Remark 4.2.1. Condition (1) is needed since ν has to satisfy Lemma 4.1.12. It is well defined since for any $\lambda \in L_0$, $\mu \in L$, $(\lambda, \mu) \in \mathbf{Z}$ by assumption and so $e^{2\pi i(\lambda, \mu)} = 1$. Moreover, it implies that $\nu(g_i, g_i) = \pm e^{\pi i(\gamma_i, \gamma_i)}$. Condition (2) is imposed for consistency reasons that will appear later. It is also well defined since L_0 is an even sublattice and $L \subset L_0^*$.

It is not hard to check that bi-multiplicative maps satisfying (1) and (2) do exist.

Lemma 4.2.2. *The map*

$$\Gamma \times \Gamma \to \mathbf{C}^*$$
$$(x_1, x_2) \mapsto e^{\pi i(\lambda_1, \lambda_2)},$$

where $L_0 + \lambda_i = x_i$, is bi-multiplicative and satisfies the above conditions (1) *and* (2).

We will construct what the physicists call a Fock space. For this we next need a central extension of the abelian group L by $\mathbf{C}^*$. So we remind the reader of some well known facts about central extensions. First set

$$B : L \times L \to \mathbf{C}^*$$
$$(\beta_i, \beta_j) \mapsto e^{-\pi i(\beta_i, \beta_j)}\nu(g_i, g_j), \quad \beta_i \in g_i.$$

Lemma 4.2.3. *The map B is*

(i) skew-symmetric: $\forall \ \alpha, \beta \in L$,

$$B(\beta, \alpha) = B(\alpha, \beta)^{-1} \quad and \quad B(\alpha, \alpha) = 1$$

(ii) and bi-multiplicative: $\forall \ \alpha, \beta, \gamma \in L$,

$$B(\alpha + \beta, \gamma) = B(\alpha, \gamma)B(\beta, \gamma) \quad and \quad B(\gamma, \alpha + \beta) = B(\gamma, \alpha)B(\gamma, \beta)$$

Proof. The map B is clearly bi-multiplicative as this is the case of ν. Furthermore by Remark 4.2.1,

$$B(\beta_i, \beta_j)B(\beta_j, \beta_i) = e^{-2\pi i(\beta_i, \beta_j)}\nu(g_i, g_j)\nu(g_j, g_i) = 1$$

since ν satisfies condition (1). In other words, B is skew-symmetric. As ν satisfies condition (2), Remark 4.2.1 implies that

$$B(\beta_i, \beta_i) = e^{-\pi i(\beta_i, \beta_i)}e^{\pi i(\beta_i, \beta_i)} = 1.$$

$\square$

Remarks 4.2.4.

(i) For $B(\alpha, \alpha) = 1$ to hold, the map ν needs to satisfy condition (2). Otherwise, we can only deduce that $B(\alpha, \alpha) = \pm 1$. As we will see further down, $B(\alpha, \alpha) = 1$ is essential for our purposes as it is necessary in order to associate a central extension of L by $\mathbf{C}^*$ to B.

(ii) Suppose that L is an integral lattice.

If L is an even lattice, then $L_0 = L$ and so $\Gamma = 1$.

If L is an odd lattice, then $\Gamma = \mathbf{Z}_2$ since for any $\alpha, \beta \in L - L_0$,

$$(\alpha + \beta, \alpha + \beta) = (\alpha, \alpha) + (\beta, \beta) + 2(\alpha, \beta) \equiv 0 \pmod 2.$$

In both cases, $\Delta = 0$ and bi-multiplicativity and condition (2) satisfied by ν implies that $\nu(\bar{a}, \bar{b}) = (-1)^{ab}$ for $a, b = 0, 1$.

Hence, for $\alpha, \beta \in L$,

$$B(\alpha, \beta) = (-1)^{(\alpha, \beta) + p(\alpha)p(\beta)} = (-1)^{(\alpha, \beta) + (\alpha, \alpha)(\beta, \beta)}.$$

The cohomology group $H^2(L, \mathbf{C}^*)$ is the group of equivalence classes of 2-cocycles of the group L with values in $\mathbf{C}^*$. These are maps $\epsilon : L \times L \to \mathbf{C}^*$ satisfying for any $\alpha\beta, \gamma \in L$

$$\epsilon(\alpha, 0) = 1 = \epsilon(0, \alpha); \quad \epsilon(\alpha, \beta)\epsilon(\alpha + \beta, \gamma) = \epsilon(\alpha, \beta + \gamma)\epsilon(\beta, \gamma).$$

Two such cocycles ϵ_1 and ϵ_2 are equivalent if there is a group homomorphism $\mu : L \to \mathbf{C}^*$ such that $\epsilon_2(\alpha, \beta) = \epsilon_1(\alpha, \beta)\mu(\alpha)\mu(\beta)\mu(a + \beta)^{-1}$.

Lemma 4.2.5.

(i) *There is a bijection between isomorphism classes of central extensions of group L by the group $\mathbf{C}^*$ and the cohomology group $H^2(L, \mathbf{C}^*)$.*

(ii) *There is a unique (up to isomorphism) central extension $\hat{L}$ of the abelian group L by the group $\mathbf{C}^*$:*

$$1 \longrightarrow \mathbf{C}^* \longrightarrow \hat{L} \longrightarrow L \longrightarrow 1$$

satisfying

$$aba^{-1}b^{-1} = B(\pi(a), \pi(b)), \quad a, b \in \hat{L}, \tag{$*$}$$

where for $x \in \hat{L}$, π is the natural projection of $\hat{L}$ onto L. For any 2-cocycle ϵ corresponding to the group $\hat{L}$,

$$\epsilon(\alpha, \beta) = B(\alpha, \beta)\epsilon(\beta, \alpha) \quad \alpha, \beta \in L.$$

Furthermore, when L is an integral lattice, if α_i, $i = 1, \cdots n$ is a basis for the lattice L, there exists a 2-cocycle corresponding to $\hat{L}$ such that

$$\epsilon(\pm\alpha, \pm\alpha) = (-1)^{(\alpha, \alpha)} \quad and \quad \epsilon(\alpha, \beta) = (-1)^{\sum_i k_i l_i} \pi_{i>j} B(\alpha_i, \alpha_j)^{k_i l_j}.$$

Proof. (i): Let $\hat{L}$ be a central extension of the abelian group L by $\mathbf{C}^*$ and π the projection of $\hat{L}$ onto L. For every $\alpha \in L$, let e^α be an element of $\hat{L}$ such that $\pi(e^\alpha) = \alpha$. Set e^0 to be the identity element of $\hat{L}$. Then $\epsilon : L \times L \to \mathbf{C}^*$ given by

$$e^\alpha e^\beta = \epsilon(\alpha, \beta)e^{\alpha+\beta} \tag{1}$$

is a 2-cocycle since e^0 is the identity and because of associativity.

Let K be another central extension of L by $\mathbf{C}^*$. For all $\alpha \in L$, fix elements $k^\alpha \in K$ mapped onto α and let μ be the corresponding 2-cocycle. The groups K and $\hat{L}$ belong to the same isomorphism class of central extensions of L if and only if there is an isomorphism $\phi : K \to \hat{L}$ such that $\phi(k^\alpha) = \tau(\alpha)e^\alpha$ for some map $\tau : L \to \mathbf{C}^*$. Thus they belong to the same isomorphism class if and only if

$$\mu(\alpha, \beta) = \tau(\alpha)\tau(\beta)\tau(\alpha + \beta)^{-1}\epsilon(\alpha, \beta) \quad \forall \alpha, \beta \in L,$$

i.e. μ is equivalent to ϵ. Hence the above gives a well defined injective map from the isomorphism classes of central extensions of L by cyclic groups of order 2 and the group $H^2(L, \mathbf{C}^*)$.

We next show that the map is surjective. Let ϵ be a 2-cocycle. We construct a central extension of L mapped onto ϵ. Consider the set

$$L \times \mathbf{C}^* = \{(\alpha, a) : \alpha \in L, a \in \mathbf{C}^*\}.$$

Define the following product on this set: $(\alpha, a).(\beta, b) = (\alpha + \beta, \epsilon(\alpha, \beta)ab)$ for all $\alpha, \beta \in L$, $a, b \in \mathbf{C}^*$. As ϵ is a 2-cocycle, the product is associative and $(0, 1)$ is the identity element. Furthermore

$$(-\alpha, \epsilon(\alpha, -\alpha)^{-1}a^{-1}) = (\alpha, a)^{-1}.$$

Hence this product gives a group structure to the set $L \times \mathbf{C}^*$.
(ii): Let $\alpha_1, \cdots, \alpha_n$ be a $\mathbf{Z}$- basis of the lattice L. Set

$$\epsilon(\alpha_i, \alpha_j) = \begin{cases} B(\alpha_i, \alpha_j) & \text{if } i > j \\ 1 & \text{if } i < j. \end{cases}$$

($\epsilon(\alpha_i, \alpha_i)$ may be arbitrarily chosen). Extend this to L by bi-multiplicativity. This gives a 2-cocycle on L with values in $\mathbf{C}^*$ since bi-multiplicativity implies that

$$\epsilon(\alpha, 0) = 1 = \epsilon(0, \alpha);$$
$$\epsilon(\alpha, \beta)\epsilon(\alpha + \beta, \gamma) = \epsilon(\alpha, \beta)\epsilon(\alpha, \gamma)\epsilon(\beta, \gamma) = \epsilon(\alpha, \beta + \gamma)\epsilon(\beta, \gamma).$$

Let $\alpha = \sum_i k_i\alpha_i$ and $\beta = \sum_i l_i\alpha_i$. Then,

$$\epsilon(\alpha, \beta) = \prod_{i,j} \epsilon(\alpha_i, \alpha_j)^{k_i l_j} = \prod_{i=1}^{n} \epsilon(\alpha_i, \alpha_i)^{k_i l_i} \prod_{i>j} B(\alpha_i, \alpha_j)^{k_i l_j} \qquad (2)$$

and

$$\epsilon(\beta, \alpha) = \prod_{i=1}^{n} \epsilon(\alpha_i, \alpha_i)^{k_i l_i} \prod_{j>i} B(\alpha_j, \alpha_i)^{k_i l_j}.$$

Hence as B is skew-symmetric (see Lemma 4.2.3 (i)),

$$\epsilon(\alpha, \beta)\epsilon(\beta, \alpha)^{-1} = \prod_{i>j} B(\alpha_i, \alpha_j)^{k_i l_j} \prod_{j>i} B(\alpha_i, \alpha_j)^{k_i l_j} \prod_{i,j} B(\alpha_i, \alpha_j)^{k_i l_j} = B(\alpha, \beta).$$

Let $\hat{L}$ be the corresponding central extension of L by $\mathbf{C}^*$ and $e^\alpha \in \hat{L}$ be as described in (i). Then, $e^\alpha e^\beta = \epsilon(\alpha, \beta)e^{\alpha+\beta}$ and $e^\beta e^\alpha = \epsilon(\beta, \alpha)e^{\alpha+\beta}$. Hence from (*) we get

$$e^\alpha e^\beta = \epsilon(\alpha, \beta)\epsilon(\beta, \alpha)^{-1}e^\beta e^\alpha = B(\alpha, \beta)e^\beta e^\alpha. \qquad (3)$$

This proves the existence of a central extension of L satisfying (*).

Conversely, if $\hat{L}$ is a central extension of L satisfying (*) and ϵ is a 2-cocycle corresponding to $\hat{L}$ via the map described in (i), then the definition of $\hat{L}$ implies that $\epsilon(\alpha, \beta) = B(\alpha, \beta)\epsilon(\beta, \alpha)$ for any $\alpha, \beta \in L$.

We next check the uniqueness of the central extension $\hat{L}$.

Equality (3) shows that, more generally, for any 2-cocycle ϵ,

$$B(\alpha, \beta) = \epsilon(\alpha, \beta)\epsilon(\beta, \alpha)^{-1}$$

is a bi-multiplicative skew symmetric map. This gives a surjective group homomorphism between $H^2(L, \mathbf{C}^*)$ and the group of bi-multiplicative skew-symmetric maps

$$L \times L \to \mathbf{C}^*$$
$$\epsilon \mapsto (B : (\alpha, \beta) \mapsto \epsilon(\alpha, \beta)\epsilon(\beta, \alpha)^{-1}).$$

This map is also injective: if $\epsilon(\alpha, \beta)\epsilon(\beta, \alpha)^{-1} = 1$ for all $\alpha, \beta \in L$, then ϵ is symmetric. Equivalently, (1) implies that the corresponding central extension is abelian and so must be isomorphic to $L \times \mathbf{C}^*$. Therefore, by (i), ϵ is equivalent to the trivial cocycle.

Hence the above map is an isomorphism, and thus the uniqueness of $\hat{L}$ follows from (i).

Finally suppose that the lattice L is integral. From the above there is a 2-cocycle ϵ corresponding to $\hat{L}$ such that

$$\epsilon(\alpha_i, \alpha_j) = \begin{cases} B(\alpha_i, \alpha_j) & \text{if } i > j \\ (-1)^{(\alpha_i, \alpha_i)} & \text{if } i = j \\ 1 & \text{if } i < j. \end{cases}$$

Then, from (2) and Remark 4.2.4 (ii) it follows that for any $\alpha \in L$,

$$\epsilon(\alpha, \alpha) = \prod_{i=1}^{n} (-1)^{(\alpha_i, \alpha_i) k_i^2} \prod_{i>j} (-1)^{\pm(k_i\alpha_i, k_j\alpha_j) + \pm k_i^2 k_j^2 (\alpha_i, \alpha_i)(\alpha_j, \alpha_j)}$$

$$= (-1)^{\frac{1}{2}((\alpha, \alpha) + (\alpha, \alpha)^2)}$$

$$= (-1)^{(\alpha, \alpha)}$$

since $(\sum_i k_i \alpha_i, \sum_i k_i \alpha_i) \equiv \sum_i k_i^2 (\alpha_i, \alpha_i)$ (mod 2). This proves the result. $\square$

In what follows, we will denote by $\hat{L}$ the extension of the free abelian group L by $\mathbf{C}^*$ given in Lemma 4.2.5 (ii). For each $\alpha \in L$, we fix an element $e^\alpha \in \hat{L}$ satisfying $\pi(e^\alpha) = \alpha$ and such that e^0 is the identity element of $\hat{L}$. Set $\epsilon : L \times L \to \mathbf{C}^*$ to be the corresponding 2-cocycle: for all $\alpha, \beta \in L$,

$$e^\alpha e^\beta = \epsilon(\alpha, \beta) e^{\alpha+\beta}.$$

The group algebra $\mathbf{C}[\hat{L}]$ is a complex vector space with basis e^α, $\alpha \in L$.

Definition 4.2.6. *Let S be the symmetric algebra of $\oplus_{n<0}(H \otimes t^n)$. The Fock space is the Γ-graded vector space defined to be the tensor product*

$$V_L = S \otimes \mathbf{C}[\hat{L}].$$

The Γ-grading is given by setting $p(v \otimes e^\alpha) = g_i$, for $v \in S$, $\alpha \in L_i$.

We remind the reader that the symmetric algebra is the quotient of the tensor algebra $T(\oplus_{n<0}(H \otimes t^n))$ by the ideal generated by the elements $ab - ba$, $a, b \in (\oplus_{n<0}(H \otimes t^n))$.

Remark 4.2.7. Basically the Fock space is the tensor product of copies of the symmetric algebra of H and of $\mathbf{C}[\hat{L}]$ and has the structure of an algebra. For $h \in H$ and $n \in \mathbf{Z}_-$, $h(n)$ will denote the element $h \otimes t^n$. The elements

$$h_1(-n_1) \cdots h_r(-n_r) \otimes e^\alpha,$$

where the elements $h_i \in H$, $n_i \in \mathbf{N}$, $\alpha \in L$, generate the vector space V_L.

Our next aim is to construct a Γ-graded vertex algebra structure on V_L. Throughout this section we keep the notation of section 4.1. For simplicity of notation we will write $p(\alpha)$ for $p(1 \otimes e^\alpha)$.

Consider H as an abelian Lie algebra and set $\tilde{H}$ to be the affinization H:

$$\tilde{H} = H \otimes \mathbf{C}[t^{-1}, t] \oplus \mathbf{C}c,$$

where c is a central element and $[h \otimes t^n, h' \otimes t^m] = \delta_{n+m,0}(h, h')c$ for $m, n \in \mathbf{Z}$, $h, h' \in H$. Then $\tilde{H}$ is a Heisenberg algebra (see Remark 2.1.16). The Fock space is a natural representation of the Lie algebra $\tilde{H}$.

For $h \in H$ and $n \in \mathbf{Z}$, $h(n)$ will denote the endomorphism on V_L corresponding to the element $h \otimes t^n$ of $\tilde{H}$. The next result also implies that when $n < 0$ it is consistent to use the same notation for $h(n)$ both as the endomorphism of V_L defined in the next Lemma and as a vector of V_L.

Lemma 4.2.8.

(i) *The symmetric algebra S is a module for the Heisenberg algebra $\tilde{H}$ with the action of $\tilde{H}$ given as follows. For $x \in S$, write $x = h'(-m)y$, where $y \in S$ and $m \in \mathbf{N}$.*

$$c.x = x;$$

$$h(n).x = \begin{cases} h(n)x & \text{if } n < 0 \\ 0 & \text{if } n = 0 \\ n\delta_{n,m}(h, h')y + h'(-m)h(n).y & \text{if } n > 0 \end{cases}$$

(ii) *The space $\mathbf{C}[\hat{L}]$ is also a $\tilde{H}$-module with the action given as follows: for $e^\alpha \in \mathbf{C}[\hat{L}]$,*

$$c.(1 \otimes e^\alpha) = 0, \qquad h(n).(1 \otimes e^\alpha) = \delta_{n,0}(h, \alpha)(1 \otimes e^\alpha).$$

Proof. For any $x \in S$, $h, h' \in H$, $n, m \in \mathbf{Z}$, $\alpha \in L$ calculations give

$$[h(n), h'(m)].x = (h, h')\delta_{n+m,0}x, \quad [h(n), h'(m)].(1 \otimes e^\alpha) = 0.$$

Hence as $c.x = x$ and $c.(1 \otimes e^\alpha) = 0$, the action described in the Lemma gives representations of the Lie algebra $\tilde{H}$ on S and on $\mathbf{R}[\hat{L}]$.   □

As the Fock space is a tensor product of two $\tilde{H}$-modules, it is a tensor product module for the Lie algebra $\tilde{H}$.

Corollary 4.2.9. *The Fock space V_L is a $\tilde{H}$-module and the action is given by Lemma 4.2.8.*

Remark 4.2.10. For $n < 0$, $h(n)$ is a "creation operator" and for $n > 0$, $h(n)$ is an "annihilation operator".

Lemma 4.2.11.

(i) *There is a natural derivation d on the Fock space defined by*

$$d(h(-m) \otimes 1) = mh(-m-1) \otimes 1, \quad d.(1 \otimes e^\alpha)$$
$$= 1\alpha(-1) \otimes e^\alpha \ h \in H, m \in \mathbf{N}, \alpha \in L.$$

(ii) $d(1 \otimes 1) = 0$

Proof. (ii) follows from the definition of the derivation d and of e^0 as the identity element of the group $\hat{L}$.   □

The following consequence of Lemma 4.2.11 (ii) is immediate.

Corollary 4.2.12. *For all $h \in H$, $n \in \mathbf{N}$, $dh(-n).(1 \otimes 1) = -nh(-n-1).(1 \otimes 1)$.*

So, the definition of the derivation d suggests setting

$$Y(h(-1) \otimes 1, z) := \sum_{n \in \mathbf{Z}} h(n)z^{-n-1}.$$

In other words,
$$h(n) = h(-1)_n \quad \forall n \in \mathbf{Z}.$$

Lemma 4.2.13. *For all* $h \in H$, $Y(h(-1) \otimes 1, z)$ *is a well defined generalized vertex operator on* V_L *satisfying the vacuum and translation axioms. In particular,*
$$[d, h(n)] = -nh(n-1)$$
for all $h \in H$, $n \in \mathbf{Z}$. *Furthermore,* $p(Y(h(-1) \otimes 1), z) = 0$.

Proof. By definition of the action of $h(n)$ on V_L given in Lemma 4.2.8, for all $v \in V_L$, $h(n).v = 0$ for $n >> 0$ and so the operator $Y(h(-1) \otimes 1, z)$ is a well defined generalized vertex operator and satisfies the vacuum axiom.

For all $h \in H$ $[d, h(n)] = -nh(n-1)$ is a direct consequence of the definition of the derivation d (see Lemma 4.2.11) for $n < 0$. For $n > 0$, we show this by induction on the length of homogeneous for all $v \in V_L$. For $v = 1 \otimes e^\alpha$, $\alpha \in L$,

$$\begin{aligned}
[d, h(n)]v &= d(\delta_{n,0}(h, \alpha)v) - h(n)(\alpha(-1) \otimes e^\alpha) \\
&= \delta_{n,0}(h, \alpha)d(\alpha(-1) \otimes e^\alpha) - (h, \alpha)\delta_{n,1}v - (h, \alpha)\delta_{n,0}(\alpha(-1) \otimes e^\alpha) \\
&= -nh(n-1)v.
\end{aligned}$$

Suppose that for $v \in V_L$, $[d, h(n)]v = -nh(n-1)v$. Then, for $h' \in H$, $m \in \mathbf{N}$,

$$\begin{aligned}
dh(n)&(h'(-m)v) \\
&= (h, h')\delta_{n,m}dv + dh'(-m)h(n)v - mh(n)h'(-m-1)v - h(n)h'(-m)dv \\
&= n(h, h')\delta_{n,m}dv + mh'(-m-1)h(n)v + h'(-m)dh(n)v \\
&\quad - m(h, h')n\delta_{n,m+1}v - mh'(-m-1)h(n)v - n(h, h')\delta_{n,m}dv - h'(-m)h(n)dv \\
&= h'(-m)[d, h(n)]v - nh(n-1)h'(-m)v + nh'(-m)h(n-1)v \\
&= -nh(n-1)h'(-m)v
\end{aligned}$$

as wanted. Therefore $[d, h(n)] = -nh(n-1)$ for all $h \in H$, $n \in \mathbf{Z}$. It follows that $[d, Y(h(-1) \otimes 1, z)] = \frac{d}{dz}Y(h(-1) \otimes 1, z)$.

By definition of the gradation, $h(-1) \otimes 1$ has parity 0. Hence This also the case for the generalized vertex operator $Y(h(-1) \otimes 1, z)$ since $h(-1)_n \neq 0$ implies that $n \in \mathbf{Z}$. This proves the Lemma. $\qquad\square$

The following property of the operators $h(n)$ follows easily from the action of the Lie algebra $\tilde{H}$ on the Fock space V_L.

Lemma 4.2.14. *For all* $h, h' \in H$, $n, m \in \mathbf{Z}$,

$$[h(n), h'(m)] = n\delta_{n+m,0}(h, h').$$

Corollary 4.2.15. *The generalized vertex operators* $Y(h(-1) \otimes 1, z)$ *and* $Y(h'(-1) \otimes 1, w)$ *are mutually local for all* $h, h' \in H$.

Proof.

$$[Y(h(-1) \otimes 1, z), Y(h'(-1) \otimes 1, w)] = \sum_{m,n \in \mathbf{Z}} [h(n), h'(m)] z^{-n-1} w^{-m-1}$$

$$= (h, h') \sum_{n \in \mathbf{Z}} n z^{n-1} w^{n-1}$$

$$= (h, h') \frac{\partial}{\partial w} \delta(z - w).$$

Since $(z - w)\delta(z - w) = 0$ (see Example 4.1.10),

$$(z - w)^2 \frac{\partial}{\partial w} \delta(z - w) = \frac{\partial}{\partial w}((z - w)^2 \delta(z - w)) + 2(z - w)\delta(z - w) = 0.$$

Hence,

$$(z - w)^2 [Y(h(-1) \otimes 1, z), Y(h'(-1) \otimes 1, w)] = 0$$

and so the generalized vertex operators $Y(h(-1) \otimes 1, z)$ and $Y(h'(-1) \otimes 1, z)$ are mutually local since they have parity 0. □

For $\alpha \in L$, define

$$e^\alpha : V_L \to V_L$$
$$v \otimes e^\beta \mapsto v \otimes e^\alpha e^\beta = \epsilon(\alpha, \beta) v \otimes e^{\alpha + \beta},$$

where $v \in S$.

We can now state the result giving a Γ-graded vertex algebra structure to the Fock space V_L as it only depends on fixing mutually local vertex operators for the elements $h(-1) \otimes 1$ of V_L, $h \in H$, satisfying the translation and vacuum axioms.

Theorem 4.2.16. *The Fock space V_L with vacuum vector $1 \otimes 1$ and translation operator d whose action is defined by:*

$$d(h(-m) \otimes 1) = m h(-m-1) \otimes 1, \quad d(1 \otimes e^\alpha) = \alpha(-1) \otimes e^\alpha \quad h \in H, m \in \mathbf{N}, \alpha \in L$$

has a unique Γ-graded vertex algebra structure satisfying

$$Y(h(-1) \otimes 1, z) = \sum_{n \in \mathbf{Z}} h(n) z^{-n-1}.$$

There is a basis e^α, $\alpha \in L$, of the group algebra $\mathbf{R}[\hat{L}]$ for which

$$Y(1 \otimes e^\alpha, z) = e^\alpha e^{-\sum_{j<0} \frac{z^{-j}}{j} \alpha(j)} e^{-\sum_{j>0} \frac{z^{-j}}{j} \alpha(j)} z^{\alpha(0)}$$

and more generally, for any $h_i \in H$, $n_i \in \mathbf{Z}_+$, $\alpha \in L$, the generalized vertex operator corresponding to the vector

$$h_1(-n_1 - 1) \dots h_r(-n_r - 1) \otimes e^\alpha$$

is

$$: \frac{1}{n_1!}Y(h_1(-n_1 - 1), z) \cdots \frac{1}{n_r!}Y(h_r(-n_r - 1), z) \otimes 1)Y(1 \otimes e^{\alpha}, z) :$$

Definition 4.2.17. *A Γ-graded vector space V is said to be a Γ-graded (bosonic) lattice vertex algebra if V is the Fock space V_L for some rational lattice L and the Γ-graded vertex algebra structure is given by Theorem 4.2.16.*

It is sometimes useful to call this vertex algebra the bosonic lattice vertex algebra in order to differentiate it from the fermionic lattice vertex algebra (see Exercise 4.2.2) since both are needed in the construction of a BKM superalgebra with non-trivial odd part. We spend the rest of section 4.2 proving Theorem 4.2.16. See [Borc2], [DL] and [Kac15] for the case $\Gamma \leq \mathbf{Z}_2$. We start by showing that the Γ-graded vertex algebra structure of V_L only depends on the generalized vertex operators $Y(h(-1) \otimes 1, z)$ and $Y(1 \otimes e^{\alpha}, z)$, $h \in H$, $n \in \mathbf{N}$, $\alpha \in L$.

Lemma 4.2.18. *Suppose that V_L is a Γ-graded vertex algebra with vacuum vector $1 \otimes 1$ and translation operator d. Then,*

(i) for $h \in H$, $n_i \in \mathbf{Z}_+$, $\alpha \in L$,

$$Y(h_1(-n_1 - 1) \cdots h_r(-n_r - 1) \otimes e^{\alpha}, z)$$
$$=: \frac{1}{n_1!}Y(h_1(-n_1 - 1), z) \cdots \frac{1}{n_r!}Y(h_r(-n_r - 1), z) \otimes 1)Y(1 \otimes e^{\alpha}, z) :$$

(ii) and for $h \in H$, $n \in \mathbf{Z}_+$, $Y(h(-n) \otimes 1) = (\frac{\partial}{\partial z})^{n-1}Y(h(-1) \otimes 1)$.

Proof. (i) follows by induction on r from Corollary 4.1.32 applied to the vectors

$$a = h_i(-n_i - 1) \otimes 1$$

and

$$b = h_{i+1}(-n_{i+1} - 1) \cdots h_r(-n_r - 1) \otimes e^{\alpha}.$$

(ii) follows from Lemmas 4.1.31 and 4.2.11 since $d^n(h(-1) \otimes 1) = nh(-n-1) \otimes 1$. $\qquad \square$

Hence we only need to associate generalized vertex operators to elements $1 \otimes e^{\alpha}$, $\alpha \in L$ and finally to check that with the generalized vertex operators as given in Theorem 4.2.16, the Fock space V_L is indeed a Γ-graded vertex algebra. We first aim towards the computation of $Y(1 \otimes e^{\alpha}, z)$.

Lemma 4.2.19. *Suppose that V_L is a Γ-graded vertex algebra with vacuum vector $1 \otimes 1$, translation operator d and $Y(h(-1) \otimes 1, z) = \sum_{n \in \mathbf{Z}} h(n)z^{-n-1}$ for all $h \in H$. For any $n \in \mathbf{Z}$, $h \in H$, $\alpha \in L$,*

$$[h(n), Y(1 \otimes e^{\alpha}, z)] = (h, \alpha)z^n Y(1 \otimes e^{\alpha}, z).$$

Proof. We apply Corollary 4.1.26 (i) to $a = h(-1) \otimes 1$ and $b = 1 \otimes e^\alpha$. Since $h(n) = (h(-1) \otimes 1)_n$ and $h(n)(1 \otimes e^\alpha) = \delta_{n,0}(h, \alpha) \otimes e^\alpha$,

$$[h(n), Y(1 \otimes e^\alpha, z)] = \sum_{j=0}^\infty \binom{n}{j} Y(h(j)(1 \otimes e^\alpha), z) z^{n-j}$$
$$= (h, \alpha) Y(1 \otimes e^\alpha, z) z^n$$

by linearity of generalized vertex operators. $\qquad\square$

The following result is an immediate consequence of Lemma 4.1.31 and Corollary 4.1.32.

Lemma 4.2.20. *Suppose that V_L is a Γ-graded vertex algebra with vacuum vector $1 \otimes 1$ and translation operator d. For any $\alpha \in L$,*

$$\frac{d}{dz} Y(1 \otimes e^\alpha z) =: Y(\alpha(-1) \otimes 1, z) Y(1 \otimes e^\alpha, z) :$$

Hence

$$\frac{d}{dz} Y(1 \otimes e^\alpha z) = \sum_{j<0} \alpha(j) z^{-j-1} Y(1 \otimes e^\alpha, z) + Y(1 \otimes e^\alpha, z) \sum_{j\geq 0} \alpha(j) z^{-j-1}.$$

Now,

$$\frac{d}{dz} e^{-\sum_{j<0} \frac{z^{-j}}{j} \alpha(j)} = \sum_{j<0} \alpha(j) z^{-j-1} e^{-\sum_{j<0} \frac{z^{-j}}{j} \alpha(j)}.$$

Similarly for $j > 0$. For $j = 0$,

$$\frac{d}{dz} e^{\alpha(0) \log z} = \alpha(0) z^{-1} e^{\alpha(0) \log z}.$$

Furthermore, the vacuum axiom implies that

$$(e^\alpha)_{-1}.1 = e^\alpha.$$

On the other hand, the constant term in the expression

$$e^{-\sum_{j<0} \frac{z^{-j}}{j} \alpha(j)} e^{-\sum_{j>0} \frac{z^{-j}}{j} \alpha(j)} z^{\alpha(0)}.1 = e^{-\sum_{j<0} \frac{z^{-j}}{j} \alpha(j)} \otimes 1$$

is 1. It thus makes sense to consider the following generalized vertex operator on the space V_L:

$$Y_\alpha(z) = e^{-\alpha} e^{\sum_{j<0} \frac{z^{-j}}{j} \alpha(j)} Y(1 \otimes e^\alpha, z) e^{\sum_{j>0} \frac{z^{-j}}{j} \alpha(j)} z^{-\alpha(0)}.$$

Lemma 4.2.21. *Suppose that V_L is a Γ-graded vertex algebra. For all $h \in H$, $n \in \mathbf{N}$,*

(i) $[h(n), e^\alpha] = \delta_{n,0}(h, \alpha) e^\alpha$ and
(ii) $[h(n), Y_\alpha(z)] = 0$.

Proof. From the action of $h(n)$ on V_L given in Lemma 4.2.8, (i) is obvious for $n \neq 0$. For any $v \in S$, $\gamma \in L$,

$$h(0)e^\alpha(ve^\gamma) = \epsilon(\alpha, \gamma)(h, \alpha + \gamma)ve^{\alpha+\gamma}$$
$$= (h, \alpha + \gamma)ve^\alpha(1 \otimes e^\gamma)$$

and

$$e^\alpha h(0)(ve^\gamma) = (h, \gamma)ve^\alpha(1 \otimes e^\gamma),$$

proving (i) for $n = 0$.
For $n, j \in \mathbf{Z}$, $j \neq 0$, by Lemma 4.2.14,

$$[h(n), e^{\frac{z^{-j}}{j}\alpha(j)}] = (h, \alpha)\delta_{n+j,0}z^n e^{-\frac{z^n}{n}\alpha(-n)} \quad \text{and} \quad [h(n), z^{\alpha(0)}] = 0.$$

Hence (ii) follows from (i) and Lemma 2.4.19. $\qquad\square$

In fact, the vertex operator $Y_\alpha(z)$ does not depend on z.

Corollary 4.2.22. *Suppose that V_L is a Γ-graded vertex algebra with vacuum vector $1 \otimes 1$, translation operator d, and $Y(h(-1) \otimes 1, z) = \sum_{n \in \mathbf{Z}} h(n)z^{-n-1}$ for all $h \in H$. For any $\alpha \in L$, the vertex operator $Y_\alpha(z) \in End(V_L)$.*

Proof. From Lemmas 4.2.14 and 4.2.21 it follows that

$$Y(1 \otimes e^\alpha, z) = e^\alpha e^{-\sum_{j<0} \frac{z^{-j}}{j}\alpha(j)} e^{\sum_{j>0} \frac{z^{-j}}{j}\alpha(j)} z^{\alpha(0)} Y_\alpha(z).$$

So, by Lemmas 4.2.20 and 4.2.21, $\frac{d}{dz}Y_\alpha(z) = 0$, which proves the result. $\qquad\square$

We will write $Y_\alpha(z) = y_\alpha$, where y_α is an endomorphism of the vector space V_L. We next compute these endomorphisms.

Lemma 4.2.23. *Suppose that V_L is a Γ-graded vertex algebra with vacuum vector $1 \otimes 1$, translation operator d, and $Y(h(-1) \otimes 1, z) = \sum_{n \in \mathbf{Z}} h(n)z^{-n-1}$ for all $h \in H$. For any $\alpha, \beta, \in L$, $v \in S$, $y_\alpha(v \otimes e^\beta) = y(\alpha, \beta)v \otimes e^\beta$ for some $y(\alpha, \beta) \in \mathbf{C}$ and $y(\alpha, 0) = 1 = y(0, \alpha)$.*

Proof. For any $v \in S$, $\beta \in L$,

$$y_\alpha(v \otimes e^\beta) = vy_\alpha(1 \otimes e^\beta)$$

since y_α commutes with all operators $h(-n)$ for all $h \in H$, $n \in \mathbf{N}$, by Lemmas 4.2.8 and 4.2.21 (ii). For any $h \in H$,

$$h(0)y_\alpha(1 \otimes e^\beta) = (h, \beta)y_\alpha(1 \otimes e^\beta)$$

by Lemma 4.2.21 (ii). Hence, since the bilinear form is non-degenerate on L, there is an element $y(\alpha, \beta) \in S$ such that

$$y_\alpha(1 \otimes e^\beta) = y(\alpha, \beta) \otimes e^\beta.$$

So by Lemma 4.2.21 (ii), $[h(n), y(\alpha, \beta) \otimes e^\beta] = 0$ for any $h \in H$, $n > 0$. Therefore, $y(\alpha, \beta) \in \mathbf{C}$.

Applying the vacuum axiom, we then get

$$1 \otimes e^\alpha = Y(1 \otimes e^\alpha, z)(1 \otimes 1)\big|_{z=0} = y(\alpha, 0)(1 \otimes e^\alpha)$$

and

$$1 \otimes e^\alpha = Y(1 \otimes 1, z)(1 \otimes e^\alpha) = y(0, \alpha)(1 \otimes e^\alpha).$$

And so $y(\alpha, 0) = 1 = y(0, \alpha)$. □

To compute the scalars $y(\alpha, \beta)$, we use the fact that the generalized vertex operators $Y(e^\alpha, z)$ and $Y(e^\beta, w)$ are mutually local. We split the proof into two parts as this will be useful later on.

Lemma 4.2.24. *Suppose that V_L is a Γ-graded vertex algebra with vacuum vector $1 \otimes 1$, translation operator d, and $Y(h(-1) \otimes 1, z) = \sum_{n \in \mathbf{Z}} h(n) z^{-n-1}$ for all $h \in H$. For any $\alpha, \beta \in L$,*

$$i_{z,w}(z-w)^{\Delta(p(\alpha), p(\beta))} Y(1 \otimes e^\alpha, z) Y(1 \otimes e^\beta, w)$$
$$- i_{w,z}(z-w)^{\Delta(p(\alpha), p(\beta))} \nu(p(\alpha), p(\beta)) Y(1 \otimes e^\beta, w) Y(1 \otimes e^\alpha, z)$$
$$= (e^\alpha y_\alpha e^\beta y_\beta - e^{\pi i \Delta(p(\alpha), p(\beta))} \nu(p(\alpha), p(\beta)) e^\beta y_\beta e^\alpha y_\alpha) Y_{\alpha, \beta}(z, w), \tag{$*$}$$

where

$$Y_{\alpha, \beta}(z, w) = z^{\alpha(0)} w^{\beta(0)} e^{-\sum_{j<0}(\frac{z^{-j}}{j}\alpha(j) + \frac{w^{-j}}{j}\beta(j))} e^{-\sum_{j>0}(\frac{z^{-j}}{j}\alpha(j) + \frac{w^{-j}}{j}\beta(j))}.$$

Proof. From Corollary 4.2.22,

$$Y(1 \otimes e^\alpha, z) Y(1 \otimes e^\beta, w)$$
$$= e^\alpha e^{-\sum_{j<0}\frac{z^{-j}}{j}\alpha(j)} e^{\sum_{j>0}\frac{z^{-j}}{j}\alpha(j)} z^{\alpha(0)} y_\alpha e^\beta e^{-\sum_{j<0}\frac{w^{-j}}{j}\beta(j)} e^{\sum_{j>0}\frac{w^{-j}}{j}\beta(j)} w^{\beta(0)} y_\beta.$$

Lemma 4.2.14 implies that $[\alpha(j), \beta(k)] = 0$ for $k \neq -j$ and $[\alpha(j), \beta(-j)]$ commutes with both $\alpha(j)$ and $\beta(-j)$. Hence,

$$e^{-\frac{z^{-j}}{j}\alpha(j)} e^{-\frac{w^{-k}}{k}\beta(-k)} = \delta_{j+k,0} e^{-(\frac{w}{z})^j \frac{1}{j}(\alpha, \beta)} e^{\frac{w^j}{j}\beta(-j)} e^{-\frac{z^{-j}}{j}\alpha(j)}.$$

And so,

$$e^{\sum_{j>0}\frac{z^{-j}}{j}\alpha(j)} e^{-\sum_{j<0}\frac{w^{-j}}{j}\beta(j)} = e^{-\sum_{j<0}\frac{w^{-j}}{j}\beta(j)} e^{\sum_{j>0}\frac{z^{-j}}{j}\alpha(j)} e^{-(\alpha,\beta)\sum_{j>0}(\frac{w}{z})^j \frac{1}{j}}.$$

Claim: $e^\beta z^{\alpha(0)} = z^{-(\alpha,\beta)} z^{\alpha(0)} e^\beta$.

For any $v \in S$, $\gamma \in L$, $e^\beta z^{\alpha(0)}(v \otimes e^\gamma) = \epsilon(\beta, \gamma) z^{(\alpha,\gamma)}(v \otimes e^{\beta+\gamma})$, where ϵ is the 2-cocycle corresponding to the central extension $\hat{L}$ (see Lemma 4.2.5 and following paragraph), and $z^{\alpha(0)} e^\beta(v e^\gamma) = \epsilon(\beta, \gamma) z^{(\alpha,\beta+\gamma)}(v e^{\beta+\gamma})$, proving the claim.

Considered as complex functions,

$$-\sum_{j>0}(\frac{w}{z})^j \frac{1}{j} = \log(1 - \frac{w}{z})$$

and

$$z^{(\alpha,\beta)} i_{z,w}(1 - \frac{w}{z})^{(\alpha,\beta)} = i_{z,w}(z - w)^{(\alpha,\beta)}$$

in the domain $|w| < |z|$. Hence as formal series,

$$e^{-(\alpha,\beta)\sum_{j>0}(\frac{w}{z})^j \frac{1}{j}} z^{(\alpha,\beta)} = i_{z,w}(z - w)^{(\alpha,\beta)}.$$

Note that by definition of Δ, $(\alpha,\beta) = -\Delta(p(\alpha), p(\beta))$. Therefore, since by Lemma 4.2.21, y_α and y_β commute with the endomorphisms $h(n)$ for all $n \in \mathbf{Z}$,

$$Y(1 \otimes e^\alpha, z)Y(1 \otimes e^\beta, w) = i_{z,w}(z - w)^{-\Delta(p(\alpha),p(\beta))} e^\alpha y_\alpha e^\beta y_\beta Y_{\alpha,\beta}(z, w), \quad (1)$$

where

$$Y_{\alpha,\beta}(z, w) = z^{\alpha(0)} w^{\beta(0)} e^{-\sum_{j<0}(\frac{z^{-j}}{j}\alpha(j) + \frac{w^{-j}}{j}\beta(j))} e^{-\sum_{j>0}(\frac{z^{-j}}{j}\alpha(j) + \frac{w^{-j}}{j}\beta(j))}.$$

Similarly,

$$Y(1 \otimes e^\beta, w)Y(1 \otimes e^\alpha, z) = i_{w,z}(w - z)^{-\Delta(p(\alpha),p(\beta))} e^\beta y_\beta e^\alpha y_\alpha Y_{\alpha,\beta}(z, w). \quad (2)$$

The result follows from (1) and (2).   □

We are now able to derive the vertex operators $Y(1 \otimes e^\alpha, z)$.

Lemma 4.2.25. *Suppose that V_L is a Γ-graded vertex algebra with vacuum vector $1 \otimes 1$, translation operator d, and $Y(h(-1) \otimes 1, z) = \sum_{n\in\mathbf{Z}} h(n)z^{-n-1}$ for all $h \in H$. There is a basis e^α, $\alpha \in L$ of the group algebra $\mathbf{C}[\hat{L}]$ satisfying $e^\alpha e^\beta (e^\alpha)^{-1} = B(\alpha, \beta)e^\beta$ with respect to which*

$$Y(1 \otimes e^\alpha, z) = e^\alpha e^{-\sum_{j<0} \frac{z^{-j}}{j}\alpha(j)} e^{-\sum_{j>0} \frac{z^{-j}}{j}\alpha(j)} z^{\alpha(0)}.$$

Proof.   We keep the same notation as in Lemma 4.2.24.   Since $Y_{\alpha,\beta}(z, w)^{-1}$ makes sense, multiplying both sides of equality (∗) in Lemma 4.2.24 by $(z - w)^m Y_{\alpha,\beta}(z, w)^{-1}$, where $m \in \mathbf{Z}$, we can deduce that for large enough $n \in \mathbf{Z} + \Delta(p(a), p(b))$ $(n = m + \Delta(p(a), p(b)))$,

$$i_{z,w}(z - w)^n Y(1 \otimes e^\alpha, z)Y(1 \otimes e^\beta, w)$$
$$= i_{w,z}(z - w)^n \nu(p(\alpha), p(\beta))Y(1 \otimes e^\beta, w)Y(1 \otimes e^\alpha, z) \quad (1)$$

if and only if

$$(z - w)^{n-\Delta(p(a),p(b))} e^\alpha y_\alpha e^\beta y_\beta = (z - w)^{n-\Delta(p(a),p(b))}$$
$$\times \nu(p(\alpha), p(\beta))e^{\pi i \Delta(p(\alpha),p(\beta))} e^\beta y_\beta e^\alpha y_\alpha. \quad (2)$$

(In particular if (1) holds, then it holds for all $n \geq \Delta(p(a), p(b))$) Applying the operators on both sides of equality (2) to the vector $1 \otimes e^\gamma$, $\gamma \in L$, we get from Lemma 4.2.23,

$$y(\beta, \gamma)\epsilon(\beta, \gamma)y(\alpha, \beta + \gamma)\epsilon(\alpha, \beta + \gamma)$$
$$= \nu(p(\alpha), p(\beta))e^{\pi i \Delta(p(\alpha),p(\beta))} y(\alpha, \gamma)\epsilon(\alpha, \gamma)y(\beta, \alpha + \gamma)\epsilon(\beta, \alpha + \gamma). \quad (3)$$

As ϵ is a 2-cocycle,

$$\epsilon(\beta,\gamma)\epsilon(\alpha,\beta+\gamma) = \epsilon(\alpha,\beta)\epsilon(\alpha+\beta,\gamma)$$

and

$$\epsilon(\alpha,\gamma)\epsilon(\beta,\alpha+\gamma) = \epsilon(\beta,\alpha)\epsilon(\alpha+\beta,\gamma).$$

Hence (3) becomes

$$y(\beta,\gamma)y(\alpha,\beta+\gamma)\epsilon(\alpha,\beta) = \nu(p(\alpha),p(\beta))e^{\pi i\Delta(p(\alpha),p(\beta))}\epsilon(\beta,\alpha)y(\alpha,\gamma)y(\beta,\alpha+\gamma).$$

By definition (see Lemmas 4.2.5 and 4.2.3),

$$\epsilon(\alpha,\beta) = B(\alpha,\beta)\epsilon(\beta,\alpha)$$

and

$$B(\alpha,\beta) = e^{-\pi i(\alpha,\beta)}\nu(p(\alpha),p(\beta)).$$

Also by definition of Δ,

$$-(\alpha,\beta) = \Delta(p(\alpha),p(\beta)).$$

Hence, it follows that

$$y(\beta,\gamma)y(\alpha,\beta+\gamma) = y(\alpha,\gamma)y(\beta,\alpha+\gamma). \tag{4}$$

In particular, setting $\gamma = 0$, by Lemma 4.2.23,

$$y(\alpha,\beta) = y(\beta,\alpha). \tag{5}$$

Using both equalities (5) and (4), it follows that

$$\begin{aligned}
y(\beta,\gamma)y(\alpha,\beta+\gamma) &= y(\gamma,\alpha)y(\beta,\gamma+\alpha)\\
&= y(\beta,\alpha)y(\gamma,\alpha+\beta)\\
&= y(\alpha,\beta)y(\alpha+\beta,\gamma).
\end{aligned}$$

Together with Lemma 4.2.23, this says that

$$y: L \times L \to \mathbf{C}$$
$$(\alpha,\beta) \mapsto y(\alpha,\beta)$$

is a 2-cocycle.

Suppose that $y(\alpha,\beta) \neq 0$ for all $\alpha,\beta \in L$. As the 2-cocycle y is symmetric, the corresponding central extension of L by $\mathbf{C}^*$ is abelian and thus isomorphic to the direct product of groups $L \times \mathbf{C}^*$. Therefore, Lemma 4.2.5 (i) implies that there exists a map $\tau: L \to \mathbf{C}^*$ such that

$$y(\alpha,\beta) = \tau(\alpha)\tau(\beta)\tau(\alpha+\beta)^{-1}.$$

For all $\alpha \in L$, set $f^\alpha = \tau(\alpha)^{-1}e^\alpha$. These elements form a basis of the group algebra $\mathbf{C}[\hat{L}]$ and

$$Y(1 \otimes f^\alpha, z)(v \otimes f^\beta) = \tau(\alpha)^{-1}\tau(\beta)^{-1}Y(1 \otimes e^\alpha, z)(v \otimes e^\beta)$$

$$= z^{\langle\alpha,\beta\rangle}e^{-\sum_{j<0}\frac{z^{-j}}{j}\alpha(j)}e^{-\sum_{j>0}\frac{z^{-j}}{j}\alpha(j)}(v \otimes f^{\alpha+\beta}).$$

Hence, rescaling the basis elements e^α if necessary, we may assume that

$$Y(1 \otimes e^\alpha, z) = e^\alpha e^{-\sum_{j<0}\frac{z^{-j}}{j}\alpha(j)}e^{-\sum_{j>0}\frac{z^{-j}}{j}\alpha(j)}z^{\alpha(0)}.$$

So to complete the proof of the Lemma, we only need to show the following claim.

Claim: $y(\alpha,\beta) \neq 0$ for all $\alpha,\beta \in L$.

Suppose that $y(\alpha,\beta) = 0$ for some $\alpha,\beta \in L$. Then, by Corollary 4.2.22 and Lemma 4.2.23,

$$Y(1 \otimes e^\alpha, z)(1 \otimes e^\beta) = e^\alpha e^{-\sum_{j<0}\frac{z^{-j}}{j}\alpha(j)}e^{-\sum_{j>0}\frac{z^{-j}}{j}\alpha(j)}z^{\alpha(0)}(y(\alpha,\beta) \otimes e^\beta) = 0.$$
(6)

Now,

$$y_\alpha(1 \otimes e^\beta) = y_\alpha e^\beta(1 \otimes 1) = Y(0,z)(1 \otimes 1). \tag{7}$$

This last equality follows from (6) and linearity in the first variable of the operators $Y(a,z)$, $a \in V_L$. Since $X(z) = Y(1 \otimes e^\alpha, z)e^\beta$ is clearly a generalized vertex operator, applying Lemma 4.1.18 to (7), we can deduce that $X(z) = 0$. Hence, $X(z)e^{-\beta} = 0$. However, for any $s \in S$, $\gamma \in L$,

$$e^\beta e^{-\beta}(s \otimes e^\gamma) = \epsilon(-\beta,\gamma)\epsilon(\beta,-\beta+\gamma)(s \otimes e^\gamma) = \epsilon(\beta,-\beta)(s \otimes e^\gamma). \tag{8}$$

In other words, the map $e^\beta e^{-\beta}$ on V_L is multiplication by $\epsilon(\beta,-\beta)$ and this scalar is non-trivial as it is in $\mathbf{C}^*$ by definition of the 2-cocycle ϵ. Therefore, (8) implies that

$$Y(1 \otimes e^\alpha, z) = 0.$$

In particular,

$$1 \otimes e^\alpha = Y(1 \otimes e^\alpha, z)(1 \otimes 1)\mid_{z=0} = 0.$$

This contradiction forces $y(\alpha,\beta) \neq 0$, proving our claim. $\square$

Remark 4.2.26. Note that because of the isomorphism between the group of equivalence classes of 2-cocycles with values in $\mathbf{C}^*$ and the group of bi-multiplicative skew-symmetric maps $L \times L \to \mathbf{C}^*$, the Γ-graded vertex algebra structure of V_L is independent, up to equivalence class, of the 2-cocycle ϵ chosen as long as $\epsilon(\alpha,\beta)\epsilon(\beta,\alpha)^{-1} = B(\alpha,\beta)$.

In order to finish the proof of Theorem 4.2.16, it only remains to verify that the generalized vertex operators $Y(1 \otimes e^\alpha, z)$ given in Lemma 4.2.25 satisfy the Γ-graded vertex algebra axioms.

Lemma 4.2.27. *Suppose that* $Y(h(-1)\otimes 1, z) = \sum_{n\in\mathbf{Z}} h(n)z^{-n-1}$ *for all* $h \in H$ *and that* d *is the derivation given in Lemma 4.2.11 (i). Let*

$$Y(1 \otimes e^\alpha, z) = e^\alpha e^{-\sum_{j<0} \frac{z^{-j}}{j}\alpha(j)} e^{-\sum_{j>0} \frac{z^{-j}}{j}\alpha(j)} z^{\alpha(0)}.$$

Then,

  (i) $[d, Y(1 \otimes e^\alpha, z)] = \frac{d}{dz} Y(1 \otimes e^\alpha, z)$,
  (ii) $Y(1 \otimes 1, z) = I_{V_L}$, $\quad Y(1 \otimes e^\alpha, z).1 \in V[[z]]$ *and* $Y(1 \otimes e^\alpha, z).1\big|_{z=0} = 1 \otimes e^\alpha$,
  (iii) $Y(1 \otimes e^\alpha, z)$ *is a generalized vertex operator of parity* $p(\alpha)$.
  (iv) *The generalized vertex operators* $Y(v, z)$ *and* $Y(1 \otimes e^\alpha, z)$ *are mutually local for* $v = h(-1) \otimes 1$ *and* $v = 1 \otimes e^\beta$ *for all* $h \in H$, $\beta \in L$.

Proof. (i): Since for all $v \in S$, $\beta \in L$,

$$d e^\alpha(v \otimes e^\beta) = \epsilon(\alpha, \beta)(d(v) + (\alpha + \beta)(-1)v) \otimes e^{\alpha+\beta}$$

and

$$e^\alpha d(v \otimes e^\beta) = \epsilon(\alpha, \beta)(d(v) + \beta(-1)v) \otimes e^{\alpha+\beta},$$

we get

$$[d, e^\alpha] = e^\alpha \alpha(-1).$$

From Lemma 4.2.13, we can deduce that

$$[d, e^{-\frac{z^{-j}}{j}\alpha(j)}] = z^{-j} j\alpha(j-1)e^{-\frac{z^{-j}}{j}\alpha(j)} \quad \forall \quad j \neq 0, \quad \text{and} \quad [d, z^{\alpha(0)}] = 0.$$

Hence,

$$[d, Y(1 \otimes e^\alpha, z)] = e^\alpha \alpha(-1) e^{-\sum_{j<0} \frac{z^{-j}}{j}\alpha(j)} e^{-\sum_{j>0} \frac{z^{-j}}{j}\alpha(j)} z^{\alpha(0)}$$

$$+ \sum_{j<0} \alpha(j-1)z^{-j} Y(1 \otimes e^\alpha, z) + Y(1 \otimes e^\alpha, z) \sum_{j>0} \alpha(j-1)z^{-j}$$

$$= \sum_{j<0} \alpha(j)z^{-j-1} Y(1 \otimes e^\alpha, z) + Y(1 \otimes e^\alpha, z) \sum_{j\geq 0} \alpha(j)z^{-j-1}$$

$$= \frac{d}{dz} Y(1 \otimes e^\alpha, z).$$

(ii) is easy to check.
(iii): For any $v \in S$,

$$Y(1 \otimes e^\alpha, z)(v \otimes e^\beta) = e^\alpha e^{-\sum_{j<0} \frac{z^{-j}}{j}\alpha(j)} e^{-\sum_{j>0} \frac{z^{-j}}{j}\alpha(j)} (v \otimes e^\beta) z^{(\alpha,\beta)}.$$

Hence for all $n \in \mathbf{Q}$, $(1 \otimes e^\alpha)_n$ has parity $p(\alpha)$ and $(1 \otimes e^\alpha)_n (v \otimes e^\beta) \neq 0$ implies that $n \in \mathbf{Z} - (\alpha, \beta)$. However, by definition of Δ, $\Delta(\alpha, \beta) = -(\alpha, \beta)$.

(iv): From Lemmas 4.2.14 and 4.2.21 (i) we can deduce that, for all $n \in \mathbf{Z}$,

$$[h(n), Y(1 \otimes e^\alpha, z)] = (h, \alpha) z^n Y(1 \otimes e^\alpha, z)$$

and so

$$[Y(h(-1) \otimes 1, z), Y(1 \otimes e^\alpha, w)] = \sum_{n \in \mathbf{Z}} (h, \alpha) Y(1 \otimes e^\alpha, w) z^{-n-1} w^n.$$

Therefore,

$$(z - w)[Y(h(-1) \otimes 1, z), Y(1 \otimes e^\alpha, w)] = 0.$$

Since the generalized vertex operator $Y(h(-1) \otimes 1, z)$ has parity 0, $Y(h(-1) \otimes 1, z)$ and $Y(1 \otimes e^\alpha, w)$ are mutually local.

Applying Lemma 4.2.24,

$$\begin{aligned}
i_{z,w}(z - w)^{\Delta(p(\alpha), p(\beta))} & Y(1 \otimes e^\alpha, z) Y(1 \otimes e^\beta, w) \\
& - i_{w,z}(z - w)^{\Delta(p(\alpha), p(\beta))} \nu(p(\alpha), p(\beta)) Y(1 \otimes e^\beta, w) Y(1 \otimes e^\alpha, z) \\
& = (\epsilon(\alpha, \beta) e^{\alpha + \beta} - e^{\pi i \Delta(p(\alpha), p(\beta))} \nu(p(\alpha), p(\beta)) \epsilon(\beta, \alpha) e^{\alpha + \beta}) Y_{\alpha, \beta}(z, w) \\
& = 0
\end{aligned}$$

since

$$\begin{aligned}
\epsilon(\alpha, \beta) &= B(\alpha, \beta) \epsilon(\beta, \alpha) \quad \text{by Remark 4.2.26} \\
&= e^{-pii(\alpha, \beta)} \nu(p(\alpha), p(\beta)) \epsilon(\beta, \alpha) \quad \text{by definition of } B \\
&= e^{-pi\Delta(p(\alpha), p(\beta))} \nu(p(\alpha), p(\beta)) \epsilon(\beta, \alpha) \quad \text{by definition of } \Delta.
\end{aligned}$$

□

We are now ready to complete the proof of Theorem 4.2.16.

Proof of Theorem 4.2.16. By Lemma 4.2.11 (ii), the translation operator d annihilates the vacuum vector. By Lemmas 4.2.13, 4.2.27 and Corollary 4.2.15, the generalized vertex operators $Y(h(-1) \otimes 1, z)$ and $Y(1 \otimes e^\alpha, z)$ satisfy the axioms of a Γ-graded vertex algebra for all $h \in H$, $\alpha \in L$. Hence by Lemma 4.2.18, the generalized vertex operators given in Theorem 4.2.16 for arbitrary generating vectors give the Fock space V_L with translation operator d and vacuum vector $1 \otimes 1$ a Γ-graded vertex algebra structure. Uniqueness follows from Lemmas 4.2.18 and 4.2.25. This completes the proof of Theorem 4.2.16.□

Exercises 4.2

1. Let L be the root lattice of a finite dimensional simple Lie algebra G of type A, D or E, i.e. all of whose roots have norm 2. Show that the lattice vertex algebra V_L is an integrable highest weight irreducible $\hat{G}$-module with highest weight vector the vacuum vector 1, where $\hat{G}$ is the extended affine Lie algebra associated to the Lie algebra G.

For a solution, see [Kac15, §5.6].

2. Let L be an integral lattice, V_L^f the exterior algebra of

$$\oplus_{-n\in\mathbf{N}}(H\otimes t^{n-\frac{1}{2}}),$$

and

$$\hat{H} = H\otimes\mathbf{C}[t^{-1},t]t^{\frac{1}{2}}\oplus\mathbf{C}c,$$

where

$$\hat{H}_{\bar{0}} = \mathbf{C}c,$$

$$\hat{H}_{\bar{1}} = H\otimes\mathbf{C}[t^{-1},t]t^{\frac{1}{2}},$$

c is a central element and $[h(n),h'(m)] = \delta_{n+m,0}(h,h')c$ for $m,n\in\mathbf{Z}$, $h,h'\in H$.

(i) Show that $\hat{H}$ is a Heisenberg superalgebra and that V_L^f is an irreducible $\hat{H}$-module.

(ii) Show that the space V_L^f with vacuum vector 1, translation operator d with action defined by:

$$d(h(-m)\otimes 1) = mh(-m-1)\otimes 1, d(1\otimes e^\alpha)$$
$$= \alpha(-1)\otimes e^\alpha h\in H, m\in\mathbf{N}, \alpha\in L,$$

has a unique vertex superalgebra structure satisfying

$$Y(h(-1-\frac{1}{2})\otimes 1, z) = \sum_{n\in\mathbf{Z}} h(n-\frac{1}{2})z^{-n-1}.$$

This is called the fermionic lattice vertex superalgebra. For a solution, see [Sch1].

4.3 From Lattice Vertex Algebras to Lie Algebras

In this subsection, we show how to construct Lie algebras from lattice vertex algebras. For details, see [Borc2], [Fren] and [Sch1]. We assume that the lattice L is non-degenerate and integral. Hence, by Remark 4.2.4 (ii), $\Gamma\leq\mathbf{Z}_2$, $\Delta = 0$, $\nu(\bar{a},\bar{b}) = (-1)^{ab}$ for $a,b = 0,1$, and

$$B(\alpha,\beta) = (-1)^{(\alpha,\beta)+(\alpha,\alpha)(\beta,\beta)}.$$

So, by definition, the Fock space V_L is graded by $\Gamma = \mathbf{Z}_2$: $v\otimes e^\alpha\in (V_L)_{\bar{0}}$ (resp. $(V_L)_{\bar{1}}$) if (α,α) is even (resp. odd).

Our aim is to construct a BKM algebra from V_L. The first thing to notice in this respect is that the quotient of V_L by the subspace generated by the action of the translation operator d has the structure of a Lie superalgebra.

Lemma 4.3.1. *The quotient space $V_L/d(V_L)$, where $d(V_L) = \langle dv, v \in V_L \rangle$, is a Lie superalgebra with the Lie super-bracket given by:*

$$[u, v] = u_0(v), \quad \forall u\, v \in V_L.$$

Proof. We check that (anti)-commutativity and the Jacobi identity hold. By Lemma 4.1.20, for all $u, v \in V_L$,

$$u_0(v) = -(-1)^{p(u)p(v)} \sum_{j=0}^{\infty} (-1)^j \frac{d^j}{j!} v_j u.$$

It follows that

$$[u, v] \equiv -(-1)^{p(v)p(u)}[v, u] \pmod{d(V_L)}.$$

By Corollary 4.1.26(ii), for any $u, v, w \in V_L$,

$$u_0(v_0 w) - (-1)^{p(u)p(v)} v_0(u_0 w) = (u_0 v)_0 w.$$

Equivalently,

$$[u[v, w]] = [[u, v]w] + (-1)^{p(u)p(v)}[v[u, w]].$$

$\square$

Remark 4.3.2. The above proof shows that the Jacobi identity holds in V_L but not the (anti)-commutativity axiom of the Lie super bracket. It is in order for this axiom to hold that to get a Lie superalgebra structure, we have to quotient out the Fock space V_L by the subspace generated by the derivation d.

Now there is a fundamental aspect of the vertex algebra V_L which we have not touched upon as yet. Recall from Lemmas 2.2.4 and 2.2.5 that an essential characteristic of a BKM superalgebra G is the existence of a non-degenerate, consistent, supersymmetric bilinear form such that the subspaces G_α and G_β, where $\alpha, \beta \in \Delta \cup \{0\}$, are orthogonal unless $\alpha + \beta = 0$. A basic aspect of the structure of lattice vertex algebras is that they too have a natural non-degenerate bilinear form. However, it is symmetric and not supersymmetric. Remember from paragraph preceding Corollary 2.2.7 that, in this book, we take Hermitian forms to be antilinear in the first argument and linear in the second, unlike usual conventions.

Theorem 4.3.3. *There is a unique, consistent, symmetric, Hermitian form $(.,.)$ on V_L such that:*

(i) $(1 \otimes e^\alpha, 1 \otimes e^\beta) = \begin{cases} 1 & if\ \alpha = \beta \\ 0 & otherwise \end{cases}$ *and*

(ii) for any $h \in H$, $n \in \mathbf{Z}$, the adjoint of the operator $h(n)$ is $h(-n)$. Furthermore, the form $(.,.)$ is non-degenerate.

Proof. Let $u = h_1(-n_1) \cdots h_r(-n_r) \otimes e^\alpha$ and $v = h'_1(-m_1) \cdots h'_s(-m_s) \otimes e^\beta$, where $h_i, h'_i \in H$, $n_i, m_i \in \mathbf{N}$ and $\alpha, \beta \in L$. By (ii),

$$(u, v) = (h_1'(m_s) \cdots h_s'(m_1)h_1(-n_1) \cdots h_r(-n_r) \otimes e^\alpha, 1 \otimes e^\beta) \qquad (*)$$

and

$$(u, 1 \otimes e^\beta) = (h_1(-n_2) \cdots h_r(-n_r) \otimes e^\alpha, h_1(n_1)(1 \otimes e^\beta)) = 0$$

unless $r = 0$. And so together with (i), this proves the existence and uniqueness.

The Hermitian form is clearly symmetric. If $p(u) = \bar{0}$ and $p(v) = \bar{1}$, then by definition of the parity, $(\alpha, \alpha) \equiv 0 \pmod 2$ and $(\beta, \beta) \equiv 0 \pmod 2$. In particular, $\alpha \neq \beta$ and so from $(*)$ and (i), $(u, v) = 0$. Hence the form is consistent.

We now show that the Hermitian form is non-degenerate. From $(*)$ and (i), all we need to show is that there are maps $h_1'(m_1)$ such that $h_1'(m_s) \cdots h_s'(m_1)s \neq 0$, where $s = h_1(-n_1) \cdots h_r(-n_r) \in S$. Let us re-write the vector s as follows:

$$s = h_{1,1}(-l_1) \cdots h_{1,k}(-l_1)h_2(-l_2) \cdots h_r(-l_t),$$

where $0 < l_1 < l_j$ for all $j > 1$. As the Hermitian form on H is non-degenerate, there is some $h \in H$ such that $(h, \sum_{i=1}^k h_{1,i}) \neq 0$ for all $1 \leq i \leq k$. The action of $\tilde{H}$ on S given in 4.2.8 (i) implies that

$$h(l_1)^k s = k l_1 (h, \sum_{i=1}^k h_{1,i})h_2(-l_2) \cdots h_r(-l_t).$$

Therefore, non-degeneracy follows by induction on t. $\square$

From now on, for reasons of simplicity, we will denote by $(.,.)$ both the bilinear form on the lattice L and the Hermitian form on the Fock space V_L defined in Theorem 4.3.3.

The Lie superalgebra $V_L/d(V_L)$ introduced in Lemma 4.3.1 is too big. The Lie superalgebra which is of importance to us is a sub-quotient of this one. This sub-quotient is derived from the natural action of the Virasoro algebra on V_L. A typical characteristic of a lattice vertex algebra is that the Virasoro algebra acts on it. It induces a natural $\mathbf{Z}$-gradation on V_L which will give us the Lie algebra we are looking for. Moreover the properties of this action will enable us to show that this Lie sub-superalgebra is indeed a BKM superalgebra when the lattice L is Lorentzian of rank at most 26. So let us first remind the reader of the definition of a Virasoro algebra.

Definition 4.3.4. *The Virasoro algebra Vir is the Lie algebra generated by elements L_n, $n \in \mathbf{Z}$ and a central element c satisfying*

$$[L_m, L_n] = (m - n)L_{m+n} + \frac{1}{12}(m^3 - m)\delta_{m+n,0}c.$$

A characteristic aspect of the lattice vertex algebras V_L is that it has a vector with the special property that the coefficients of the corresponding vertex operator generate a Virasoro algebra.

Definition 4.3.5. *A vertex algebra V with an even vector v such that*

$$Y(v, z) = \sum_{n \in \mathbf{Z}} L_n z^{-n-2},$$

where $\langle L_n, c \rangle \cong Vir$, c being a scalar, is called a conformal vertex algebra. The vector v is said to be a conformal vector with central charge c.

Theorem 4.3.6. *The vector*

$$\zeta = \frac{1}{2} \sum_i \alpha_i(-1)\alpha_i'(-1) \otimes 1,$$

where the vectors (α_i) run through a basis of L and (α_i') through a dual basis in H is a conformal vector of the lattice vertex algebra V_L with central charge $\dim L$. For all $j \in \mathbf{Z}$,

$$L_j = \frac{1}{2} \sum_i \left(\sum_{n<0} \alpha_i(n)\alpha_i'(j-n) + \sum_{n \geq 0} \alpha_i'(j-n)\alpha_i(n) \right).$$

In particular, $L_{-1} = d$ and L_0 is diagonalizable derivation on V_L:

$$L_0(1 \otimes e^\alpha) = \frac{1}{2}(\alpha, \alpha)(1 \otimes e^\alpha), \quad L_0(h(-m)v) = mh(-m)v + h(-m)L_0v.$$

Furthermore, the operators L_n and L_{-n} are adjoint with respect to the Hermitian form $(., .)$ on V_L.

Proof. By Corollary 4.1.26 (i), since $L_m = \zeta_{m+1}$,

$$[L_m, L_n] = \sum_{j=0}^{\infty} \binom{m+1}{j} (L_{j-1}\zeta)_{m+n+2-j}. \tag{1}$$

By Corollary 4.1.32, since $h(-1) = (h(-1) \otimes 1)_{-1}$ for all $h \in H$,

$$
\begin{aligned}
Y(\omega, z) &= \frac{1}{2} \sum_i \; : Y(\alpha_i(-1) \otimes 1, z) Y(\alpha_i'(-1) \otimes 1, z) : \\
&= \frac{1}{2} \sum_i \left(\sum_{n<0} \alpha_i(n) z^{-n-1} \sum_{m \in \mathbf{Z}} \alpha_i'(m) z^{-m-1} \right. \\
&\quad \left. + \sum_{m \in \mathbf{Z}} \alpha_i'(m) z^{-m-1} \sum_{n \geq 0} \alpha_i(n) z^{-n-1} \right) \\
&= \frac{1}{2} \sum_j z^{-j-2} \sum_i \left(\sum_{n<0} \alpha_i(n)\alpha_i'(j-n) + \sum_{n \geq 0} \alpha_i'(j-n)\alpha_i(n) \right).
\end{aligned}
\tag{2}
$$

Hence,

$$L_j = \frac{1}{2} \sum_i \left(\sum_{n<0} \alpha_i(n)\alpha_i'(j-n) + \sum_{n \geq 0} \alpha_i'(j-n)\alpha_i(n) \right). \tag{3}$$

Hence, from the action of $\tilde{H}$ on V_L given in Lemma 4.2.8, we get, for any $h \in H$, $m \in \mathbf{Z}$, $\alpha \in L$, $v \in V_L$,

$$
L_{-1}h(-m)v = \frac{1}{2}\sum_i (\alpha_i(-m-1)m(\alpha_i', h)v
$$

$$
+ m(\alpha_i, h)\alpha_i'(-1-m)) + h(-m)L_{-1}v
$$
$$
= mh(-m-1)v + h(-m)L_{-1}v
$$

and

$$
L_{-1}(1 \otimes e^\alpha) = \frac{1}{2}\sum_i ((\alpha_i', \alpha)\alpha_i(-1) + (\alpha_i, \alpha)\alpha_i'(-1)) \otimes e^\alpha
$$
$$
= \alpha(-1) \otimes e^\alpha
$$

and so $L_{-1} = d$.

$$
L_0(h(-m)v) = \frac{1}{2}\sum_i (\sum_{n<0} \alpha_i(n)\alpha_i'(-n) + \sum_{n \geq 0} \alpha_i'(-n)\alpha_i(n))h(-m)v
$$
$$
= \frac{1}{2}\sum_i (\alpha_i(-m)m(\alpha_i', h)v + m(\alpha_i, h)\alpha_i'(-m)) + h(-m)L_0v
$$
$$
= mh(-m)v + h(-m)L_0v
$$

and

$$
L_0(1 \otimes e^\alpha) = \frac{1}{2}\sum_i (\alpha_i, \alpha)\alpha_i'(0)(1 \otimes e^\alpha)
$$
$$
= \frac{1}{2}\alpha(0)(1 \otimes e^\alpha)
$$
$$
= \frac{1}{2}(\alpha, \alpha)(1 \otimes e^\alpha).
$$

We next evaluate the right hand side of (1). From the above,

$$
L_0\omega = 2\zeta. \tag{4}
$$

For $j > 0$, since $j - n > 1$ when $n < 0$, (3) gives

$$
L_j\zeta = \frac{1}{4}\sum_{i,k} \alpha_i'(j-1)\alpha_i(1)(\alpha_k(-1)\alpha_k'(-1) \otimes 1)
$$
$$
= \frac{1}{4}(\sum_{i,k}(\alpha_i, \alpha_k)\alpha_i'(j-1)\alpha_k'(-1) + \sum_i \alpha_i'(j-1)\alpha_i(-1)) \otimes 1
$$
$$
= \frac{1}{2}(\sum_i \alpha_i'(j-1)\alpha_i(-1)) \otimes 1
$$
$$
= \begin{cases} \frac{1}{2}\dim L(1 \otimes 1) & \text{if } j = 2 \\ 0 & \text{otherwise.} \end{cases}
$$

So, setting I_{V_L} to be the identity homomorphism on the space V_L, (1) becomes

$$[L_m, L_n]$$
$$= (L_{-1}\zeta)_{m+n+2} + (m+1)(L_0\zeta)_{m+n+1} + \frac{(m+1)m(m-1)}{6}(L_2\zeta)_{m+n-1}$$
$$= (d\zeta)_{m+n+2} + 2(m+1)\zeta_{m+n+1} + \frac{m^3-m}{12}(\dim L)I_{V_L}$$

by (4) and the vacuum axiom

$$= -(m+n+2)\zeta_{m+n+1} + 2(m+1)\zeta_{m+n+1} + \frac{m^3-m}{12}(\dim L)I_{V_L}$$

by Lemma 4.1.31

$$= (m-n)L_{m+n} + \frac{m^3-m}{12}(\dim L)I_{V_L}$$

as expected.

Finally, we show that

$$(L_j u, v) = (u, L_{-j}v).$$

For all $u, v \in V_L$,

$$(L_j u, v)$$
$$= \frac{1}{2}\sum_i \left(\left(\sum_{n<0}\alpha_i(n)\alpha_i'(j-n) + \sum_{n\geq 0}\alpha_i'(j-n)\alpha_i(n)\right)u, v\right) \quad \text{by} \quad (2)$$
$$= \frac{1}{2}\sum_i \left(u, \left(\sum_{n<0}\alpha_i'(-j-n)\alpha_i(-n) + \sum_{n\geq 0}\alpha_i(-n)\alpha_i'(-j-n)\right)v\right)$$

by Theorem 4.3.3

$$= \frac{1}{2}\sum_i \left(u, \left(\sum_{n\leq 0}\alpha_i(n)\alpha_i'(-j+n) + \sum_{n>0}\alpha_i'(-j+n)\alpha_i(n)\right)v\right)$$

replacing n by $-n$

$$= \frac{1}{2}\sum_i \left(u, \left(\sum_{n<0}\alpha_i(n)\alpha_i'(-j+n) + \sum_{n\geq 0}\alpha_i'(-j+n)\alpha_i(n)\right)v\right)$$

by Lemma 4.2.14

$$= (u, L_{-j}v).$$

$\square$

Remark 4.3.7. **(i).** The definition of the operator L_0 implies that for any eigenvector $v \otimes e^\alpha$ of L_0, $v \in S$, $\alpha \in L$, the corresponding eigenvalue $x \in \mathbf{C}$ is such that $x - \frac{1}{2}(\alpha, \alpha) > 0$.

(ii). Note that vertex operator algebras were defined by Frenkel, Lepowsky and Meurman [FrenLM2] to reflect some crucial features of the moonshine module. They are conformal vertex algebras with the added condition that all the eigenspaces of L_0 on V are finite dimensional. However not all conformal vertex algebras are vertex operator algebras. In particular, if the lattice L is not positive definite, then the lattice vertex algebra V_L is not a vertex operator algebra. Indeed, for example, the eigenspace $V_1 = \{v \in V_L : L_0 v = v\}$ is then infinite dimensional:

$$\{h(\frac{1}{2}(\alpha, \alpha) - 1) \otimes e^\alpha : h \in H, \alpha \in L, (\alpha, \alpha) \leq 0\} \leq V_1$$

since for any $\alpha \in L$ of non-positive even norm, as $1 - \frac{1}{2}(\alpha, \alpha) \in \mathbf{N}$, for all $h \in H$, $h(\frac{1}{2}(\alpha, \alpha) - 1) \otimes e^\alpha \in V_L$; and the definition of the operator L_0 given in Theorem 4.3.6 implies that

$$L_0(h(\frac{1}{2}(\alpha, \alpha) - 1) \otimes e^\alpha) = h(\frac{1}{2}(\alpha, \alpha) - 1) \otimes e^\alpha.$$

Corollary 4.3.8. *Distinct eigenspaces for the operator L_0 are orthogonal with respect to the Hermitian form $(.,.)$. In particular, the restrictions of the Hermitian form to the eigenspaces of the operator L_0 are non-degenerate.*

Proof. Let $u, v \in V_L$ be eigenvectors for L_0 with eigenvalues k, j respectively. By Theorem 4.3.6, L_0 is self dual. Hence,

$$k(u, v) = (L_0 u, v) = (u, L_0 v) = j(u, v).$$

The result follows. $\qquad\qquad\qquad\qquad\qquad\qquad\qquad\qquad\qquad\qquad\square$

For any subspace U of V_L on which the operator L_0 acts, for $n \in \frac{1}{2}\mathbf{Z}$, we will write

$$U^n = \{u \in U : L_0 u = nu\}.$$

As mentioned earlier, the Lie algebra $V_L/d(V_L)$ is too large. This was realized independently by Goddard-Thorn [GodT] and Frenkel [Fren] who showed that the space we should consider is not the whole lattice vertex algebra V_L, nor is it a full eigenspace of the operator L_0 but the subspace P^1 of V_L, where we set

$$P^i = \{v \in V | L_0 v = iv, L_i v = 0, \forall i > 0\}.$$

We next give two technical but useful results.

Lemma 4.3.9. *The linear derivation d on V_L is a monomorphism.*

Proof. Suppose that $du = 0$ for some $u \in V_L$. So by Lemma 4.1.31,

$$\frac{d}{dz} Y(u, z) = 0$$

or equivalently

$$u_n = 0 \quad \forall n \neq -1.$$

Hence, by Corollary 4.1.26 (ii), for all $v \in V_L$, $n \in \mathbf{Z}$, $[u_{-1}, v_n] = 0$. In particular, from the vacuum axiom, we get

$$v_n u = v_n u_{-1}(1 \otimes 1) = 0 \quad \forall n \geq 0. \tag{$*$}$$

Now, there exists $r \in \mathbf{N}$, $s_j \in S$ and distinct elements $\beta_j \in L$, $1 \leq j \leq r$, such that $u = \sum_{j=1}^{r} s_j \otimes e^{\beta_j}$. From $(*)$, we can deduce that for all $h \in H$, taking $v = h(-1) \otimes 1$ in the above,

$$0 = h(0)u = \sum_j (h, \beta_j) s_j \otimes e^{\beta_j}$$

since $v_0 = h(0)$. As the bilinear form on L is non-degenerate, there is some $h \in H$ such that $(h, \beta_j) \neq 0$ for any $1 \leq j \leq r$ for which $\beta_j \neq 0$. Therefore, as the vectors $s_j \otimes e^{\beta_j}$ are linearly independent, this implies that $j = 1$ and $\beta_1 = 0$ and so

$$u = s_1 \otimes 1.$$

Taking $v = h(-1) \otimes 1$ in $(*)$, for all $h \in H$, $n \in \mathbf{N}$, $h(n)s_1 = 0$. As the bilinear form remains non-degenerate on H, we can conclude that $s_1 = 0$ or equivalently $u = 0$, proving the Lemma. $\qquad\square$

Lemma 4.3.10. *For all $u, v \in P^1$, $u_0(v) \in P^1$.*

Proof. For any $n \in \mathbf{Z}_+$, by Corollary 4.1.26 (i), for all $u \in P^1$,

$$
\begin{aligned}
[L_n, u_0] &= [\zeta_{n+1}, u_0] \\
&= \sum_{j=0}^{n+1} \binom{n+1}{j} (\zeta_j u)_{n+1-j} \\
&= \sum_{j=0}^{n+1} \binom{n+1}{j} (L_{j-1}u)_{n+1-j} \\
&= (L_{-1}u)_{n+1} + (n+1)u_n \quad \text{since } u \in P^1 \\
&= (du)_{n+1} + (n+1)u_n \quad \text{by Theorem 4.3.6} \\
&= -(n+1)u_n + (n+1)u_n \quad \text{by Lemma 4.1.31} \\
&= 0
\end{aligned}
$$

Hence, for all $v \in P^1$, $0 = L_n u_0 v$ for $n > 0$ and $L_0 u_0 v = u_0 v$. In other words, $u_0(v) \in P^1$. $\qquad\square$

Corollary 4.3.11. *The vector space P^1/dP^0 is a Lie subalgebra of the Lie superalgebra $V_L/d(V_L)$.*

Proof. We first show that

$$P^1 \cap d(V_L) = dP^0.$$

Let $v \in V_L$ be such that $dv \in dV \cap P^1$. Hence, $L_0 dv = dv$. By Theorem 4.3.6, $d = L_{-1}$ and $[L_0, L_{-1}] = L_{-1}$. This implies that $dv - dL_0 v = dv$ and so

$$dL_0 v = 0.$$

Applying Lemma 4.3.9, we then get $L_0 v = 0$.

For $n > 1$, again by Theorem 4.3.6,

$$0 = L_n L_{-1} v = L_{-1} L_n v + (n+1) L_{n-1} v$$

and so by induction on n, $L_{-1} L_n v = 0$. Thus by Lemma 4.3.9, $L_n v = 0$. Hence, $v \in P^0$ as expected.

All we now need to check for P^1 / dP^0 to be a Lie superalgebra is that for all $u, v \in P^1$, $u_0(v) \in P^1$. This follows from Lemma 4.3.10.

Finally, by Theorem 4.3.6, the derivation L_0 has non-integral eigenvalues only on the odd part of the Fock space V_L. Therefore the subspace P^1 is even and so P^1 / dP^0 is a Lie algebra.   □

Remark 4.3.12. We will see in section 5.5 that the quotient of this Lie algebra by the kernel of the Hermitian form is a BKM algebra when the lattice L is even, Lorentzian of rank at most 26. It is important to note that the subspace P^1 / dP^0 is a Lie algebra, i.e. that its odd part is trivial. Therefore, even when the lattice L is odd, which is the case for the root lattice of the fake monster superalgebras [Sch2], and $\Gamma = \mathbf{Z}_2$, the above construction though it starts from a lattice vertex superalgebra (i.e. $\mathbf{Z}_2$-graded) only leads to a BKM algebra.

The Lie algebra derived in Exercises 4.3.1 and 4.3.2 is also a BKM algebra modulo the kernel of the bilinear form (see Exercise 5.5.2) when the lattice L is Lorentzian of rank at most 10, not necessarily even.

Exercises 4.3

1. Assume that the lattice L is integral. Remember the definition of the fermionic lattice vertex superalgebra V_L^f from Exercise 4.2.2.

(i) Show that the quotient space $V_L^f / d(V_L^f)$ is a Lie superalgebra.

(ii) Show that there is a unique, consistent, symmetric, natural Hermitian form on V_L^f such that: for any $h \in H$, $n \in \mathbf{Z}$, the adjoint of the operator $h(n + \frac{1}{2})$ is $h(-n + \frac{1}{2})$. Deduce that the form is non-degenerate.

(iii) Show that the vector

$$\zeta_f = \frac{1}{2} \sum_i \alpha_i(-\frac{3}{2}) \alpha_i'(-\frac{1}{2}) \otimes 1,$$

where the vectors (α_i) run through a basis of L and (α_i') through a dual basis in H is a conformal vector of the lattice vertex algebra V_L^f with central charge $\dim L$. Check that for all $j \in \mathbf{Z}$,

$$L_j = \frac{1}{2} \sum_i \left(\sum_{\substack{n \in \mathbf{Z} - \frac{1}{2} \\ n < 0}} (-n + \frac{j}{2}) \alpha_i(n) \alpha_i'(j-n) - \sum_{\substack{n \in \mathbf{Z} - \frac{1}{2} \\ n \geq 0}} (-n + \frac{j}{2}) \alpha_i'(j-n) \alpha_i(n) \right).$$

Deduce that $L_{-1} = d$ and L_0 is diagonalizable derivation on V_L^f:

$$L_0(1) = \frac{1}{2}(\alpha, \alpha)(1 \otimes e^\alpha), \quad L_0(v) = (m_1 + \cdots m_r) v,$$

for $v = h_1(-m_1) \cdots h_r(-m_r)$, $h_i \in H$, $m_i > 0$. Furthermore, show that the operators L_n and L_{-n} are adjoint with respect to the above Hermitian form on V_L^f.

2. Assume that the lattice L is integral. Let V_L^b (resp. V_L^f) be the bosonic (resp. fermionic) lattice vertex superalgebra defined in Definition 4.2.17 (resp. Exercise 1) and ζ_b (resp. ζ_f) its conformal vector given in Theorem 4.3.6 (resp. Exercise 1). Set $V_L = V_L^f \otimes V_L^b$.

(i) Show that the vector $\zeta = \zeta_f \otimes 1 + 1 \otimes \zeta_b$ is a conformal vector of the lattice vertex superalgebra V_L and that the Virasoro algebra acts on it.

(ii) Let P^i be the generalized eigenspaces for the operator L_0 (corresponding to ζ instead of ζ_b as defined in section 4.3). Show that the quotient space V_L/dV_L has the structure of a Lie superalgebra and P^1/dP^0 is a Lie subalgebra of V_L/dV_L.

(iii) Let $\tau = a(-\frac{1}{2}) \otimes a(-1) \in V_{\bar{1}}$. Show that $\zeta = \frac{1}{2}\tau_0\tau$.

(iv) Deduce that the operators $L_n = \zeta_{n+1}$ and $G_m = \tau_{m+\frac{1}{2}}$ generate a Neveu-Schwarz algebra with central charge $\frac{3}{2} \dim L$, i.e. for all $m, n \in \mathbf{Z}$, $r, s \in \mathbf{Z} + \frac{1}{2}$,

$$[c, L_m] = 0 = [c, G_r],$$

$$[L_m, L_n] = (m - n)L_{m+n} + \frac{1}{12}(m^3 - m)\delta_{m+n,0}c,$$

$$[L_n, G_r] = (\frac{1}{2}n - r)G_{n+r},$$

$$[G_r, G_s] = 2L_{r+s} + \frac{1}{3}(r^2 - \frac{1}{4})\delta_{r+s,0}c.$$

(iv) Set

$$P^{\frac{1}{2}} = \{v \in V_L : L_0v = \frac{1}{2}v, \ L_nv = 0 = G_mv, \ \forall n, m > 0\}.$$

Show that the subspace $G_{-\frac{1}{2}}P^{\frac{1}{2}}/dP^0$ is a Lie subalgebra of P^1/dP^0.

For a solution of Exercises 1 and 2, see [Sch1].

Chapter 5

Lorentzian BKM Algebras

5.1 Introduction

The aim of the chapter is to give the reader the tools to understand the classification of the "interesting" BKM superalgebras. This work is still in progress today and has not been completed. We have seen in section 2.3 that a given BKM superalgebra is fully known if we have a means of describing its root system Δ, a base of Δ or equivalently a set of generators for G, and the multiplicities of its roots of infinite type. Given any abstract generalized Cartan matrix, one can construct an associated BKM superalgebra. However, it is usually impossible to find all this information about its structure. By "interesting", we mean a BKM superalgebra for which this is feasible and which can be constructed in some concrete manner. The root lattices of these BKM superalgebras is either Lorentzian, semi-positive definite of corank 1 or positive definite. The latter two cases correspond respectively to affine and finite dimensional semisimple Lie algebras (see Exercise 2.3.10). These are Kac-Moody Lie algebras. Based on what is known so far, there is good reason to believe that the "interesting" BKM superalgebras that are not Kac-Moody Lie superalgebras have Lorentzian root lattice. This is for example the case for the two explicitly known BKM superalgebras at the origin of the general theory of BKM superalgebras, namely the Monster and Fake Monster Lie algebras (Examples 2.3.11 (i), 2.6.39, 2.6.40): their root lattices are the even unimodular Lorentzian lattices $II_{1,1}$ and $II_{1,25}$ of rank 2 and 26 respectively.

Definition 5.1.1. *A BKM superalgebra with Lorentzian root lattice is a Lorentzian BKM (LBKM) superalgebra.*

In this chapter we consider the roots as elements of the Cartan subalgebra or of its dual. This is more natural. As the example of the Monster Lie algebra (Example 2.3.11.1) shows, in general there are infinitely many simple roots of infinite type and so the formal root lattice Q is infinite dimensional with an artificial structure due to the fact that the simple roots are forced to be linearly independent.

There is a related project whose aim is the classification and construction of hyperbolic Kac-Moody Lie algebras or equivalently hyperbolic Weyl (reflection) groups. These are Kac-Moody Lie algebras with Lorentzian root lattice. Equivalently, this is the classification of Lorentzian BKM superalgebras whose root lattice can be generated by its simple roots of positive norm. The root lattice of an "interesting" Lorentzian BKM superalgebra is not necessarily generated by its simple roots of positive norm and the simple roots of non-positive norm may have to be taken into account to construct a basis for the root lattice. In fact, it may not have roots of positive norm. In other words, its Weyl group may be uninteresting or even trivial. As we saw in Example 2.3.11.1, this is the case for the Monster Lie algebra: its root lattice is the even unimodular lattice $II_{1,1}$ of rank 2 and its Weyl group has order 2. The "interesting" Lorentzian BKM superalgebras do not necessarily have hyperbolic reflection groups. If the classification problem of hyperbolic reflection group can be completely solved, then it would therefore give a subclass of "interesting" Lorentzian BKM superalgebras.

Attempts to tackle this classification problem directly from information about the root lattice or equivalently the generalized Cartan matrix do not lead far.

In sections 5.2 and 5.3, we will introduce the reader to the classification project of LBKM algebras to which we can associate a vector valued modular form satisfying certain properties and whose Fourier coefficients give the root multiplicities. Therefore the classification of these LBKM algebras should be equivalent – though this is not proved yet – to the classification of these modular forms. The classification of adequate modular forms should be an easier problem to study. The only literature in this subject is in the form of research papers. We follow the methodology of [Borc9,11] where these results were originally proved.

In section 5.4 we will give evidence showing that this classification problem should be a finite one. Namely, the existence of the Weyl vector (see Definition 2.3.47) gives a strong upper bound on the rank of the root lattices of LBKM algebras. To do this, we need to consider the converse problem, namely that of associating vector valued modular forms to automorphic forms on Grassmannians in such a way that the coefficients of its product expansion are given by some of the coefficients of the modular form. For further study, the reader should consult [Bar] and [Br], where this problem has been solved under some technical restrictions.

The classification part of the project is only a first step. We need to find a way of constructing them concretely and naturally. This we do in section 5.5 from lattice vertex algebras. See [Borc2] and [Fren] for details.

In the classification we sometimes get BKM superalgebras with non-trivial odd parts and we have thus given in chapter 2 the theory in its most general form rather than restricting ourselves to the Lie algebra setup. In chapter 4 we could have kept to the more classical theory of ordinary lattice vertex superalgebras. They are all that is needed to construct LBKM algebras. Though in this book we do not explain how to construct LBKM superalgebras with non-trivial odd parts as it would make the exposition too advanced and too technical for a first approach, the theory of Γ-graded lattice vertex algebras gives the reader interested in pursuing work in this direction the tools to understand this construction.

5.2 Automorphic forms on Grassmannians

In this section, we study how the definition of automorphic forms on the upper half plane for a Fuschian group can be generalized to the context of Lorentzian lattices to functions transforming under the action of a discrete subgroup of a real Lie group. For this we need to construct from a Lorentzian lattice a space that has the same typical geometric properties as $\mathcal{H}$. In other words, it should be a Hermitian symmetric space and its group of isometries should be a semisimple Lie group (see Lemma B.2.25). Indeed $\mathcal{H}$ is isomorphic to $SL(2, \mathbf{R})/SO(2, \mathbf{R})$ since $SO(2, \mathbf{R})$ is the stabilizer of the point $i \in \mathcal{H}$.

Remark 5.2.1. In the context of BKM superalgebras the Lorentzian lattices we encounter have signature $(n, 1)$ as simple roots of finite type have been assumed to have non-negative norm. However in sections 5.2-5.4, we develop the theory for Lorentzian lattices having signature $(1, n)$ as this is more natural in the framework of modular functions. However, the main result given in Chapter 5 on product expansions of automorphic forms (Theorem 5.2.5) can "symmetrically" be adapted to the case of a Lorentzian lattice with signature $(n, 1)$.

In sections 5.2-5.4, we will assume that L is a Lorentzian lattice with signature $(1, b^- - 1)$ and we set

$$M = L \oplus II_{1,1}.$$

Hence the lattice M has signature $(2, b^-)$. We let $z \in M$ be a primitive element of norm 0 in the second factor $II_{1,1}$ of M. Hence z' is also in $II_{1,1}$ and has norm 0.

The bilinear form on $M \otimes_{\mathbf{Z}} \mathbf{C}$ will be written $(.,.)$. Consider the Grassmannian $G(M)$ consisting of all the 2-dimensional positive definite vector subspace of the real vector space $M \otimes_{\mathbf{Z}} \mathbf{R}$.

Choose an orientation of maximal positive definite subspaces (see Appendix A). Set P to be the set of elements $Z_M = X_M + iY_M \in M \otimes_{\mathbf{Z}} \mathbf{C}$ with norm 0 and for which X_M, Y_M form an oriented basis for a maximal positive definite subspace of $M \otimes_{\mathbf{Z}} \mathbf{R}$.

Note that $Z_M^2 = 0$ if and only if $X_M^2 = Y_M^2$ and $(X_M, Y_M) = 0$.

Lemma 5.2.2. *The Grassmannian $G(M)$ is a Hermitian symmetric space isomorphic to the submanifold of the projective space $\mathbf{P}(M \otimes_{\mathbf{C}} \mathbf{C})$ consisting of points represented by vectors in P and its isometry group is isomorphic to $O(M, \mathbf{R})^+/SO(2, \mathbf{R})$.*

Proof. We choose an orientation (see Appendix A) on the 2-dimensional positive definite subspaces of $M \otimes_{\mathbf{Z}} \mathbf{R}$. Consider the map:

$$G(M) \to \mathbf{P}(M \otimes_{\mathbf{Z}} \mathbf{C})$$
$$v^+ \mapsto X_M + iY_M,$$

where X_M, Y_M is a oriented orthogonal basis of the 2-dimensional positive definite subspace v^+ of $M \otimes_{\mathbf{Z}} \mathbf{R}$ such that $X_M^2 = Y_M^2$, and $X_M + iY_M$ is a representative of a point in the complex projective space $\mathbf{P}(M \otimes_{\mathbf{Z}} \mathbf{C})$. We first check that this map is well defined. Let X_{1M} and Y_{1M} be another oriented orthogonal basis of the subspace v^+ such that $X_{1M}^2 = Y_{1M}^2$. Since for any $\lambda \in \mathbf{C}$, $\lambda(X_M + iY_M)$ and $X_M + iY_M$ represent the same point in $\mathbf{P}(M \otimes_{\mathbf{Z}} \mathbf{C})$, we may assume that $X_{1M}^2 = X_M^2$. Then, $X_{1M} = aX_M + bY_M$ and $Y_{1M} = cX_M + dY_M$, where $a, b, c, d \in \mathbf{R}$. So $X_{1M} \wedge Y_{1M} = (ad - bc)X_M \wedge Y_M$, which implies that $ad - bc > 0$. Since $\begin{pmatrix} a & c \\ b & d \end{pmatrix} \in O(2, \mathbf{R})$, we therefore get $c = -b$ and $a = d$. Hence,

$$X_{1M} + iY_{1M} = (aX_M + bY_M) + i(-bX_M + aY_M) = (a - ib)(X_M + iY_M)$$

and so $X_{1M} + iY_{1M}$ and $X_M + iY_M$ represent the same point in $\mathbf{P}(M \otimes_{\mathbf{Z}} \mathbf{C})$. These calculations also show that the map is injective. Hence by Lemma B.2.14, the Grassmannian $G(M)$ is a Hermitian manifold and the group $O(M \otimes_{\mathbf{Z}} \mathbf{R})^+$ of the isometries of the real inner product space $M \otimes_{\mathbf{Z}} \mathbf{R}$ keeping the orientation fixed is its isometry group. The latter clearly acts transitively on $G(M)$.

It only remains to check that the group $O(M \otimes_{\mathbf{Z}} \mathbf{R})^+$ contains an involution with an isolated fixed point (see Definition B.2.17). Let X_M, Y_M be the oriented basis of a 2-dimensional positive definite subspace v^+ of $M \otimes_{\mathbf{Z}} \mathbf{R}$. Then, its orthogonal complement $v^{+\perp}$ is a positive definite subspace. Consider the linear map ϕ on $M \otimes_{\mathbf{Z}} \mathbf{R}$ defined by:

$$X_M \mapsto Y_M; \ Y_M \mapsto -X_M; \ v \mapsto v, \ \forall v \in v^{+\perp}.$$

This map belongs to $O(M \otimes_{\mathbf{Z}} \mathbf{R})^+$ as $X_M \wedge Y_M = Y_M \wedge (-X_M)$ (see Appendix A) and simple calculations show that v^+ is its unique fixed point in $G(M)$.

We can therefore conclude that the Grassmannian $G(M)$ is a Hermitian symmetric space. Moreover, the stabilizer of the subspace v^+ is the isometry group of a 2-dimensional Euclidean vector space. $\square$

Lemma 5.2.3. *There is a natural projection* $\pi : P \to G(M)$ *and* $(P, G(M), \pi)$ *is a principal* $\mathbf{C}^*$ *bundle.*

Proof. Let $V \in G(M)$. Then, $\pi^{-1}(V)$ is the set of norm 0 vectors $X_M + iY_M \in M \otimes_{\mathbf{Z}} \mathbf{C}$ such that X_M, Y_M form an oriented basis of the positive definite subspace v^+ of $M \otimes_{\mathbf{Z}} \mathbf{R}$. Since the manifold structure on the projective space $\mathbf{P}(M \otimes_{\mathbf{Z}} \mathbf{C})$ is transferred from the manifold $M \otimes_{\mathbf{Z}} \mathbf{C}$ via the projection π, it follows that the triple $(P, G(M), \pi)$ is a fiber bundle (see Definition B.3.1).

To show that P is a principal $\mathbf{C}^*$-fiber over $G(M)$, we need to check that the fibers $\pi^{-1}(V)$ are the orbits of the action of $\mathbf{C}^*$ on P and that this action is free and transitive on the fibers (see Definition B.3.3).

Let $X_M + iY_M \in \pi^{-1}(V)$ and $a + ib \in \mathbf{C}$. The vector

$$(X_M + iY_M)(a + ib) = (aX_M - bY_M) + i(bX_M + aY_M)$$

has norm 0 and $aX_M - bY_M \in V$ and so has positive norm. Moreover the matrix $\begin{pmatrix} a & -b \\ b & a \end{pmatrix}$ has positive determinant. It follows that $(X_M + iY_M)(a + ib) \in \pi^{-1}(V)$.

If $(X_M + iY_M)(a + ib) = X + iY$, then $aX_M - bY_M = X_M$ and $bX_M + aY_M = Y_M$, in which case $b = 0$ and $a = 1$.

Let $X_{1M} + iY_{1M} \in \pi^{-1}(V)$. There exists $A \in O(2, \mathbf{R})^+$ such that

$$A \begin{pmatrix} X_M \\ Y_M \end{pmatrix} = \begin{pmatrix} X_{1M} \\ Y_{1M} \end{pmatrix}.$$

Since $O(2, \mathbf{R})^+ = \{ \begin{pmatrix} a & b \\ -b & d \end{pmatrix} : a, b \in \mathbf{R}, a^2 + b^2 \neq 0 \}$,

$$(X_M + iY_M)(a - bi) = X_{1M} + iY_{1M}$$

for some $a - bi \in \mathbf{C}^*$.   □

Definition 5.2.4. *An automorphic form of weight k and character χ on the Grassmannian $G(M)$ is a function Ψ_M on P satisfying:*

(i) Ψ_M *is holomorphic on P;*

(ii) Ψ_M *is homogeneous of degree $-k$, i.e.* $\Psi_M(v\tau) = \tau^{-k}\Psi_M(v)$, $v \in P$, $\tau \in \mathbf{C}^*$;

(iii) *There is a subgroup Γ of finite index in $O(M)^+$ and a one dimensional character χ of Γ such that $\Psi_M(\sigma v) = \chi(\sigma)\Psi_M(v)$.*

We next give a different description of the Grassmannian $G(M)$. For this, let us consider the positive norm vectors in $L \otimes_{\mathbf{Z}} \mathbf{R}$.

Lemma 5.2.5. *There are two cones of positive norm vectors in the Lorentzian vector space $L \otimes_{\mathbf{Z}} \mathbf{R}$. They are the sets $\pm\{dx + y : d > 0, (dx + y)^2 > 0, (x, y) = 0\}$, where $x \in L \otimes_{\mathbf{Z}} \mathbf{R}$ and $x^2 > 0$.*

Proof. Let $\{e_i, i = 1, \cdots, n\}$, where e_i has positive norm for $i = 1, \cdots, n - 1$, be an orthogonal basis of $L \otimes_{\mathbf{Z}} \mathbf{R}$. With respect to this basis, $(a_1, \cdots, a_n)$ is a positive norm vector if and only if $-a_n < \sqrt{\sum_{i=1}^{n-1} a_i^2} < a_n$ and $a_n > 0$ or $a_n < \sqrt{\sum_{i=1}^{n-1} a_i^2} < -a_n$ and $a_n < 0$.   □

We next explain how to choose one of the cones of positive norm vectors in $L \otimes_{\mathbf{Z}} \mathbf{R}$.

Any vector in the complex vector space $M \otimes_{\mathbf{Z}} \mathbf{C}$ is of the form $Z + mz' + nz$, where $Z \in L \otimes_{\mathbf{Z}} \mathbf{C}$, $m, n \in \mathbf{C}$. We will write these vectors as (Z, m, n).

Lemma 5.2.6. *If X_M, Y_M is an oriented basis of a 2-dimensional positive definite subspace of $M \otimes_{\mathbf{Z}} \mathbf{R}$ satisfying $(X_M, z) > 0$ and $(Y_M, z) = 0$, then set $Y \in L \otimes_{\mathbf{Z}} \mathbf{R}$ to be the vector satisfying $Y_M - Y \in < z, z' > \otimes_{\mathbf{Z}} \mathbf{R}$. The set of these vectors Y forms a cone of positive norm vectors in $K \otimes_{\mathbf{Z}} \mathbf{R}$.*

Proof. First note that for any oriented basis X_M, Y_M with the above property, Y has positive norm since $(Y, z) = 0$ implies that $Y_M = (Y, 0, n)$ for some $n \in \mathbf{R}$.

We now show that for any two oriented bases X_M, Y_M and X_{1M}, Y_{1M} with the above property, Y and Y_1 belong to the same cone of positive norm vectors.

By Lemma A.5, without loss of generality, we may take the positive norm vector Y_M in $L \otimes_{\mathbf{Z}} \mathbf{R}$. Then, $(Y_M, z) = 0$. Let $X_M \in M \otimes_{\mathbf{Z}} \mathbf{R}$ be a positive norm vector such that $(X_M, z) > 0$ and X_M, Y_M form an orthogonal basis for a 2-dimensional positive definite subspace v^+ of $M \otimes_{\mathbf{Z}} \mathbf{R}$. Then, either X_M, Y_M or $X_M, -Y_M$ is an oriented basis. So, without loss of generality, we may assume that X_M, Y_M is an oriented basis.

Let X_{1M}, Y_{1M} be another oriented basis of a 2-dimensional positive definite subspace of $M \otimes_{\mathbf{Z}} \mathbf{R}$ satisfying $(X_{1M}, z) > 0$ and $(Y_{1M}, z) = 0$. Let π be the projection of $M \otimes_{\mathbf{Z}} \mathbf{R}$ onto v^+. Then, there exist $a, b, c, d \in \mathbf{R}$ such that

$$ad - bc > 0 \tag{1}$$

and

$$\pi(X_{1M}) = aX_M bY_M;\ \pi(Y_{1M}) = cX_M + dY_M.$$

Since $0 = (Y_{1M}, z)$, $c = 0$. Hence, $(Y_{1M}, Y_M) = (\pi(Y_{1M}), Y_M) = d$ and so $(Y_{1M}, Y_M) > 0$ by (1). Equivalently, extending Y_M to an orthogonal basis of the subspace $K \otimes_{\mathbf{Z}} \mathbf{R}$, the Y_M-coefficient of Y_1 is positive.

It only remains to check that all positive norm vectors in the cone $\mathcal{C}$ containing Y_M can be obtained in this manner. Applying the above argument for Y_M to any positive norm vector Y_{1M} in $\mathcal{C}$, it follows that there is an oriented basis X_{1M}, Y_{1M} or $X_{1M}, -Y_{1M}$ with the correct properties. In the latter case, what precedes implies that $-Y_{1M} \in \mathcal{C}$, contradicting assumption. Hence, X_{1M}, Y_{1M} is an oriented basis. $\qquad\Box$

Set $\mathcal{C}$ to be the cone of all positive norm vectors Y in $L \otimes_{\mathbf{Z}} \mathbf{R}$ as defined in Lemma 5.2.6. Set

$$P_1 = \{Z_M = X_M + iY_M \in M \otimes_{\mathbf{Z}} \mathbf{C} : X_M^2 = Y_M^2 > 0, (X_M, Y_M) = 0, (Z_M, z) = 1\}.$$

Clearly $P_1 \leq P$.

Lemma 5.2.7.

(i)

$$P_1 = \{(Z, 1, -\frac{Z^2}{2}) : Z \in (L \otimes_{\mathbf{Z}} \mathbf{R}) + i\mathcal{C}\}.$$

(ii) There is an injection from $L \otimes_{\mathbf{Z}} \mathbf{R} + i\mathcal{C}$ to P_1.

Proof. We leave (i) for the reader to check. The map

$$L \otimes \mathbf{R} + i\mathcal{C} \to P_1$$
$$Z \mapsto Z - z' - \frac{Z^2}{2}z$$

is injective.   □

Note that the action of $O(M, \mathbf{R})^+$ on P does not restrict to $L \otimes \mathbf{R} + i\mathcal{C}$.

Corollary 5.2.8. *The Grassmannian $G(M)$ is isomorphic to the manifold $L \otimes_{\mathbf{Z}} \mathbf{R} \oplus i\mathcal{C}$. In particular it is a hyperbolic space of dimension b^-.*

Proof. There is a natural injective map from P_1 to $G(M)$. Hence, by Lemma 5.2.7, there is an injection from $L \otimes_{\mathbf{Z}} \mathbf{R} + i\mathcal{C}$ to $G(M)$. This map is surjective by Lemma 5.2.9. The second statement is a direct consequence of the definition of a hyperbolic space (see Definition B.2.24).   □

Lemma 5.2.9. *Let v^+ be a 2-dimensional positive definite subspace of $M \otimes_{\mathbf{Z}} \mathbf{R}$. There is a unique oriented orthogonal basis X_M, Y_M of V satisfying $(X_M + iY_M, z) = 1$ and $(X_M + iY_M)^2 = 0$.*

Proof. Since $\dim v^+ > 1$, the subspace v^+ contains a non-zero vector Y orthogonal to z. Let X_M, Y_M be an orthogonal oriented basis for v^+. Then, $(X_M, z) \neq 0$. Otherwise $(X_M, z) = 0 = (Y_M, z)$. However the maximal dimension of a positive definite subspace in the complement of z in $M \otimes_{\mathbf{Z}} \mathbf{R}$ is 1. Hence, we may choose $X_M \in v^+$ so that

$$(X_M + iY_M, z) = 1 \quad \text{and} \quad (X_M, Y_M) = 0.$$

Furthermore, $X_M^2 = \lambda Y_M^2$ for some $\lambda > 0$. Hence $X_M, \sqrt{\lambda}Y_M$ forms a basis for V with the right characteristics. Without loss of generality, we may thus assume that $\lambda = 1$.

Suppose that X_{1M}, Y_{1M} is another oriented orthogonal basis for v^+ satisfying

$$(X_{1M} + iY_{1M}, z) = 1 \quad \text{and} \quad X_{1M}^2 = Y_{1M}^2.$$

Let $X_{1M} = aX_M + bY_M$ and $Y_{1M} = cX_M + dY_M$. Then,

$$0 = (cX_M + dY_M, z) = c, \, 1 = (aX_M + bY_M, z) = a, \quad \text{and}$$

$$0 = (X_{1M}, Y_{1M}) = bd.$$

Since $ad - bc > 0$, it follows that $b = 0$. Moreover

$$X_M^2 = X_{1M}^2 = Y_{1M}^2 = d^2 Y_M^2 = d^2 X_M^2.$$

So, $d = 1$. As a result

$$X_{1M} = X_M \quad \text{and} \quad Y_{1M} = Y_M.$$

□

For an automorphic form Ψ_M on $G(M)$, set Ψ_z to be the function on $L \otimes_{\mathbf{Z}} \mathbf{R} + i\mathcal{C}$ defined as follows:

$$\Psi_z(Z) = \Psi_M((Z, 1, Z^2/2)).$$

This is well defined by Lemma 5.2.7. Since an automorphic form is homogeneous, the following result is immediate.

Lemma 5.2.10. *The automorphic form Ψ_M is fully determined by the function Ψ_z on $L \otimes_{\mathbf{Z}} \mathbf{R} + i\mathcal{C}$.*

By Corollary 5.2.8, the positive definite 2-dimensional vector space v^+ is given by a vector in $L \otimes_{\mathbf{Z}} \mathbf{R} \oplus i\mathcal{C}$ and hence by a vector

$$Z_M = (Z, 1, -Z^2/2) \in P_1$$

(see Lemma 5.2.7), where

$$Z = X + iY \in L \otimes_{\mathbf{Z}} \mathbf{R} \oplus i\mathcal{C}$$

in such a way that

$$Z_M = X_M + iY_M$$

and X_M, Y_M form an oriented orthogonal basis for v^+.

Lemma 5.2.11.
  (i) $X_M = (X, 1, (Y^2 - X^2)/2)$ *and* $Y_M = (Y, 0, -(X, Y))$;
  (ii) $X_M^2 = Y^2 = Y_M^2$;
  (iii) $z_{v^+} = X_M/Y^2$;
  (iv) $z_{v^+}^2 = 1/Y^2$;
  (v) w^+ *is the one dimensional subspace of $L \otimes_{\mathbf{Z}} \mathbf{R}$ generated by Y;*
  (vi) $\lambda_{w^+} = (\lambda, Y)Y/Y^2$ *for any $\lambda \in L$.*
  (vii) $\mu = X$.

Proof. Since

$$z_{v^+} = (z_{v^+}, X_M)X_M/X_M^2 + (z_{v^+}, Y_M)Y_M/Y_M^2$$

$$= (z, X_M)X_M/X_M^2 + (z, Y_M)Y_M/Y_M^2$$

as $z = z_{v^+} + z_{v^-}$ and $(z_{v^-}, X_M) = 0 = (z_{v^-}, Y_M)$, (iii) follows from the fact that $(z, X_M) = 1$, $(z, Y_M) = 0$, and $X_M^2 = Y_M^2$ (see Lemma 5.2.7).
By definition (see Theorem 3.3.11),

$$\mu = -z' + (z', z_{v^-})z_{v^+}/z_{v^+}^2 + (z', z_{v^+})z_{v^-}/z_{v^-}^2 - z' + (z_{v^+} - (z', z_{v^+})z)/z_{v^+}^2$$

$$= (0, -1, 0) + (X, 1, (Y^2 - X^2)/2)/Y^4 - (0, 0, (Y^2 - X^2)/2)/Y^4$$

$$= (X, 0, 0).$$

□

Before closing this section, we give an example illustrating why we have to be careful when dealing with actions on $L \otimes_{\mathbf{Z}} \mathbf{R}$ and actions on P_1. The automorphisms on $L \otimes_{\mathbf{Z}} \mathbf{R} \oplus +i\mathcal{C}$ do not act in the "same way" on P_1. Namely, we consider translations by elements in L^*. Let us find how translation by the element $u \in L^*$ acts on P_1. For any element $Z \in L \otimes_{\mathbf{Z}} \mathbf{R} \oplus +i\mathcal{C}$, it maps

$$Z \mapsto Z + u,$$

and hence it take the corresponding element in P_1

$$Z_M \mapsto (Z + u)_M = (Z + u, 1, -(Z + u)^2/2).$$

So the translation map by u corresponds to an automorphism of M whose extended action on $M \otimes_{\mathbf{Z}} \mathbf{C}$ takes Z_M to $(Z + u)_M$.

Definition 5.2.12. *For any $u \in L^*$, define the automorphism t_u on M as follows:*

$$t_u(v) = v + (v, z)u - ((u, v) + (v, z)(u, u)/2)z.$$

Lemma 5.2.13. $t_u(Z_M) = (Z + u)_M$ *for any $Z \in K \otimes_{\mathbf{Z}} \mathbf{R} \oplus +i\mathcal{C}$.*

Proof.

$$t_u(Z_M) = (Z, 1, -Z^2/2) + (u, 0, 0) - (0, 0, (Z, u) + u^2/2) = (Z+u, 1, -(Z+u)^2/2).$$

$\square$

Exercises 5.2

1. Check that the denominator formula for the Fake Monster Lie superalgebra F (see Example 2.6.40) is given by au automorphic form on the Grassmannian $G(M)$, where M is a unimodular even lattice of rank $(26, 2)$. Determine the group of isometries fixing this automorphic form.

2. The simple roots of the Monster Lie algebra were given in Example 2.3.11.1. Show that its roots are the elements $(m, n) \in II_{1,1}$ with multiplicity $c(mn)$, where $J(q) = \sum_n c(n)q(n)$ is the normalized modular.

Hint: Consider the logarithm of the product side of the denominator formula (see Example 2.6.39) and use Hecke operators $T(m)$, $m \in \mathbf{Z}$, where

$$T(m)f(\tau) = m^{k-1} \sum_{\substack{ad=n \\ a \geq 1 \\ 0 \leq b < d}} d^{-k} f(\frac{a\tau + b}{d}),$$

f being an (ordinary) modular function of weight k. For a solution, see [Borc7, Lemma 7.1 and Theorem 7.2].

5.3   Vector Valued Modular forms and LBKM Algebras

We reconsider the examples of the Monster Lie algebra within the framework of modular and automorphic forms.

Example 5.3.1. The Monster Lie algebra M.

The denominator formula gives a product expansion for the function $j(\tau) - j(\sigma)$ on the product space $\mathcal{H} \times \mathcal{H}$, where $\mathcal{H} = \{x + iy : x, y \in \mathbf{C}, y > 0\}$ is the upper half plane, with the following properties.

1. The exponents of the product expansion of the function $j(\tau) - j(\sigma)$ are given by the coefficients of the modular function j.

2.

    (i) The function $j(\tau) - j(\sigma)$ is holomorphic on $\mathcal{H} \times \mathcal{H}$ and meromorphic at the cusps.

    (ii) For any $g \in SL(2, \mathbf{Z}) \times SL(2, \mathbf{Z})$, $g(j(\tau) - j(\sigma)) = j(\tau) - j(\sigma)$

    (iii) For any $g \in \mathbf{Z}_2$, $g(j(\tau) - j(\sigma)) = -(j(\tau) - j(\sigma))$.

3. The hyperbolic space $\mathcal{H} \times \mathcal{H}$ is a Hermitian symmetric space (see Lemma B.2.23) isomorphic to G/K, where $G = (SL(2, \mathbf{R}) \times SL(2, \mathbf{R})).\mathbf{Z}_2$ is a real Lie group and the stabilizer K of the point $(i, i) \in \mathcal{H} \times \mathcal{H}$ is isomorphic to $(SO(2, \mathbf{R}) \times SO(2, \mathbf{R})).\mathbf{Z}_2$, where $SO(2, \mathbf{R}) = \{g \in SL_2(\mathbf{R}) : g^t g = 1\}$.

4. The root lattice of the Monster Lie algebra is the even unimodular lattice $L = II_{1,1}$ of rank 2 and the hyperbolic space $\mathcal{H} \times \mathcal{H}$ is isomorphic to the Grassmannian $G(M)$ of all negative definite 2-dimensional vector subspaces of the real vector space $M \otimes_{\mathbf{Z}} \mathbf{R}$, where $M = L \oplus II_{1,1}$. This can be seen in the following manner. Consider the basis u_1, u_2 of L with respect to which the bilinear form is given by the matrix $\begin{pmatrix} 0 & -1 \\ -1 & 0 \end{pmatrix}$. Therefore a vector $u = y_1 u_1 + y_2 u_2$ has negative norm if and only if $y_1 > 0$ and $y_2 > 0$ or $y_1 < 0$ and $y_2 < 0$. So, there are two cones of negative norm vectors in L. We choose any one of them. Set $\mathcal{C} = \{(y_1, y_2) : y_1 > 0, y_2\}$. Then,

$$L \otimes_{\mathbf{Z}} \mathbf{R} + i\mathcal{C} = \{((x_1, y_1), (x_2, y_2)) : y_1 > 0, y_2 > 0\}$$

and so the hyperbolic space $\mathcal{H} \times \mathcal{H}$ is isomorphic to $L \otimes \mathbf{R} + i\mathcal{C}$.

Set z, z' to be the basis the second factor $II_{1,1}$ of M satisfying $z^2 = 0 = z'^2$. The bilinear form on M will be written $(.,.)$.

The map

$$L \otimes_{\mathbf{Z}} \mathbf{R} + i\mathcal{C} \to P_1 = \{Z = X + iY \in M \otimes_{\mathbf{Z}} \mathbf{C} :$$
$$X^2 = Y^2 < 0, (X, Y) = 0, (Z, z') = 1\}$$
$$Z \mapsto Z - z - \frac{Z^2}{2} z'$$

is injective.

There is an obvious projection from the above set to the Grassmannian $G(M)$. We therefore get a map from $L \otimes \mathbf{R} + i\mathcal{C}$ to $G(L)$, which is also clearly injective. It is bijective since any 2-dimensional negative definite vector subspace of $M \otimes_{\mathbf{Z}} \mathbf{R}$ has a unique ordered orthogonal basis X, Y such that $(X+iY, z') = 1$ (see Lemma 3.2.11).

5. The group $(SL(2, \mathbf{Z}) \times SL(2, \mathbf{Z})).\mathbf{Z}_2$ is isomorphic to the discrete group $O(M)^+$. Indeed, the Lie group $O(M \otimes_{\mathbf{Z}} \mathbf{R})^+$ of isometries of the real vector space $M \otimes_{\mathbf{Z}} \mathbf{R}$ keeping a chosen orientation invariant (see Appendix A) acts as a transitive group of isometries of $G(M)$ (see Lemmas 3.2.7 and B.2.24). Moreover, the stabilizer of a point in the Grassmannian $G(M)$ is isomorphic to the above subgroup K since $SO(2, \mathbf{R})$ is the subgroup of isometries of a 2-dimensional Euclidean space (see Lemma 3.2.7).

In other words, one can associate to the Monster Lie algebra a modular function, namely j, whose Fourier coefficients are the exponents of a function, namely $j(\tau) - j(\mu)$, on the Grassmannian $G(M)$, satisfying the relations (i)-(iii) under the action of discrete group $O(M)^+$ – an automorphic form on $G(M)$. Remember from Definition 2.3.10 that j is a vector valued modular form.

As illustrated in this example, our aim is to characterize a BKM Lorentzian algebras by an automorphic form on the Grassmannian $G(M)$ having a product expansion whose exponents are given by the Fourier coefficients of a vector valued modular form. The main result of this section gives a class of vector valued modular forms to which automorphic forms on $G(M)$ can be associated in this manner.

In sections 5.3 and 5.4, we assume that the Lorentzian lattice L is even and contains the unimodular lattice $II_{1,1}$ of rank 2 as a direct summand as this holds for the root lattices of all known interesting cases of LBKM algebras.

As $b^+ = 2$, the modular function F, holomorphic on $\mathcal{H}$, and meromorphic at the cusps with Laurent series $\sum_{n \in \mathbf{Z}} c(n)q^n$ at infinity satisfying $c(n) \in \mathbf{Z}$ for $n \leq 0$, considered in section 3.4, has weight $1 - \frac{b^-}{2}$. We also found the singularities of the theta transform Φ_M of F. We start this section by taking our investigations further in the case that interests us: namely when $b^+ = 2$. The zeros and poles of the automorphic forms on $G(M)$ that can be associated with a BKM algebra are elements of sets called rational quadratic divisors.

Definition 5.3.2. A *rational quadratic divisor is a set of points* $Z \in L \otimes_{\mathbf{Z}} \mathbf{R} + i\mathcal{C}$ *satisfying* $aZ^2 + (b, Z) + c = 0$ *for some* $a, c \in \mathbf{Z}$, $b \in L$ *such that* $b^2 - 4ac > 0$.

Considering the image of $L \otimes_{\mathbf{Z}} \mathbf{R} + i\mathcal{C}$ in P, they can be described as follows.

Lemma 5.3.3. *The rational quadratic divisor are the sets of points in* P_1 *orthogonal to a given negative norm vector in* M. *Two rational quadratic divisors are equal if they correspond to negative norm vectors in* M *that are rational multiples of each other.*

Proof. Let $(Z, 1, -Z^2/2) \in P_1$, where $Z \in L \otimes_{\mathbf{Z}} \mathbf{R} + i\mathcal{C}$. Let $(b, -a, c) \in M$ be a negative norm vector. So

$$(b, -a, c)^2 = b^2 + 2ac > 0$$

and

$$((Z, 1, -Z^2/2), (b, -a, c)) = (b, Z) + c + aZ^2/2.$$

The result is now immediate.   □

Since the lattice L is Lorentzian, we show that the real vector space $L \otimes_{\mathbf{Z}} \mathbf{R}$ can be subdivided into connected regions on which the theta transform Φ_L is analytic. These regions generalize the notion of Weyl chambers for a reflection group. Applying Theorem 3.4.13 and Corollary 5.2.8 to the lattice L, we can deduce the following.

Lemma 5.3.4. *The singular points of $\Phi_L(w^+, F)$ in the hyperbolic space $G(L)$ of dimension b^- lie on hyperplanes.*

Therefore the domains on which the function $\Phi_L(w^+, F)$ is analytic are disconnected. Note that any 1-dimensional positive definite subspace $w^+ \in G(L)$ contains precisely two norm 1 vectors $\pm Y_1$ and only one of them is in the fixed positive cone $\mathcal{C}$ (see section 5.2).

Definition 5.3.5. *A Weyl chamber determined by the function $\Phi_L(w^+, F)$ is the positive cone generated by*

$$\{Y/|Y| : Y \in R \cap \mathcal{C}\},$$

where R is a maximal connected domain in the hyperbolic space $G(K)$ on which the function $\Phi_L(w^+, F)$ is analytic.

If for some $\lambda \in L \otimes_{\mathbf{Q}} \mathbf{R}$, $(\lambda, Y) > 0$ for all vectors Y in some Weyl chamber W, we will write $(\lambda, W) > 0$.

We are now ready to state the principal theorem of this section. It gives conditions enabling a vector valued modular form to be associated to an automorphic form on a Grassmannian in such a way that the coefficients of the Laurent series of the modular form are exponents of the product expansion of the automorphic form.

Theorem 5.3.6. *Let F be an almost holomorphic modular form on $\mathcal{H}$ of weight $1 - \frac{b^-}{2}$ and type ρ_M such that $F = \sum_{\gamma \in M^*/M} f_\gamma$, where for all $\gamma \in M^*/M$, f_γ has Fourier series $\sum_{n \in \mathbf{Q}} c_\gamma(n) q^n$ at infinity satisfying $c_\gamma(n) \in \mathbf{Z}$ for $n \leq 0$ and $c_\gamma(n) = 0$ for $n << 0$. Then, there is an automorphic form Ψ_M of weight $c(0)/2$ for the group $\mathrm{Aut}(M, F)$ with the following properties:*

(i) For each primitive vector $z \in M$ of norm 0 and Weyl chamber W of L, there is a uniquely defined vector $\rho = \rho(L, W, F) \in L \otimes_{\mathbf{Z}} \mathbf{R}$ such that for $Z \in L \otimes_{\mathbf{Z}} \mathbf{R} + i\mathcal{C}$,

$$\Psi_z(Z) = e^{-2\pi i(\rho, Z)} \prod_{\substack{\lambda \in L^* \\ (\lambda, W) > 0}} \left(1 - e^{2\pi i(\lambda, Z)}\right)^{c_\lambda(\lambda^2/2)}$$

and the product is convergent for $Y \in W$ and Z in the neighbourhood of the cusp of z.

(ii) The zeros and poles of Ψ_M lie on rational quadratic divisors $\lambda^\perp$, where $\lambda \in M$ and $\lambda^2 < 0$. On $\lambda^\perp$, where $\lambda \in M$ has negative norm, they have order

$$\sum_{\substack{n \in \mathbf{N} \\ n\lambda \in M^*}} c_{n\lambda}(n^2 \lambda^2/2).$$

If the order is positive (resp. negative), it corresponds to a zero (resp. pole). If this number is always positive, then the automorphic form is holomorphic.

Remark 5.3.7. Let e_i, $1 \leq i \leq n$, be an orthogonal basis for $L \otimes_{\mathbf{Z}} \mathbf{R}$ such that $e_i^2 = -1$ for $i \leq n-1$, $e_n^2 = 1$ and $e_n \in \mathcal{C}$. Then, for any $Y \in \mathcal{C}$, $Y = \sum_i a_i e_i$ such that $a_n > 0$. Remember that the positive cone $\mathcal{C}$ is determined by the vector z (see section 5.2). By the cusp of z we mean $Y \in \mathcal{C}$ such that $Y^2 >> 0$, i.e roughly speaking, Y is near "infinity". This generalizes the notion of the cusp at infinity in the upper half plane, namely the hyperbolic space of dimension 2, to the hyperbolic space $G(M) \cong (L \otimes_{\mathbf{Z}} \mathbf{R}) + i\mathcal{C}$ of dimension b^- (see Corollary 5.2.8).

We spend most of this section proving this Theorem. Basically we show that the Theta transform of the modular form F is closely related to the required automorphic form. More precisely, we know the singularities of the Theta transform and hence the idea is to consider its exponential.

The first step is therefore to compute the Theta transform $\Phi_M(v^+, F)$ for $v^+ \in G(M)$ such that Φ_M is analytic at v^+. To do this, we apply Theorem 3.4.14.

In the rest of this section, F will denote a modular form with the properties described in Theorem 5.3.6. Set v^+ to be a point in the Hermitian symmetric space $G(M)$ to be a non-singular point of the function Φ_M.

We split the calculation into several steps. We first evaluate the terms in the series given for $G(s, v^+)$ in Theorem 3.4.14 corresponding to $\lambda_{w^+} = 0$.

Lemma 5.3.8. *If $\lambda \in L$ such that $\lambda_{w^+} = 0$, then*

$$\int_{y>0} c_{\lambda, \lambda^2/2}(y) \exp(-\pi n^2 / 2 y z_{v+}^2 - \pi y (\lambda_{w+}^2 - \lambda_{w-}^2)) y^{-s-5/2} dy$$

$$= c_\lambda(\lambda^2/2)(\frac{\pi n^2}{2 z_{v+}^2})^{-s+b^+/2-\frac{3}{2}} \Gamma(s - b^+/2 + 3/2).$$

Proof. Let $\lambda \in L$ be such that $\lambda_{w+} = 0$. Then, $\lambda = \lambda_{w-}$. Since

$$\lambda_{w+}^2 - \lambda_{w-}^2 = 2\lambda_{w+}^2 - \lambda^2$$

for $\lambda \in L$ and $c_{\lambda, \lambda^2/2}(y) = c_\lambda(\lambda^2/2) y^{b^+/2} \exp(-2\pi y \lambda^2/2)$,

$$\int_{y>0} c_{\lambda, \lambda^2/2}(y) \exp(-\pi n^2 / 2 y z_{v+}^2 - \pi y (\lambda_{w+}^2 - \lambda_{w-}^2)) y^{-s-5/2} dy$$

$$= c_\lambda(\lambda^2/2) \int_0^\infty \exp(-\pi n^2 / 2 y z_{v+}^2) y^{-s+b^+/2-5/2} dy.$$

Set $t = \pi n^2 / 2 y z_{v+}^2$. Then,

$$dt = -\frac{\pi n^2}{2 y^2 z_{v+}^2} dy$$

and

$$y^{-s+\frac{b^+-1}{2}} = (\frac{\pi n^2}{2z_{v+}^2})^{-s+\frac{b^+-1}{2}} t^{s-\frac{b^+-1}{2}}.$$

So, the above integral becomes

$$c_\lambda(\lambda^2/2)(\frac{\pi n^2}{2z_{v+}^2})^{-s+\frac{b^+-3}{2}} \int_0^\infty e^{-t} t^{s-\frac{b^+-3}{2}-1} dt$$

The result now follows from Lemma C.4.1.   □

Since Φ_M is analytic at v^+, so is Φ_L at w^+. Set W to be the Weyl chamber containing a generator for the 1-dimensional subspace w^+.

Corollary 5.3.9. *Let $v^+ \in G(M)$ with oriented basis X_M and Y_M. Assume that the function $\Phi_M(v^+, F)$ does not have any singularity at v^+. Then, for Y^2 small enough,*

$$\Phi_M(v^+, F) = \frac{|Y|}{\sqrt{2}} \Phi_L(w^+, F)$$

$$+ 4 \sum_{\substack{\lambda \in L^* \\ (\lambda, W) > 0}} c_\lambda(\lambda^2/2) \log|1 - \mathbf{e}((\lambda, X + iY))|$$

$$+ c_0(0)(-\log(Y^2) - \log 2\pi - \Gamma'(1)).$$

Proof. Note that $\lambda_{w+}^2 - \lambda_{w-}^2 = 2\lambda_{w+}^2 - \lambda^2$ for $\lambda \in L$ and the lattice M having signature $(2, b^-)$, $c_{\lambda, \lambda^2/2}(y) = c_\lambda(\lambda^2/2)y \exp(-2\pi y \lambda^2/2)$. Hence, Theorem 3.4.14 becomes

$$G_M(s; v^+) = \frac{1}{\sqrt{2}|z_{v+}|} G_L(s; w^+) + \frac{\sqrt{2}}{|z_{v+}|} \sum_{\lambda \in L^*} \sum_{n>0} \mathbf{e}((n\lambda, \mu)) c_\lambda(\lambda^2/2)$$

$$\times \int_{y>0} \exp(-\pi n^2/2yz_{v+}^2 - 2\pi y \lambda_{w+}^2) y^{-s-3/2} dy. \tag{1}$$

We first calculate the sum of terms corresponding to $\lambda \in L^*$ for which $\lambda_{w+} = 0$. If $\lambda \in L^*$ and $\lambda_{w+} = 0$ then as $v^+ \cap L \otimes_{\mathbf{Z}} \mathbf{R} = \mathbf{w}^+$, $\lambda_{v+} = 0$. Hence in this case $\lambda = 0$ by Theorem 3.4.13 since the theta transform does not have any singularity at v^+. Therefore applying Lemma 5.3.8 we find that

$$\frac{\sqrt{2}}{|z_{v+}|} \sum_{\substack{\lambda \in L^* \\ \lambda_{w+}=0}} \sum_{n>0} \mathbf{e}((n\lambda, \mu)) c_\lambda(\lambda^2/2) \int_{y>0} \exp(-\pi n^2/2yz_{v+}^2 - 2\pi y \lambda_{w+}^2) y^{-s-3/2} dy$$

$$= \frac{\sqrt{2}}{|z_{v+}|} c_0(0) \sum_{n>0} (\frac{\pi n^2}{2z_{v+}^2})^{-s-\frac{1}{2}} \Gamma(s+1/2)$$

$$= c_0(0)\pi^{-s-\frac{1}{2}}(2z_{v+}^2)^s \Gamma(s+1/2) \sum_{n>0} \frac{1}{n^{2s+1}}. \tag{2}$$

Let us compute the constant term of the Laurent series at $s = 0$ of the right hand side of this equality. By Corollary C.4.4, the function $\pi^{-s-\frac{1}{2}}(2z_{v+}^2)^s\Gamma(s+1/2)$ is analytic at $s = 0$ whereas the function $\sum_{n>0}\frac{1}{n^{2s+1}} = \zeta(2s+1)$ has a simple pole at $s = 0$ (by Corollary C.4.14). Therefore to find the desired constant term, we only need to calculate the first two terms of the Taylor series (see Definition C.2.6) for the functions $\pi^{-s-\frac{1}{2}}$, $(2z_{v+}^2)^s$, $\Gamma(s+1/2)$, and the first two terms of the Laurent series (see Corollary C.2.7) for the function $\zeta(2s+1)$. The sum of the first two terms of the Taylor series for $\pi^{-s-\frac{1}{2}}$ at $s = 0$ is:

$$\pi^{-\frac{1}{2}} - \log(\pi)\pi^{-\frac{1}{2}}s$$

since $\pi^{-s-\frac{1}{2}} = \exp^{(-s-\frac{1}{2})\log(\pi)}$ and so $-\log(\pi)\pi^{-s-\frac{1}{2}} = \frac{d}{ds}\pi^{-s-\frac{1}{2}}$. Similarly, the sum of the first two terms of the Taylor series for $(2z_{v+}^2)^s$ is

$$1 + \log(2z_{v+}^2)s$$

and that of $\Gamma(s+1/2)$ is

$$\Gamma(1/2) + \Gamma'(1/2)s. \tag{3}$$

Applying Corollary C.4.7 to the case $s = 1/2$ and using Corollary C.4.3 and Lemma C.4.5 (ii), the expression in (3) is equal to

$$\pi^{\frac{1}{2}} + \pi^{\frac{1}{2}}(\Gamma'(1) - 2\log 2)s.$$

Corollary C.4.14 shows that the sum of the first two terms of the Laurent series for $\zeta(2s+1)$ is

$$\frac{1}{2s} - \Gamma'(1).$$

As a result, the constant coefficient of the Laurent series at $s = 0$ of the right hand side of equality (2) is the constant coefficient of

$$c_0(0)(\pi^{-\frac{1}{2}} - \pi^{-\frac{1}{2}}\log(\pi)s)(1 + \log(2z_{v+}^2)s)$$
$$\times (\pi^{\frac{1}{2}} + \pi^{\frac{1}{2}}(\Gamma'(1) - 2\log 2)s)(\frac{1}{2}s^{-1} - \Gamma'(1))$$
$$= c_0(0)(1 - \log(\pi)s)(1 + \log(2z_{v+}^2)s)$$
$$\times (1 + (\Gamma'(1) - 2\log 2)s)(\frac{1}{2}s^{-1} - \Gamma'(1)).$$

So it is equal to

$$c_0(0)(-\Gamma'(1) + (\Gamma'(1) - 2\log 2)/2 + \log(2z_{v+}^2)/2 - \frac{1}{2}\log(\pi)) \tag{4}$$
$$= c_0(0)(\log(|z_{v+}|) - \log(\sqrt{2\pi}) - \Gamma'(1)/2).$$

We next calculate the constant term of the Laurent series at $s = 0$ of the sum

$$\frac{\sqrt{2}}{|z_{v+}|}\sum_{\substack{\lambda \in L^* \\ \lambda_{w+} \neq 0}}\sum_{n>0}e((n\lambda, \mu))c_\lambda(\lambda^2/2)\int_{y>0}\exp(-\pi n^2/2yz_{v+}^2 - 2\pi y\lambda_{w+}^2)y^{-s-3/2}dy.$$

In order to do this, we first compute the integral

$$\int_{y>0} \exp(-\pi n^2/2yz_{v+}^2 - 2\pi y\lambda_{w+}^2)y^{-s-3/2}dy. \tag{5}$$

It is equal to

$$\exp(-2\pi n|\lambda_{w+}|/|z_{v+}|)\int_0^\infty \exp(-\frac{\pi}{y}(\sqrt{2}y|\lambda_{w+}| - n/\sqrt{2}|z_{v+}|)^2)y^{-s-3/2}dy.$$

Set

$$t = -\frac{\pi}{y}(\sqrt{2}y|\lambda_{w+}| - \frac{n}{\sqrt{2}|z_{v+}|})^2.$$

Then,

$$dt = [-\frac{\pi}{y^2}(\sqrt{2}y|\lambda_{w+}| - \frac{n}{\sqrt{2}|z_{v+}|})^2 + \frac{\pi\sqrt{2}|\lambda_{w+}|}{y}(\sqrt{2}y|\lambda_{w+}| - \frac{n}{\sqrt{2}|z_{v+}|})]dy$$

$$= [-\pi(\sqrt{2}|\lambda_{w+}| - \frac{n}{\sqrt{2}|z_{v+}|y})]dy$$

$$= \pi(\sqrt{2}|\lambda_{w+}| - \frac{n}{\sqrt{2}|z_{v+}|y})\frac{n}{\sqrt{2}|z_{v+}|y}dy$$

$$= \pi^{1/2}t^{1/2}\frac{n}{\sqrt{2}|z_{v+}|}dy/y^{3/2}$$

since $\sqrt{2}|\lambda_{w+}| - \frac{n}{\sqrt{2}|z_{v+}|y} = \sqrt{\frac{t}{\pi y}}$. Hence the integral (5) is equal to

$$\exp(-2\pi n|\lambda_{w+}|/|z_{v+}|)\frac{\sqrt{2}|z_{v+}|}{\pi^{1/2}n}\int_0^\infty e^{-t}t^{-1/2}y^{-s}dt.$$

Since we want to find the constant term of its Laurent series at $s = 0$ and the previous expression is analytic at $s = 0$, the integral part of it being equal to $\Gamma(1/2) = \sqrt{\pi}$ (see Lemmas C.4.1 and C.4.5), at $s = 0$ it is equal to

$$\exp(-2\pi n|\lambda_{w+}|/|z_{v+}|)\frac{\sqrt{2}|z_{v+}|}{n} = \exp(-2\pi n|(\lambda, Y)|)\frac{\sqrt{2}|z_{v+}|}{n}. \tag{6}$$

This equality follows from Lemma 5.2.11 (iv) and (vi). Therefore, from (1), (4), (6) and Lemma 5.2.11 (iv),(vi) and (vii), we can deduce that

$$\Phi_M(v^+, F) = \frac{|Y|}{\sqrt{2}}\Phi_L(w^+, F)$$

$$- 2\sum_{\substack{\lambda\in L^* \\ \lambda_{w+}\neq 0}} c_\lambda(\lambda^2/2)\sum_{n>0} \mathbf{e}((n\lambda, X))\exp(-2\pi n|(\lambda, Y)|)/n$$

$$+ c_0(0)(-\log(Y^2) - \log 2\pi - \Gamma'(1)).$$

Since
$$|\mathbf{e}((n\lambda, X)) \exp(-2\pi n|(\lambda, Y)|)| = \exp(-2\pi n|(\lambda, Y)|) < 1,$$

by Lemma C.4.13, the series

$$\sum_{n>0} \mathbf{e}((n\lambda, X)) \exp(-2\pi n|(\lambda, Y)|)/n$$

converges and is equal to

$$-\log(1 - \exp(2\pi i(\lambda, X) - 2\pi|(\lambda, Y)|)) = -\log(1 - \mathbf{e}((\lambda, X) + i|(\lambda, Y)|)).$$

Moreover, Lemma 5.2.11 (vi) implies that $\lambda_{w^+} \neq 0$ if and only if $(\lambda, Y) \neq 0$. Hence, as

$$\log(1 - \mathbf{e}((\lambda, X) + i|(\lambda, Y)|)) + \log(1 - \mathbf{e}(-(\lambda, X) + i|(-\lambda, Y)|))$$
$$= \log((1 - \mathbf{e}((\lambda, X) + i|(\lambda, Y)|))(1 - \mathbf{e}(-(\lambda, X) + i|(\lambda, Y)|)))$$
$$= \log|1 - \mathbf{e}((\lambda, X) + i|(\lambda, Y)|)|,$$

$(\lambda, Y) = |(\lambda, Y)||$ if $(\lambda, Y) > 0$, and by Lemma 3.4.1, $c_\lambda(\lambda^2/2) = c_{-\lambda}(\lambda^2/2)$,

$$\Phi_M(v^+, F) = \frac{|Y|}{\sqrt{2}} \Phi_L(w^+, F) + 4 \sum_{\substack{\lambda \in L^* \\ (\lambda, Y) > 0}} c_\lambda(\lambda^2/2) \log|1 - \mathbf{e}((\lambda, X + iY))|$$

$$+ c_0(0)(-\log(Y^2) - \log 2\pi - \Gamma'(1)).$$

Finally, note that $(\lambda, Y) > 0$ if and only if $(\lambda, W) > 0$: As Φ_M is analytic at v^+, so is Φ_L at w^+. From Lemma 5.2.11 (v) we know that this latter subspace is generated by Y. By assumption, $Y \in \mathcal{C}$. Hence, $Y/|Y| \in W$ (see Definition 5.3.5 of the Weyl chamber W). Then, by Theorem 3.4.13 on the singularities of Φ_K and by definition of W, $(\lambda, Y) \neq 0$ if and only if $(\lambda, W) \neq 0$. Since by definition W is generated by vectors in $\mathcal{C}$, $(\lambda, Y') > 0$ for all $Y' \in W$. The result now follows. $\qquad\square$

To complete the calculation of the function $\Phi_M(v^+, F)$, we need to compute $\Phi_L(w^+, F)$. In fact we will show that each to each Weyl chamber corresponds a Weyl vector and that $\Phi_L(w^+, F)$ is given by the inner product of this vector with the norm 1 generator for w^+ in $\mathcal{C}$.

By definition, the Weyl chambers are precisely the regions on which the function $\Phi_L(w^+, F)$ is analytic and they are disconnected. So the expression we get for $\Phi_L(w^+, F)$ is only valid on a Weyl chamber and it may differ from one chamber to the next. Set $z_L \in K$ to be a primitive norm 0 vector such that

$$(z_L, Y) > 0.$$

By assumption of the lattice L, there is a negative definite lattice K such that

$$L = K \oplus K^\perp,$$

where $K^\perp \cong II_{1,1}$ and $z_L \in K^\perp$. Set $z'_L \in K^\perp$ to be such that $(z_L, z'_L) = 1$. We will write the element $\lambda + mz' + nz$, where $\lambda \in K$, $m, n \in \mathbf{R}$ as (λ, m, n).

Let $Y_1 = \dfrac{Y}{|Y|} = (Y_K, m, n)$, where $m, n \in \mathbf{R}$. Then, Y_1 is the unique norm 1 vector in $\mathcal{C}$ generating w^+. Let $\mu_L \in K \otimes_{\mathbf{Z}} \mathbf{R}$ be the vector μ for the lattice L as defined in Theorem 3.3.10.

Lemma 5.3.10. $(z_L)_{w^+} = mY_1$ and $\mu_L = Y_1/m$.

Proof.

$$(z_L)_{w^+} = (z_L, Y_1)Y_1 = mY_1$$

and by definition,

$$\mu_L = -z_1 + (z', (z_L)_{w^-})(z_L)_{w^+}/(z_L)^2_{w^+} + (z', (z_L)_{w^+})(z_L)_{w^-}/(z_L)^2_{w^-}$$

$$= -z_1 + \frac{1}{(z_L)^2_{w^+}}((z_L)_{w^+} - (z', (z_L)_{w^+})z)$$

$$= \frac{1}{m^2}((0, -m^2, 0) + (mY_L, m^2, mn) - (0, 0, mn))$$

$$= Y_L/m.$$

$\square$

The lattice K being negative definite, $G(K) = \emptyset$ and so,

$$\Theta_K(\tau; u^+) = \theta_K(\tau) = \sum_{\lambda \in K} e^{\pi i \bar{\tau} \lambda^2}$$

is the ordinary theta function (see Definition 3.2.15) of the lattice K. Hence,

$$\Phi_K(\tau, F) = \int_{SL_2(\mathbf{Z}) \backslash \mathcal{H}} F(\tau) \bar{\theta}_K(\tau) \, dx dy / y^2.$$

Corollary 5.3.11. *For the element $w^+ \in G(L)$ generated by the vector Y in the Weyl chamber W,*

$$\Phi_L(w^+, F) = \frac{1}{\sqrt{2}(z_L, Y_1)} \Phi_K(F) + 2\sqrt{2}\pi(z_L, Y_1) \sum_{\lambda \in K^*} c_\lambda(\lambda^2/2) B\left(\frac{(\lambda, Y_1)}{(z_L, Y_1)}\right),$$

where $B(x)$ is a polynomial of degree 2 in x and $B(x) = x^2 - |x| + \frac{1}{6}$ if $-1 \leq x \leq 1$.

Proof. Since $G(K) = \emptyset$ and by Lemma 5.3.10, $(z_L)^2_{w^+} = m^2 = (z_L, Y_1)^2$, we can deduce from Theorem 3.4.14 and Lemma 5.3.8 that $\Phi_L(v^+, F)$ is equal to the constant term at $s = 0$ of the Laurent expansion of

$$\frac{1}{\sqrt{2}|(Y_1, z_L)|} \Phi_K(F) - \frac{\sqrt{2}}{|(Y_1, z_L)|} \sum_{\lambda \in K^*} \sum_{n > 0} c_\lambda(\lambda^2/2) \mathbf{e}((n\lambda, \mu_L))$$

$$\times(\frac{\pi n^2}{2(z_L, Y_1)^2})^{-s-1}\Gamma(s+1).$$

Remember that by assumption, $|(Y, z_L)| = (Y, z_L)$.

Since $\Gamma(1) = 1$ by Corollary C.4.3, from Corollary D.4 we can deduce that the series $\sum_{n>0} \mathbf{e}((n\lambda, \mu_L))/n^2$ converges and is equal to a polynomial $B((\lambda, \mu))$ of degree 2 in (λ, μ_L) as given. By Lemma 5.3.10,

$$(\lambda, \mu_L) = \frac{(\lambda, Y_1)}{|(z_L)_{w+}|} = (\lambda, Y_1)/(z_L, Y_1).$$

The result follows.   □

Note that for $\lambda^2 < 0$, only finitely many $c_\lambda(\lambda^2/2) \neq 0$ (see Lemma 3.4.11). Hence the sum in the previous Lemma is finite.

We want to find for each Weyl chamber W a vector $\rho_W \in M \otimes_{\mathbf{Z}} \mathbf{R}$ satisfying

$$(\rho_W, Y_1) = \Phi_L(w^+, F),$$

where Y_1 is the unique vector of norm 1 in the positive cone $\mathcal{C}$ generating the vector space w^+. For any $Y \in W \cap w^+$, this implies that

$$(\rho_W, Y) = |Y|\Phi_L(w^+, F). \tag{III}$$

Hence, we want to define a vector ρ_W such that equality (III) holds for all $Y \in W$. As a function of Y, (ρ, Y) is linear in Y. Note that because of the isomorphism between norm 1 vectors in $\mathcal{C}$ and $G(L)$, we can consider Φ_L as a function on Y_1. Hence, the following conditions are sufficient for the existence of such a vector.

Lemma 5.3.12. *A vector ρ_W satisfying equality (III) exists for the Weyl chamber W if the function Φ_L is linear in Y_1.*

Lemma 5.3.13. *The function Φ_L is a function of Y_1 is linear.*

Proof. Since the polynomial B is of degree 2, from Corollary 5.2.10, we get

$$\Phi_L(w^+, F) = \frac{1}{\sqrt{2}(z_L, Y_1)}\Phi_K(F) + \sqrt{2}\pi \sum_{\lambda \in K^*} c_\lambda(\lambda^2/2)(a\frac{(\lambda, Y_1)^2}{(z_L, Y_1)}$$
$$+ b(\lambda, Y_1) + c(z_L, Y_1)), \tag{1}$$

for some $a, b, c \in \mathbf{R}$. Hence we have to show that some of the terms cancel out, in particular those having (z_L, Y_1) in the denominator.

For any vector $Y' \in \mathcal{C}$ generating w^+, $Y_1 = Y'/|Y'|$. Let $z_1 \neq z$ be another primitive vector of norm 0 such that $(z_1, Y) > 0$ for $Y \in \mathcal{C}$ (note that no norm 0 vector is orthogonal to a vector of positive norm in $L \otimes_{\mathbf{Z}} \mathbf{R}$). Let W_1 be another Weyl chamber. Let Φ and Φ_1 be the restrictions of Φ_K to W and W_1 respectively. By Lemma 5.3.14, $\Phi - \Phi_1$ is a linear function of Y_1 defined on the closure of $W \cup W_1$. Furthermore, the common poles of Φ and Φ_1 belong to

$\overline{W} \cap \overline{W}_1$. Hence they have positive norm. If $Y \in \overline{W}$ is a pole of Φ, then by Corollary 5.3.11, $(Y, z_L) = 0$. Hence, Y cannot have positive norm and so Φ and Φ_1 have no common poles. As a result, Φ is a polynomial in Y_1. Hence, the result follows from (1). □

Lemma 5.3.14. *Let W_i, $i = 1, 2$, be two Weyl chambers and Φ_i the restriction of Φ_L to W_i. Then, Φ_i can be continued to an analytic function on $\overline{W}_1 \cup \overline{W}_2$ and*

$$\Phi_1(w^+) - \Phi_2(w^+) = -8\sqrt{2}\pi \sum_{\substack{\lambda \in L^* \cap (\overline{W}_1 \cap \overline{W}_2)^\perp \\ (\lambda, W_1) > 0}} c_\lambda(\lambda^2/2) |(\lambda, Y_1)|,$$

where Y_1 is the unique vector of norm 1 in w^+.

Proof. Since $\Gamma(-1/2) = -2\sqrt{\pi}$ by Corollary C.4.4 and Lemma C.4.5 (ii), applying Theorem 3.4.13 to the case $b^+ = 1$, we can deduce that on $\overline{W}_1 \cap \overline{W}_2$, Φ_L has singularities of type

$$-2\sqrt{2}\pi \sum_{\substack{\lambda \in L^* \cap (\overline{W}_1 \cap \overline{W}_2)^\perp \\ \lambda \neq 0}} c_\lambda(\lambda^2/2) |(\lambda, Y_1)|$$

Hence Φ_i can be analytically continued on $\overline{W}_1 \cap \overline{W}_2$ and

$$\Phi_1(w^+) - \Phi_2(w^+) = -4\sqrt{2}\pi \sum_{\substack{\lambda \in L^* \cap (\overline{W}_1 \cap \overline{W}_2)^\perp \\ \lambda \neq 0}} c_\lambda(\lambda^2/2) |(\lambda, Y_1)|.$$

Indeed if a function $g(\tau) = f(\tau) - |\tau|$ is analytic, then it has analytic continuations $f(\tau) - \tau$ and $f(\tau) + \tau$. Hence the result follows since by Lemma 3.4.1, $c_\lambda(\lambda^2/2) = c_{-\lambda}(\lambda^2/2)$. □

The next definition thus makes sense.

Definition/Lemma 5.3.15. *The Weyl vector for the Weyl chamber W is the vector $\rho(L, W, F) \in L \otimes_{\mathbf{Z}} \mathbf{R}$ defined by*

$$8\sqrt{2}\pi(\rho(L, W, F), Y) = |Y| \Phi_L(w^+, F),$$

where $Y \in W$ is a vector generating $w^+ \in G(L)$.

The factor $8\sqrt{2}\pi$ is only a normalization factor due to technical considerations coming from Lemma 5.3.14. Using Definition 5.2.14, We can now rewrite the theta transform Φ_M for the lattice M given in Corollary 5.3.9 in terms of the Weyl vector.

Corollary 5.3.16. *For $v^+ \in G(M)$ such that the generators in $\mathcal{C}$ of w^+ are in the Weyl chamber W,*

$$\Phi_M(v^+, F) = 8\pi(\rho(L, W, F), Y) - 4 \sum_{\substack{\lambda \in L^* \\ (\lambda, W) > 0}} c_\lambda(\lambda^2/2) \log|1 - \mathbf{e}((\lambda, X + iY))|$$

$$+ c_0(0)(-\log(Y^2) - \log 2\pi - \Gamma'(1)).$$

Corollary 5.3.17. *The infinite product expansion of the function $\Psi_z(Z)$ given in Theorem 5.3.6 converges for $Y \in W$ and Z near the cusp of z. Moreover, for $v^+ \in G(M)$,*

$$\Phi_M(v^+, F) = -4\log|\Psi_z(Z, F)| - 4\frac{c_0(0)}{2}(\log|Y| + \Gamma'(1)/2 + \log\sqrt{2\pi}).$$

Proof. Taking the exponential of both sides of the equality in Corollary 5.3.16, we get

$$\exp(\Phi_M(v^+, F)) = \exp(8\pi(\rho(L, W, F), Y))(Y^2)^{-c_0(0)}(2\pi)^{-c_0(0)}$$
$$\times \exp(-c_0(0)\Gamma'(1)) \prod_{\substack{\lambda \in L^* \\ (\lambda, W) > 0}} |1 - e((\lambda, X + iY))|^{-4c_\lambda(\lambda^2/2)}. \tag{1}$$

Define

$$\Psi_z(Z, F) = e^{-2\pi i(\rho, Z)} \prod_{\substack{\lambda \in L \\ (\lambda, W) > 0}} (1 - e^{2\pi i(\lambda, Z)})^{c_\lambda(\lambda^2/2)}.$$

The convergence follows since the absolute value of this product, namely

$$e^{2\pi(\rho, Y)} \prod_{\substack{\lambda \in L \\ (\lambda, W) > 0}} |1 - e^{2\pi i(\lambda, Z)}|)^{c_\lambda(\lambda^2/2)},$$

is convergent from Theorem 3.4.14 and the definition of the Weyl chamber (see Definition 5.3.5). Substituting into (1), we can deduce that

$$\exp(\Phi_M(v^+, F)) = |\Psi_z(Z, F)|^{-4}(Y^2)^{-c_0(0)}(2\pi)^{-c_0(0)}\exp(-c_0(0)\Gamma'(1)).$$

The result now follows by the taking the logarithm of both sides of this equality and noticing that

$$c_0(0)(\log Y^2 + \log(2\pi) + \Gamma'(1)) = c_0(0)(2\log|Y| + 2\log(\sqrt{2\pi}) + 2(\Gamma'(1)/2)).$$

$\square$

We next show that the function Ψ_z is the restriction of an automorphic form on the Grassmannian $G(M)$.

Lemma 5.3.18. *Define the function Ψ_M on M by setting*

$$\Psi_M(Z_M, F) = \Psi_z(Z, F).$$

Then, Ψ_M is an automorphic form Ψ_M holomorphic on P invariant under the action of the group $\text{Aut}(M, F)$ of weight $c_0(0)/2$ and character χ.

Proof. As the space P is a principal fiber bundle over $G(M)$ together with the action of $\mathbf{C}^*$ and the Hermitian symmetric domain is isomorphic to the subset

P_1 of P and to the space $K \otimes_{\mathbf{Z}} \mathbf{R} + i\mathcal{C}$ (see section 5.2 and Definition B32), the function Ψ_z can be extended to a function Ψ_M on P defined by:

$$\Psi_M(Z_M, F) = \Psi_z(Z, F)$$

for $Z \in K \otimes_{\mathbf{Z}} \mathbf{R} + i\mathcal{C}$ and

$$\Psi_M(Z_M\tau, F) = \tau^{-c_0(0)/2}\Psi_M(\tau)$$

for all $\tau \in \mathbf{C}^*$. Since Ψ_z is holomorphic on $K \otimes_{\mathbf{Z}} \mathbf{R} + i\mathcal{C}$, so is Ψ_M on P.

To show that Ψ_M is an automorphic form (see Definition 5.3.5), it remains to check that for all $\sigma \in Aut(M, F)$ (see Definition 3.4.15), $\Psi_M(\sigma Z_M\tau) = \chi(\sigma)\Psi_M(Z_M\tau)$ for some 1-dimensional character χ of $Aut(M, F)$. Now, $\Phi_M(\sigma v^+, \sigma F) = \Phi_M(v^+, F)$ by definition of $Aut(M, F)$. Hence by Corollary 5.3.16, σ fixes the function Ψ_z. Consider the function

$$\Psi(Z_M) = \left|\Psi_M(Z_M)\right|\left|Y_M\right|^{c_0(0)/2}.$$

Then,

$$\Psi(Z_M\tau) = \left|\Psi_M(Z_M)\right|\left|\tau\right|^{-c_0(0)/2}\left|bX_M + aY_M\right|^{c_0(0)/2}$$

for $\tau = a + ib \in \mathbf{C}^*$. Now,

$$(bX_M + aY_M, bX_M + aY_M) = (b^2 + a^2)Y_M^2 = \tau^2 Y_M^2$$

since $(X_M, Y_M) = 0$ and $X_M^2 = Y_M^2$ (see section 5.2). Hence,

$$\Psi(Z_M\tau) = \Psi(Z_M)$$

for all $\tau \in \mathbf{C}^*$. Since $Aut(M, F)$ stabilizes Ψ_z, it stabilizes Ψ_M as a function on P_1 and hence it stabilizes Ψ_M as a function on P since the norm of the vector Y_M and that of the vector σY_M are equal for all $\sigma \in Aut(M, F)$.

Hence

$$\left|\Psi_M(Z_M)/\Psi_M(\sigma Z_M)\right| = \left|Y_M\right|^{c_0(0)/2}/\left|\sigma(Y_M)\right|^{c_0(0)/2} = 1$$

for all $\sigma \in Aut(M, F)$. So

$$\Psi_M(\sigma Z_M) = \chi(\sigma)\Psi_M(Z_M)$$

for some map $\chi : Aut(M, F) \to \mathbf{C}^*$ such that $\left|\chi(\sigma)\right| = 1$ for all $\sigma \in Aut(M, F)$. The definition of χ implies that it is a group homomorphism. Hence χ is a 1-dimensional character of $Aut(M, F)$. $\square$

To prove Theorem 5.3.6, it only remains to check that the orders of the zeros and singularities are as stated.

Lemma 5.3.19. *The singularities of the automorphic form Φ_M are on rational quadratics $\lambda^\perp$ for $\lambda \in M$ with $\lambda^2 < 0$ and are of order*

$$\sum_{\substack{n \in \mathbf{Z} \\ n\lambda \in M^*}} c_{n\lambda}(n^2\lambda^2/2).$$

Proof. The singularities of Ψ_M on P_1 are those of Ψ_z on $L\otimes_{\mathbf{Z}}\mathbf{R}+i\mathcal{C}$. Moreover, if $Z_M \in P_1$ is a singular point of Ψ_M, then so are $Z_M\tau$ for all $\tau \in \mathbf{C}^*$ by definition of the action of C^*. By Corollary 5.3.17 and Theorem 3.4.13, the singularities of $-4\log|\Psi_M(Z_M)|$ near the point $Z_{0M} = X_{0M} + iY_{0M}$, where Z_0 corresponds to the point $v_0^+ \in G(M)$, are of type

$$- \sum_{\substack{\lambda \in M^* \cap Z_{0M}^{\perp} \\ \lambda \neq 0}} c_\lambda(\lambda^2/2) \log(\lambda_{v^+}^2)$$

since $(u, Z_{0M}) = 0$ for $u \in M^*$ if and only if $(u, X_{0M}) = 0 = (u, Y_{0M})$ and this is equivalent to $u \in (v_0^+)^{\perp} = v_0^-$. Since X_M, Y_M form an orthogonal basis for v^+ such that $X_M^2 = Y_M^2$,

$$\lambda_{v^+} = (\lambda, X_M)X_M/X_M^2 + (\lambda, Y_M)Y_M/Y_M^2$$

and so

$$\log \lambda_{v^+}^2 = \log((\lambda, X_M)^2/X_M^2 + (\lambda, Y_M)^2/Y_M^2)$$
$$= \log(\frac{1}{Y_M^2}|(\lambda, Z_M)|^2)$$
$$= 2\log(|(\lambda, Z_M)|) - 2\log|Y_M|.$$

Hence the singularities of $\log(|\Psi_M(Z_M)|^2)$ near the point Z_{0M} are of type

$$(\log(|(\lambda, Z_M)|) - \log|Y_M|) \sum_{\substack{\lambda \in M^* \cap Z_{0M}^{\perp} \\ \lambda \neq 0}} c_\lambda(\lambda^2/2).$$

Equivalently,

$$\log(|\Psi_M(Z_M)|^2) - (\log(|(\lambda, Z_M)|) - \log|Y_M|) \sum_{\substack{\lambda \in M^* \cap Z_{0M}^{\perp} \\ \lambda \neq 0}} c_\lambda(\lambda^2/2) = \Upsilon(Z_M)$$

is analytic near the point Z_{0M}. Therefore, taking the exponential of both sides of this equality, we get

$$|\Psi_M(Z_M)|^2 = \exp(\Upsilon(Z_M)) \prod_{\substack{\lambda \in M^* \cap Z_{0M}^{\perp} \\ \lambda \neq 0}} \left|\frac{(\lambda, Z_M)}{Y_M}\right|^{c_\lambda(\lambda^2/2)}.$$

Hence the singularities lie on $\lambda^{\perp}$, where $\lambda \in M$ such that $\lambda^2 < 0$. By Lemma 5.3.3, they have order

$$\sum_{\substack{0 < x \in \mathbf{R} \\ x\lambda \in M^*}} c_{x\lambda}(x^2\lambda^2/2).$$

Note that this sum is finite by Lemma 3.4.11. Since by assumption, for all $\gamma \in M^*/M$ and $m \leq 0$, $c_\gamma(m) \neq 0$ only if $m \in \mathbf{Z}$, the result follows. $\square$

Remark 5.3.20. (i) In sections 5.2–5.4 the Lorentzian lattice L has been assumed to have rank $(1, n)$ in order to remain compatible with usual conventions about modular forms, in particular about the theta function. However, we saw in Chapter 2 that in the theory of BKM algebras, the non-diagonal entries of the generalized Cartan matrix A are assumed to be non-positive and as a consequence the roots of finite type have positive norm. So usually LBKM algebras have Lorentzian root lattices of rank $(n, 1)$ and so sections 5.2–5.4 are not compatible with this assumption, but with the opposite one, i.e. with taking non-negative non-diagonal entries in A and thus negative norm roots of finite type. The material in sections 5.2–5.4 can be adapted to the usual conventions about BKM algebras by taking as our definition of our theta function, the function

$$\theta_\gamma(\tau) = \sum_{\lambda \in M+\gamma} \mathbf{e}(-\tau \lambda_{v-}^2/2 - \overline{\tau} \lambda_{v+}^2/2)$$

and for the weight of our modular form F,

$$\frac{b^- - b^+}{2}$$

instead of $\frac{b^+ - b^-}{2}$. This also changes the representation ρ_M. The representation space $\mathbf{C}[M^*/M]$ remains the same but gives the complex conjugate of ρ_M. In Theorem 5.3.6 the product expansion then becomes

$$\Psi_z(Z) = e^{-2\pi i(\rho, Z)} \prod_{\substack{\lambda \in K \\ (\lambda, W) > 0}} (1 - e^{2\pi i(\lambda, Z)})^{c_\lambda(-\lambda^2/2)}$$

and the zeros and poles of Ψ_M lie on rational quadratic divisors $\lambda^\perp$, where $\lambda \in M$ and $\lambda^2 > 0$. On $\lambda^\perp$, where $\lambda \in M$ has positive norm, they have order

$$\sum_{\substack{n \in \mathbf{N} \\ n\lambda \in M^*}} c_{n\lambda}(-n^2\lambda^2/2).$$

(ii) When F is an ordinary complex valued modular function, the full isometry group $O(M)$ of the lattice M is equal to the group $Aut(F, M)$ fixing F.

Theorem 5.3.6 gives a class of modular forms whose coefficents are the exponents of the product expansion of automorphic forms. As these automorphic forms have a Fourier expansion, their product expansion is in many cases equivalent to the denominator formula of a BKM algebra. We have seen in Example 5.3.1 this to be the case for the normalized modular invariant J. This is however not always the case. A fundamental question yet unanswered is therefore the following one:

Open Question 1. Find necessary and sufficient conditions for the vector valued modular form F to be associated to a BKM algebra. In other words, the product expansion of the automorphic form associated to F in the manner described in Theorem 5.3.6 should be the denominator formula of a BKM algebra. Hence the exponents of this product, namely some of the coefficients of the

Fourier series for F at infinity should give the root multiplicities for this BKM algebra.

In order to be able to fully describe the BKM algebra we need to know not only the roots but also which among them are simple. Equivalently, to get the full denominator formula, we need to compute the Fourier series of the automorphic form Ψ_M. In general this is not easy to do. However it is worth noticing that the weight of the automorphic form related to the most typical LBKM algebra, namely the Fake Monster Lie algebra (see Example 5.3.33) is equal to minus half the signature of the lattice M (in the convention $b^+ = 2$). This is also true for the Monster Lie algebra.

Definition 5.3.21. *The automorphic form is said to be of singular weight if its weight is equal to $-s/2$, where s is the signature of the Lorentzian lattice L.*

Indeed if the automorphic form has singular weight then it is possible to find the Fourier coefficients of Ψ_M when $Aut(F, M) = O(M)$ and the character χ is trivial, i.e. when the automorphic form Ψ_M is fixed by the full isometry group of the lattice M. We assume this to be the case in the rest of this section.

We briefly describe this process without giving all the proofs. The details can be found in [Borc9, Chapter 3]. We assume that the automorphic form Ψ_M corresponds to a modular form F satisfying the conditions stated in Theorem 5.3.6.

The lattice U generated by the orthogonal primitive norm 0 vectors z and z_L has rank 2. By assumption $(z_L, Y) > 0$, and so the basis z, z_L of the lattice U is positively oriented since $z_{v^+} \wedge (z_L)_{v^+} = (z_L, Y) X_M \wedge Y_M / Y^4$ by Lemma 5.2.11 (ii) and (iii). Let J be the subgroup of the isometry group $O(M)$ of M fixing this sublattice.

Definition 5.3.22. *A subgroup fixing a rank 2 null sublattice is a Jacobi group.*

For any $r \in \mathbf{R}$, set

$$t_r(\lambda) = \lambda + r(\lambda, z)z_L - (\lambda, z_L)z$$

for $\lambda \in M \otimes_{\mathbf{Z}} \mathbf{C}$. In other words, t_r is the translation by the vector $rz_L \in L \otimes_{\mathbf{Z}} \mathbf{R}$ on the vector space $M \otimes_{\mathbf{Z}} \mathbf{C}$ (see Definition 5.2.12). By Exercise 5.3.4, the automorphism t_r does not depend on the positively oriented basis of primitive vectors of U.

Lemma 5.3.23. *For any $r \in \mathbf{R}$, $t_r(P) = P$ and $t_u(P) = P$ for all $u \in L \otimes_{\mathbf{Z}} \mathbf{R}$.*

Proof. Since $t_r(Z_M) = Z_M + rz_L - r(Z_M, z_L)z$ and $r \in \mathbf{R}$,

$$t_r(Z_M) = X_M + rz_L - r(X_M, z_L)z + i(Y_M - r(Y_M, z_L)z)$$

and

$$(X_M + rz_L - r(X_M, z_L)z)^2 = X_M^2 = Y_M^2 = (Y_M - r(Y_M, z_L)z)^2.$$

Moreover,

$$(X_M + rz_L - r(X_M, z_L)z) \wedge (Y_M - r(Y_M, z_L)z) = X_M \wedge Y_M + \cdots,$$

where none of the other summands are in v^+. Hence the basis

$$X_M + rz_L - r(X_M, z_L)z, Y_M - r(Y_M, z_L)z$$

is positively oriented. Therefore $t_r(Z_M) \in P$. The second equality follows from similar arguments. $\qquad\square$

When n is an integer, $t_n \in O(M)$, and hence since an automorphic form is homogeneous, by assumption

$$\Psi_M(t_n v) = \Psi_M(v) \quad v \in P. \tag{a}$$

Hence, the next definition makes sense. For $m \in \mathbf{Z}$, set

$$\psi_m(v) = \int_{r \in \mathbf{R}/\mathbf{Z}} \Psi_M(t_r(v))e^{-2\pi i m r} dr \quad v \in P.$$

The automorphic form Ψ_M can be expanded as a Jacobi series.

Lemma/Definition 5.3.24. *On P, the automorphic form Ψ_M is equal to its Jacobi series:*
$$\Psi_M = \sum_{m \in \mathbf{Z}} \psi_m.$$

Proof. Considering the function $g(r) = \Phi_M(t_r v)$ on $\mathbf{R}$ for fixed $v \in P$, the Fourier expansion of g is $g(r) = \sum_{m \in \mathbf{Z}} \psi_m(v)e^{2\pi i m r}$ (see Lemma D.2). The result follows from equality (a) by taking $r \in \mathbf{Z}$. $\qquad\square$

Again by assumption and Lemma 5.3.23, the next integral is well defined. For $m \in L^*$, set

$$A_\lambda(v) = \int_{u \in L \otimes_{\mathbf{Z}} \mathbf{R}/L} \Psi_M(t_u(v))e^{-2\pi i(\lambda, u)} du \quad v \in P.$$

Lemma/Definition 5.3.25. *On P, the automorphic form Ψ_M is equal to its Fourier series:*
$$\Psi_M = \sum_{\lambda \in L^*} A_\lambda.$$

The same holds form the functions ψ_m, $m \in \mathbf{Z}$.

Set F to be the subgroup of the isometry group $O(M)$ of the lattice M fixing the vector z.

Definition 5.3.26.

(i) *A subgroup of $O(M)$ fixing a primitive null vector is a Fourier group.*

(ii) *The automorphic form Ψ_M or the function ψ_m is said to be holomorphic at $\mathcal{F}$ if $A_\lambda = 0$ unless λ is in the closure of the positive cone $\mathcal{C}$.*

From Exercise 5.3.5, we know that the functions ψ_m satisfy the following properties.

Lemma 5.3.27. *For all $m \in \mathbf{Z}$,*

(i) ψ_m has weight $c(0)/2$ (i.e. is homogeneous of degree $-c(0)/2$);

(ii) ψ_m is holomorphic on P;

(iii) $\psi_m(t_r v) = e^{2\pi i m r} \phi(v)$, $v \in P$;

(iv) $\psi_m(\sigma v) = \psi_m(v)$ for all $\sigma \in J^+$, $v \in P$.

(v) ψ_m is holomorphic at the cusps, i.e. at every Fourier group corresponding to a rank 1 null sublattice of U.

This result in fact says that the functions ψ_m are Jacobi forms (see Exercise 5.3.5). So let us give the generalization of Jacobi forms (see Definition 3.2.17) to higher dimensions.

Definition 5.3.28. *A Jacobi form of weight k and index m $(k \in \mathbf{Z}_+, m \in \mathbf{Z})$ for the modular group $SL_2(\mathbf{Z})$ is a holomorphic function $\phi : K \otimes_{\mathbf{Z}} \mathbf{C} \times \mathcal{H} \to \mathbf{C}$ satisfying*

(i)

$$\phi(\frac{Z}{c\tau + \delta}, \frac{a\tau + b}{c\tau + d}) = (c\tau + d)^k \mathbf{e}^m(\frac{c(Z^2/2)}{c\tau + d})\phi(Z, \tau),$$

for all $\begin{pmatrix} a & b \\ c & d \end{pmatrix} \in SL_2(\mathbf{Z})$;

(ii)

$$\psi(Z + \lambda\tau + \mu, \tau) = \mathbf{e}^{-m}((Z, \lambda) + \tau\lambda^2/2)\phi(Z, \tau),$$

for all $\lambda, \mu \in \mathbf{Z}$.

Corollary 5.3.29. *For every $m \in \mathbf{Z}$, the functions ψ_m are Jacobi forms.*

We next define a theta function on two variables as follows.

Definition 5.3.30. *A Theta function of index m is a linear combination of functions on $K \otimes_{\mathbf{Z}} \mathbf{C} \times \mathcal{H}$ of the following type:*

$$\theta_{K+\gamma}(Z, \tau) = \sum_{\lambda \in K+\gamma} e^{-\pi i \lambda^2 \tau} e^{2\pi i m(Z, \lambda)}$$

for $m \in \mathbf{Z}$, $\gamma \in K \otimes_{\mathbf{Z}} \mathbf{Q}$.

We can now deduce the result on the Fourier coefficients of the automorphic form Ψ_M in the holomorphic case from Exercises 5.3.5 and 5.3.6.

Theorem 5.3.31. *Suppose that for all $m \leq 0$, the coefficients $c(m) \geq 0$ and that $c(0) = -s$. Then, the only non-zero coefficients of the Fourier series of Ψ_z correspond to vectors of norm 0.*

Hence the following questions arise:

Open Question 2. Are the BKM algebras that can be fully described, i.e. whose root system can be explicitly given, those associated to an automorphic form of singular weight?

Open Question 3. Classify all holomorphic automorphic forms on $G(M)$ of singular weight that are automorphic products. If the answer to Question 2 is in the affirmative, then this should give all the "interesting" BKM algebras by taking their Fourier expansions at the cusps.

Note that, as in the case of the Monster Lie algebra, these BKM algebras may not have a hyperbolic Weyl (reflection) group but a fairly trivial one.

Open Question 4. Find a method for calculating the Fourier coefficients of the automorphic forms having non-singular weight. This would, in particular, allow a description of the BKM algebras with hyperbolic reflection group. If this is possible, then the answer to Question 2 is negative.

Let us now investigate the Weyl vector ρ given in Lemma 5.3.15 further.

Theorem 5.3.32. *The Weyl vector for the Weyl chamber W is*

$$\rho(K, W, F) = (-\frac{1}{2} \sum_{\substack{l \in K^* \\ (\lambda, W) > 0}} c_\lambda(\lambda^2/2)\lambda, c, \frac{1}{24} \sum_{\lambda \in K^*} c_\lambda(\lambda^2/2)),$$

where c is the constant term of $F_L(\tau)\overline{\Theta}_L(\tau)E_2(\tau)$.

Proof. We apply Corollary 5.3.11. We want a vector Y_1' of norm 1 in the Weyl chamber W of Y_1 such that

$$0 \le \frac{(\lambda, Y_1')}{(z, Y_1')} < 1 \tag{1}$$

so that $B(\frac{(\lambda, Y_1')}{(z, Y_1')}) = (\frac{(\lambda, Y_1')}{(z, Y_1')})^2 - \frac{(\lambda, Y_1')}{(z, Y_1')} + 1/6$. We want (1) to hold for $\lambda \in K^*$, i.e with non-positive norm, satisfying

$$c_\lambda(\lambda^2/2) \ne 0 \quad \text{and} \quad (\lambda, W) \ge 0 \tag{2}$$

By Lemma 3.4.11, there are only finitely many possibilities for vectors $\lambda \in K^*$ with this property. Now, by lemma 5.3.10, $(\lambda, \mu_L) = \frac{(\lambda, Y_1')}{(z, Y_1')}$, where $Y_1 = (m\mu_L, m, n)$. So we can choose the vector $\mu_L \in K \otimes_{\mathbf{Z}} \mathbf{R}$ so that inequality (1) holds for the all vectors $\lambda \in K^*$ satisfying (2). Note that since $m \ne 0$ and the singular points are on hyperplanes of codimension 1 in $G(L)$ given by $\lambda \in L$ satisfying (2), the corresponding vector Y_1' is in W. For $\rho(K, W, F) = (\rho, c, d)$ and Y_1 given by (2), the definition of the Weyl vector (see Lemma 5.3.15) says that

$$8\sqrt{2}\pi((\rho, c, d), (m\mu_L, m, n)) = \Phi_L((m\mu_L, m, 1/2m - m\mu_L^2/2), F).$$

Hence, applying Corollary 5.3.11, we get

$$8\sqrt{2}\pi(m(\rho,\mu_L) + c(\frac{1}{2m} - m\mu_L^2/2) + dm)$$

$$= \frac{1}{\sqrt{2m}}\Phi_K(F) + \sqrt{2}\pi m(2c_0(0)/6 + 4\sum_{\substack{\lambda\in K^* \\ (\lambda,W)>0}} c_\lambda(\lambda^2/2)((\lambda,\mu_L)^2 - (\lambda,\mu_L) + 1/6))$$

since the norm of Y_1' being equal to 1, $n = (1 - m\mu_L^2)/2m$.

As this holds for all $m > 0$,

$$c = \frac{1}{8\pi}\Phi_L(F)$$

and

$$(\rho,\mu_L) - c\mu_L^2/2 + d = c_0(0)/24 + \frac{1}{2}\sum_{\substack{\lambda\in K^* \\ (\lambda,W)>0}} c_\lambda(\lambda^2/2)((\lambda,\mu_L)^2 - (\lambda,\mu_L) + 1/6).$$

This in turn holds for all vectors $\mu_L \in K \otimes_{\mathbf{Z}} \mathbf{R}$ such that $0 \leq (\lambda,\mu_L) \leq 1$ for all $\lambda \in K^*$ satisfying conditions (2). Hence,

$$\rho = -\frac{1}{2}\sum_{\substack{\lambda\in K^* \\ (\lambda,W)>0}} c_\lambda(\lambda^2/2)\lambda,$$

and

$$d = c_0(0)/24 + \frac{1}{12}\sum_{\substack{\lambda\in K^* \\ (\lambda,W)>0}} c_\lambda(\lambda^2/2) = \frac{1}{24}\sum_{\lambda\in K^*} c_\lambda(\lambda^2/2).$$

Hence the result follows from Exercise 5.3.8. $\qquad\qquad\square$

We have already studied the example of the Monster Lie algebra in the context of automorphic forms (see Example 5.3.1). We now apply Theorem 5.3.6 to the typical example of a LBKM algebra, namely the Fake monster Lie algebra. We have already computed its denominator formula using properties of its root system. We give an alternative method for finding this product expansion by using Theorem 5.3.6.

Example 5.3.33. For our modular form we take $F(\tau) = 1/\Delta(\tau)$. This is an ordinary modular function of weight -12. Its Laurent series at infinity is

$$f(\tau) = \sum_n p_{24}(n+1)q^n = q^{-1} + 24 + \cdots.$$

Therefore, $\Phi_z(Z) = e^{-2\pi i(\rho,Z)}\prod_{\substack{\lambda\in K \\ (\lambda,W)>0}}(1 - e^{2\pi i(\lambda,Z)})^{p_{24}(1-\lambda^2/2)}$ is an automorphic form of weight 12 for the group $O_{2,26} = O(M)$, where $M = K + II_{1,1}$ and K is a unimodular even Lorentzian lattice of rank $(25,1)$. Applying Theorem 5.3.32, we get that the Weyl vector is $\rho = (0,0,1)$. Theorem 5.3.5 also implies that Φ_z has singular weight and since $c(-1) > 0$ it is holomorphic. So by Theorem 5.3.31, the only non-trivial coefficients of the Fourier expansion of

Φ_z correspond to norm 0 vectors. They can then be calculated from the above product and we find

$$\Phi_z(Z) = \sum_{w \in W} \det(w)\Delta((Z, w(\rho))).$$

In this section, we have studied the question of associating to a modular form the product expansion of an automorphic form corresponding to the denominator formula of a LBKM algebra.

What about the converse problem? Do all LBKM algebras have a vector valued modular form associated to them in the manner described in Theorem 5.3.6? The answer is no for most of them. The next section gives very strong evidence that this can be done only when the lattice has rank at most 26. So the problem of classifying the LBKM algebras for which this is possible should be a finite one. Note that even when the rank is at most 26, most LBKM algebras have no associated vector valued modular form. The reason is that in general the denominator formula does not give a product expansion for an automorphic form. When this is the case, the answer to the above question is yes in the case considered here of a Lorentzian root lattice L containing the lattice $II_{1,1}$ as a direct summand. This has been proved by Bruinier and later by Barnard in a different context. The interested reader should read [Br, Theorems 5.11 and 5.12] and [Bar]. Their method is briefly exposed in the next section, though we do not give detailed proofs.

Exercises 5.3

1. Prove Theorem 5.3.6 in the case of an arbitrary even Lorentzian lattice L.

2. Find the multiplicities of the roots of Monster Lie algebra by applying Theorem 5.3.6.

3. Show that the automorphism t_r ($r \in \mathbf{R}$) defined in section 5.3 is independent of the positively oriented basis chosen for the sublattice U.

4. (i) Construct a homomorphism from the Jacobi group $\mathcal{J}$ to the group $O(K) \times SL_2(\mathbf{Z})$. Show that the centre of the connected component of its kernel group consists of the automorphisms t_r, $r \in \mathbf{R}$.

(ii) Construct a homomorphism from the Fourier group $\mathcal{F}$ to the group $O(L)$ containing the automorphisms t_v, $v \in L$, in its kernel.

(iii) Show that $\psi_m(t_r v) = e^{2\pi i m r} \psi_m(v)$ for all $v \in P$, $r \in \mathbf{R}$, $m \in \mathbf{Z}$.

5. For $r \in \mathbf{C}$, does $t_r(P) = P$ hold? Give an analytic continuation of the function ψ_m on the set of all vectors v in $M \otimes_{\mathbf{Z}} \mathbf{C}$ of norm 0 satisfying $\mathcal{I}(\frac{(v,z)}{(v,z_L)}) > 0$. Show that the function

$$\phi_{z,m}(z,\tau) = \phi_m(Z + \tau(z_L)' + z' - (Z^2/2)z) \qquad (a)$$

defined on $(K \otimes C) \times \mathcal{H}$ is a Jacobi form (according to Definition 5.3.28). Conversely, show that given a Jacobi form on $(K \otimes_{\mathbf{Z}} \mathbf{C}) \times \mathcal{H}$, there is a function on P satisfying Lemma 5.3.27 satisfying the above equality (a). Remember that $z' \in M$ (resp. $(z_L)' \in L$) is the primitive norm 0 vector orthogonal to the lattice L (resp. K) having inner product 1 with z (resp. z_L).

6. (i) Show that a Theta function $\theta_{K+\gamma}$ of index $m \in \mathbf{N}$ on $K \otimes_{\mathbf{Z}} \mathbf{C} \times \mathcal{H}$ is a Jacobi form of weight $-s/2$ and index m if $\gamma \in K^*$.

  (ii) Show that the Jacobi forms ψ_m in the Jacobi series for the automorphic form Ψ_M have positive index.

 (iii) Show that if $m > 0$, then the Jacobi form $\psi_{z,m}$ has a Fourier series of the form $\sum_{\lambda \in K+\gamma} e^{-\pi i \lambda^2 \tau} e^{2\pi i m (Z, \lambda)}$ for $m \in \mathbf{Q}$, $\gamma \in K \otimes_{\mathbf{Z}} \mathbf{Q}$.

 (iv) Deduce that when the coefficients $c(n) \geq 0$ for $n \leq 0$, the automorphic form Ψ_M is a linear combination of products of modular functions and Jacobi forms. If moreover Ψ_M has weight $-s/2$, where s is the signature of the lattice L, deduce that it is a sum of Jacobi forms.

For a solution, see [EicZ, Theorem 5.1].

  (v) Deduce that if the Fourier coefficient A_λ of a holomorphic automorphic form Ψ_M of weight $-s/2$ is non-zero, then $\lambda^2 = 0$.

5.4 An Upper Bound for the Rank of the Root Lattices of LBKM Algebras?

In this section, we present some strong evidence that the classification problem of LBKM algebras associated to vector valued modular forms is essentially a finite problem. At least it is so if we assume a Conjecture of Burger, Li and Sarnak about the eigenvalues of the hyperbolic Laplacian to be true.

We do not prove all the material presented in this section but aim to present some of the basic ideas developed in [Bar] and [Br]. The details can be found in these two papers. The assumptions and notation in this section are the same as in the previous one.

Our aim is to show that if F is a vector valued modular form satisfying the conditions of Theorem 5.3.6 and the corresponding automorphic form Ψ_M has a product expansion giving the denominator formula for a LBKM algebra, then the rank of the root lattice L is bounded above.

The particularity of a root lattice L is the existence of a Weyl vector in $L \otimes_{\mathbf{Z}} \mathbf{Q}$ for a Weyl chamber, i.e. they are associated to a reflection group. Therefore the Weyl chambers determined by the theta transform of F should be the Weyl chambers for reflection (Weyl) groups and the vector $\rho(L, W, F)$ given in Lemma 5.3.15 should be a Weyl vector for the root system of the reflection group. We remind the reader that the Weyl chambers for the reflection group $\mathcal{G}$ of a BKM algebra are the open cones $w\mathcal{C}$, where $w \in \mathcal{G}$ and

$$W = \{(\lambda, \alpha) > 0; \alpha \in \Pi\},$$

where Π is a fixed base for the root system. The singularity points are in the hyperplanes

$$H_{w\alpha} = \{\lambda \in L \otimes_{\mathbf{Z}} \mathbf{R} : (w\alpha, \lambda) = 0\},$$

$\alpha \in \Pi$, $w \in \mathcal{G}$. The Weyl vector $\rho(L, W, F)$ is a Weyl vector for the root system, i.e.

$$(\rho, \alpha) = \frac{1}{2}\alpha^2 \quad \forall \alpha \in \Pi. \tag{I}$$

As we will see this is a very strong condition to impose. It leads (assuming the above mentioned conjecture to be true) to a very low upper on the possible rank of the root lattice.

Remark 5.4.1. From Theorem 5.3.32 we know that the Weyl vector $\rho(K, W, F)$ belongs to the rational vector space $L \otimes_{\mathbf{Z}} \mathbf{Q}$ since by assumption the Fourier coefficients $c_\lambda(m)$ of the components of the modular form F are integers for all $m \leq 0$.

We assume that the modular form F corresponds to a BKM algebra and thus that H_α and $\mathcal{G}$ are as above and that the vector $\rho(L, W, F)$ satisfies equalities (I). The only coefficients $c_\lambda(m)$ with $m < 0$ of the modular form F which appear in the theta correspondence are the ones given by $m = \lambda^2/2$ for some $\lambda \in K^*$ of negative norm. However there may well be other non trivial coefficients $c(m)$ in the Fourier series of the modular form F with $m < 0$ and $m \neq \lambda^2/2$ for any $\lambda \in K^*$. Hence as they do not play any role in the theory developed in section 5.3, it is unclear how to find conditions these coefficients should satisfy from the requirement that $\rho(L, W, F)$ be a Weyl vector for a reflection group.

To find the upper bound on the rank of the lattice L, we explain how under certain assumptions a vector valued modular form F_1 with the property that the only singularities it has are of the form q^m where $m = \lambda^2/2$ for some primitive vector $\lambda \in L \otimes_{\mathbf{Z}} \mathbf{Q}$ of negative norm, can be associated to the vector $\rho(K, W, F)$ in the manner described in Theorem 5.3.6. This gives a low bound on the terms q^m, $m < 0$ in the Fourier series for F_1. The upper bound for the rank is a consequence of this result. So we first construct a modular form F_1. However, a priori, it has terms containing Whittaker functions and so does not have a Fourier series as given in Lemma 3.3.7 and the desired singularities. The aim is to show that the unwanted terms are equal to zero so that the singularities of F_1 are only of the above form corresponding to primitive roots of L^*. In order to show this, we find its theta transform Φ_1 and show that it is the sum of a piecewise linear function ψ_1 on $G(L)$ and a real analytic function ξ_1 on $G(L)$, the latter coming from the undesired terms of F_1. So all that needs to be shown is that the real analytic part is equal to 0. This will follow from the fact that $\Phi_K - \Phi_1$ is an eigenvector for the hyperbolic Laplacian for then, using an as yet unproved conjecture of Burger, Li and Sarnak about the exceptional eigenvalues for the hyperbolic Laplacian [BuLS], [BuS], it can be shown that $\Phi_1 = \Phi_K$. In particular, Φ_1 is piecewise linear and $\xi_1 = 0$.

The next result is only a restatement of Lemmas 5.3.13 and 5.3.14.

Lemma 5.4.2. *The automorphic form Φ_L is a piecewise linear map on the Grassmannian $G(L)$, i.e. it is linear on each Weyl chamber.*

In order to construct the modular form F_1, we need to introduce the hyperbolic laplacian.

The ordinary Laplacian in Euclidean n-space is the operator defined in cartesian coordinates as follows:

$$\nabla^2 = \frac{\delta^2}{\delta x_1^2} + \cdots + \frac{\delta^2}{\delta x_n^2}.$$

The hyperbolic Laplacian generalizes this operator to the 2-dimensional hyperbolic space, namely the upper half plane $\mathcal{H}$.

Definition 5.4.3. *The hyperbolic Laplacian of weight (m^+, m^-) of functions on the upper half place is the operator defined as follows:*

$$\nabla^2_{m^+, m^-} = -y^2\left(\frac{\delta^2}{\delta x^2} + \frac{\delta^2}{\delta y^2}\right) + iy(m^+ - m^-)\frac{\delta}{\delta x} + i(m^+ + m^-)\frac{\delta}{\delta y}.$$

One of the most fundamental equation in mathematics and physics – namely the potential equation – aims at finding the eigenfunction of the ordinary Laplacian with eigenvalue 0. More generally, finding the eigenvectors and eigenvalues of ∇^2 is of basic importance and can be done by using Green's functions. What about the eigenvectors and eigenvalues of the hyperbolic Laplacian? In section 3.3 we defined vector valued modular forms.

Definition 5.4.4. *A vector valued modular form F of weight (m^+, m^-) is said to be of eigenvalue $\lambda \in \mathbf{C}$ if*

$$\nabla^2_{m^+, m^-}(F) = \lambda F.$$

Using the Fourier expansion of vector valued modular forms and the definition of Whittaker functions $M_{k,m}$ and $W_{k,m}$, the vector valued modular form F of eigenvalue λ can be computed. These functions are the solutions of the Whittaker differential equation:

$$W'' + \left(\frac{k}{z} - \frac{1}{4} + \frac{\frac{1}{4} - m^2}{z^2}\right)W = 0.$$

For details, see [AS] and [WhiW].

Set $m^- = b^- - 1$ to be the rank of the root lattice L. From Lemma 3.3.7, we know that the components F_γ, $\gamma \in M^*/M$, of the modular form F have Fourier expansion of type

$$\sum_{m \in \mathbf{Z}} c_{m,\gamma}(y) q^{m + \gamma^2/2}.$$

Exercise 5.4.1 gives the solutions for the coefficients $c_{m,\gamma}(y)$ if F is of eigenvalue $\lambda = s(s-1) - \frac{1+m^-}{2}(1 - \frac{1+m^-}{2})$. As this exercise also shows, the essential part of the eigenvalue is $s(s-1)$. This is why it is more useful to write λ in this form

Lemma 5.4.5. *If the vector valued modular form F of type ρ_L is of eigenvalue $\lambda = s(s-1) - \frac{1+m^-}{2}(1 - \frac{1+m^-}{2})$ then the coefficients $c_{m,\gamma}(y)$ are a linear combination of the functions:*

$$\mathcal{M}_{n,s} = y^{-(1+m^-)/2}M_{sgn(n)(1-m^-)/2,\,s-1/2}(4\pi|n|y)$$

and

$$\mathcal{W}_{n,s} = y^{-(1+m^-)/2}W_{sgn(n)(1-m^-)/2,\,s-1/2}(4\pi|n|y).$$

A vector valued modular form of weight $(1, m^-)$ and type ρ_L can then be constructed using the definition (see Definition 3.3.5) as follows:

Definition/Theorem 5.4.6. *For $\delta \in L^*/L$, set*

$$F(\tau, \delta) = \frac{1}{2}\sum_{A \in <T>\backslash Mp_2(\mathbf{Z})}(\mathbf{e}(x\delta^2/2)\mathcal{M}_{\delta^2/2,1}(y)e_\delta)|_A,$$

where for any $A \in Mp_2(\mathbf{Z})$,

$$(F|_A)(\tau) = \rho_K(A)^{-1}F(A\tau)(c\tau + d)^{-1}(c\tau + d)^{-m^-}.$$

This is the Mass-Poincaré series at $s = 1$. Then, $F(\tau, \delta)$ is a modular form of weight $(1, m^-)$, type ρ_L and eigenvalue $\frac{1+m^-}{2}(1 - \frac{1+m^-}{2})$ with singularity only at infinity of type

$$\mathbf{e}(\frac{\delta^2}{2}x)\mathcal{M}_{\frac{\delta^2}{2},1}(y)e_\delta + \mathbf{e}(\frac{\delta^2}{2}x)\mathcal{M}_{\frac{\delta^2}{2},1}(y)e_{-\delta}.$$

For $\delta \in L^*/L$ having negative norm, set

$$F_\delta(\tau) = y^{(1+m^-)/2}F(\tau, \delta).$$

Its Fourier series is:

$$F_\delta(\tau) = q^{\delta^2/2}e_\delta + q^{\delta^2/2}e_{-\delta} + \sum_{\substack{\gamma \in L^*/L \\ n \in \mathbf{Z}+\gamma^2/2 \\ n \geq 0}} a_{n,\gamma}q^n e_\gamma + \sum_{\substack{\gamma \in L^*/L \\ n \in \mathbf{Z}+\gamma^2/2 \\ n < 0}} a_{n,\gamma}e^{2\pi i n x}W(4\pi ny)e_\gamma.$$

The problem is that F_δ does not necessarily have the same type of Fourier expansion as the vector valued modular forms used in sections 3 and 5. Therefore to derive its theta transform Φ_δ, a modified version of section 3.4 is needed, namely the Rankin-Selberg method. Remember the description of the singularity sets in Theorem 3.4.13.

Definition 5.4.7. *For $\delta \in L^*/L$ having negative norm, $H(\delta) = \bigcup_{\substack{\lambda \in L^* \cap L - \delta \\ \lambda^2 = \delta^2}}\lambda^\perp$ is the δ-Heegner divisor.*

As in Theorem 3.4.13, the singularities of the theta transform Φ_δ can be worked out.

Lemma 5.4.8. *The vector $v_0^+ \in G(L)$ is a singularity point of Φ_δ if $v_0^+ \in H(\delta)$, in which case it has singularity of type*

$$-4\pi\sqrt{2} \sum_{\substack{\lambda \in L^* \cap v_0^- \\ \lambda \neq 0}} |\lambda_{v^+}|.$$

An immediate consequence of Theorem 3.4.13 applied to the theta transform Φ_L and of the assumption (I) is a description of the singularities in terms of Heegner divisors.

Corollary 5.4.9. *The singularity points of Φ_K belong to divisors of type $2H(\delta)$ where δ is a primitive root in L^*.*

Let $\mathcal{P}$ be the set of primitive roots in L^*/L. The last two results suggest that we consider

$$F_1 = 2 \sum_{\delta \in \mathcal{P}} c(\delta^2/2) F_\delta.$$

The next result is immediate from the above.

Corollary 5.4.10. *The function $\Phi_L - \Phi_1$ is real analytic on the Grassmannian $G(L)$.*

However we cannot immediately conclude that $\Phi_L = \Phi_1$ nor can we as yet deduce anything about the singularities of F_1 because the less simpler form of the Fourier series of F_1 compared to the modular forms of previous sections implies that Φ_1 is the sum of a piecewise linear function ψ_1 and a real analytic function ξ_1. What we need to show is that $\xi_1 = 0$. Since $\Phi_K - \psi_1$ is piecewise linear, the next result follows from Corollary 5.4.10.

Corollary 5.4.11. *The function $\Phi_L - \psi_1$ is linear on the Grassmannian $G(L)$.*

Any linear function can be shown (see Exercise 5.4.4) to be an eigenvector for the hyperbolic Laplacian with eigenvalue $-m^-$ and from Exercise 5.4.3, we know that ψ_1 is an eigenvector for the hyperbolic Laplacian with eigenvalue $-m^-$. Hence:

Corollary 5.4.12. *The function $\Phi_L - \Phi_1$ is an eigenvector for the hyperbolic Laplacian with eigenvalue $-m^-$.*

Conjecture (Berger, Li, Sarnak). *Set $\omega = b^-1/2 - 1$ and let $s^2 - \omega^2$ be the eigenvalues of the hyperbolic Laplacian Δ_{1,m^-}. Then, the exceptional eigenvalues (i.e. the real ones) of this Laplacian are equal to $\omega, \omega - 1, \cdots, -\omega$.*

We then get the desired result from Exercise 5.4.4.

Corollary 5.4.13. *If $m^- \geq 7$, then $\Phi_L = \Phi_1$.*

Corollary 5.4.14. *If $m^- \geq 7$, then the singularities of the modular form F_1 are of type $q^{-1/n}$, where n is a positive integer.*

This result is a consequence of the following Lemma.

Lemma 5.4.15. *Let $\alpha \in \Pi$ be of negative norm and m be minimal in $\mathbf{Q}$ such that $\delta = m\alpha \in L^*$. Then, δ has norm $-2/n$ for some positive integer n.*

Proof. Since α has negative norm, it is a root of finite type (remember that we are working in the opposite convention: the non-diagonal entries of the Cartan matrix are non-negative; see Remark 5.3.20). Since it is simple, it is a primitive root in L^*. Hence, δ is such that $l\delta \in L^*$ implies that $\lambda \in \mathbf{Z}$. Therefore there exists $\delta_1 \in L$ such that $(\delta, \delta_1) = 1$. Since α is a root, so is δ and so $r_\delta(\delta_1) \in L$, where r_δ is the reflection induced by δ. The result then follows.   □

We are finally ready to compute the bound on the rank of L.

Corollary 5.4.16. *Assuming the above Conjecture to be true, $\dim L \leq 26$.*

Proof. We assume that $m^- \geq 7$. By Corollary 5.4.14, the modular form $F_1(\tau)\Delta(\tau)$ is holomorphic since $\Delta(\tau)$ has a simple pole. Hence it must have weight $12 + 1 - \dim K/2$ and as it is holomorphic its weight is non-negative. The result then follows.   □

There is very strong evidence that the Conjecture given above is true (see [Ar]). It would nevertheless be obviously satisfactory to find a proof of Corollary 5.4.16 which is not based on it.

Open Question 5. Find an alternative proof of Corollary 5.4.16.

There is good evidence that the upper bound of 26 holds. It is indirectly supported by a result of Goddard and Thorn from theoretical physics, namely the No-Ghost Theorem. Indeed as we will see in the next section, this result is crucial in the only (as yet known) method for constructing LBKM algebras and implies that the Lie algebras we can construct for higher ranks are not BKM algebras.

As the last two sections show, the classification project is far from complete and there are many questions yet to be addressed. As we have seen in this section, the principal problem of finding the vector valued modular forms associated to a BKM algebra, can be addressed by answering the following question.

Open Question 6. When are the Weyl chambers determined by the theta transform of F associated to a reflection group and when is the vector $\rho(K, W, F)$ a Weyl vector for a reflection group?

The reader more interested in the classification of hyperbolic reflection groups or equivalently BKM algebras whose simple roots of finite type generate a Lorentzian root lattice should in particular study the works of V.A. Gritsenko and V.V. Nikulin [Grit], [GritN1-5], [Nik1-3], together with [Borc3,12].

Exercises 5.4

1. Let F be the vector valued modular form of type ρ_L and eigenvalue λ with Fourier series

$$\sum_{m \in \mathbf{Z}} c_{m,\gamma}(y) q^{m + \gamma^2/2}.$$

(i) Show that

$$c''_{m,\gamma}(y) + (\frac{1+m^-}{2})c'_{m,\gamma}(y) + (\frac{\lambda}{y^2} + \frac{2\pi m(1-m^-)}{y} - 4\pi^2 m^2)c_{m,\gamma}(y) = 0.$$

(ii) Write $\lambda = s(1-s) - \frac{1+m^-}{2}(1 - \frac{1+m^-}{2})$. Check that the modular form yF has weight $(0, m^- - 1)$. Show that yF is an eigenvector for the Laplacian of this weight of eigenvalue $s(1-s) - \frac{m^-}{2}(1 - \frac{m^-}{2})$.

(iii) Find the solutions of the above differential equation in terms of the Whittaker functions (see Lemma 5.4.5).

Hint: change variables to $d_m(y) = y^{\frac{1+m^-}{2}}c_{m,\gamma}(y)$. For a solution see [Bar, Lemma 1.3.2].

2. (i) Show that the series for the function $F(\tau; \delta, s)$ given in Theorem 5.4.6 converges uniformly for $\mathcal{R}(s) > 1 + \frac{1+m^-}{2}$. Deduce that F is a real analytic modular form of type ρ_L and eigenvalue $\lambda = s(1-s) - \frac{1+m^-}{2}(1 - \frac{1+m^-}{2})$ and that its only singularity is at the cusp $i\infty$ and of type given in Theorem 5.4.6.

For a solution see [Bar, Theorem 1.3.7].

(ii) Deduce that the Fourier series for F_δ is as given in section 5.4.

3. (i) Find the Fourier series of the function Φ_δ using the Rankin-Selberg method.

(ii) Show that the function Φ_δ on the Grassmannian $G(L)$ is the sum of a function $\epsilon(v)$ real analytic on $G(L)$ and a function $\psi(v)$ piecewise linear on $G(L)$.

(iii) Show that ψ_1 is an eigenvector for the hyperbolic Laplacian on $G(L)$ with eigenvalue $-m^-$. Deduce that the function $\Phi_L - \Phi_1$ is an eigenvector for the Laplacian with eigenvalue $-m^-$.

For a solution, see [Br, Theorem 5.11], [Borc11, §7], [Bar, §3].

4. Since $\Phi_L - \Phi_1$ is an eigenvector of the Laplacian with real eigenvalue, use the Conjecture of Berger, Li, Sarnak to show that $\Phi_L = \Phi_1$ if $m^- \geq 7$. Deduce that $\dim L \leq 26$.

For a solution, see [Bar, lemma 3.4.2, Theorem 4.1.4].

5.5 A Construction of LBKM Algebras from Lattice Vertex Algebras

The aim of this section is to explain how some LBKM algebras can be constructed from (bosonic) lattice vertex algebras. The lattice L is only assumed to be integral and non-degenerate. The Fock space V_L has a vertex superalgebra structure (see Chapter 4). Let us reconsider Theorem 2.5.4 characterizing BKM algebras according to root space properties.

We need a self-centralizing Lie subalgebra acting semisimply on the Lie algebra in such a way that the eigenspaces are finite dimensional. It seems natural to consider the image

$$H^1 = \{h(-1) \otimes 1 : h \in H\}$$

of H in the Lie algebra P^1/dP^0.

Lemma 5.5.1. *The vector space H^1 is a self-centralizing Lie subalgebra of the Lie algebra P^1/dP^0.*

Proof. By definition of the Lie bracket (see Lemma 4.3.1), for any $h_1, h_2 \in H$,

$$[h_1(-1) \otimes 1, h_2(-1) \otimes 1] = h_1(0)(h_2(-1) \otimes 1) = 0.$$

Hence the vector space H^1 is an abelian Lie subalgebra of P^1/dP^0. For any $h \in H$, $\alpha \in L$, and $s \in S$, applying Lemmas 4.2.8 and 4.3.1,

$$[h(-1) \otimes 1, s \otimes e^\alpha] = h(0)(s \otimes e^\alpha) = (h, \alpha)(s \otimes e^\alpha).$$

By the nature of the action of the L_0 operator (see Theorem 4.3.6), the eigenspace $\dim P_\alpha^1$ of H^1 is finite dimensional since $\dim H < \infty$.

It only remains to check that the vector space $\{h(-1) \otimes 1 : h \in H\}$ is self-centralizing. From the above, $s \otimes e^\alpha$ is in its centralizer for $\alpha \in L$, $s \in S$ if and only if $(h, \alpha) = 0$ for all $h \in H$. Hence, as the bilinear form is non-degenerate and the action of the vector space in consideration is semisimple on P^1/dP^0, the result follows. $\qquad\square$

This result shows that the eigenspaces of the abelian Lie subalgebra H^1 are the subspaces P_α^1, $\alpha \in L$, where for any subspace U of V_L, we write

$$U_\alpha = \{v \in U | \deg(v) = \alpha\}.$$

Note that the Lie algebra V_L/dV_L is too large to be a BKM algebra because there is no upper bound on the norms of the roots $\alpha \in L$ with $(V_L)_\alpha \neq 0$, whereas the norms of roots is bounded above for the Lie algebra P^1/dP_0. This follows from the definition of the subspace P^1.

Lemma 5.5.2. *For any $\alpha \in L$, $P_\alpha^1 \neq 0$ implies that $\alpha^2 \leq 2$.*

Proof. Let $s \in S$ be such that $s \otimes 1$ is an eigenvector for L_0. By definition, the corresponding eigenvalue m is non-negative. For $\alpha \in H$, $s \otimes e^\alpha \in P^1$ is an eigenvector for L_0 with eigenvalue $1 = \frac{1}{2}(\alpha, \alpha) + m$. Hence, $(\alpha, \alpha) = 2(1 - m) \leq 2$. $\qquad\square$

We need to define a non-degenerate symmetric invariant bilinear form on the Lie algebra. Let us return to the bilinear form $(.,.)$ on V_L. We first state an obvious property satisfied by it.

Lemma 5.5.3. *For any $\alpha \in L$, the bilinear form is non-degenerate on $(V_L)_\alpha$ and the operator L_0 is diagonalizable on $(V_L)_\alpha$.*

Proof. Clearly, the vectors $1 \otimes e^\alpha$ together with $h_1(-n_1) \cdots h_r(-n_r) \otimes e^\alpha$, where the elements h_i run through H and n_i and r through $\mathbf{N}$, generate the subspace $(V_L)_\alpha$. As these are eigenvectors for L_0, L_0 is diagonalizable on $(V_L)_\alpha$. For any $s \in S$,

$$(h_1(-n_1) \cdots h_r(-n_r) \otimes e^\alpha, s \otimes e^\beta) = (1 \otimes e^\alpha, (h_1(n_r) \cdots h_r(n_1)s \otimes e^\beta).$$

This is equal to 0 if $\alpha \neq \beta$ since by definition $(1 \otimes e^\alpha, 1 \otimes e^\beta) = 0$ in that case. The Lemma follows from the action of the Heisenberg $\tilde{H}$ on V_L given in Lemma 4.2.8. $\qquad\square$

To show that it induces a bilinear form on the quotient P^1/dP^0, we need to show that $(dP^0, P^1) = 0$. Set

$$N = \{v \in P^1 : (v, P^1) = 0\}$$

to be the kernel of the bilinear form on P^1.

Lemma 5.5.4. $dP^0 \subseteq N$.

Proof. Since by definition of the bilinear form on V_L (see Theorem 4.3.3), the operator L_1 is the adjoint of the operator L_{-1} and $L_{-1} = d$ by Lemma 4.3.9,

$$(dP^0, P^1) = (P^0, L_1 P^1) = 0$$

by definition of the subspace P^1. $\qquad\square$

The non-degeneracy condition leads us to consider the vector space P^1/N instead of the Lie algebra P^1/dP^0 as a likely candidate for the BKM algebra we want to construct. However, we do not yet know that P^1/N has the structure of a Lie algebra. Lemma 5.5.4 implies that the vector space P^1/N will have a Lie algebra structure, or more precisely that it will be the quotient Lie algebra $(P^1/dP^0)/(N/dP^0)$ if N/dP^0 is an ideal of the Lie algebra P^1/dP^0. This is not at all obvious. This will follow if for all $u \in N$, $v \in P^1$, $u_0(v) \in P^1$, i.e. that for all $v' \in P^1$, $(u_0(v), v') = 0$. To show this, the definition of the bilinear form given in Theorem 4.3.3 suggests that it could be useful to find the adjoint $(u_0)'$ of the linear map u_0. Then,

$$(u_0(v), v') = ([u, v], v') = (v, (u_0)'v').$$

This indicates that we should try to prove $(u_0)' = (\omega(u))_0$ for some involution ω on V_L, in other words contravariance of the bilinear form. We next study this question. One of the consequences will be that the subspace N/dP^0 is an ideal of the Lie algebra P^1/dP^0 since the above shows that proving contravariance amounts to finding the adjoint of the operator u_0 for $u \in P^1$. For the notion of contravariance, we need an involution with adequate properties.

Remark 5.5.5. Since the vertex algebra structure of V_L is independent of the cocycle ϵ chosen as long as $e(\alpha, \beta)e(\beta, \alpha)^{-1} = B(\alpha, \beta) = (-1)^{(\alpha,\beta)+(\alpha,\alpha)(\beta,\beta)}$ (see Remark 4.2.26), Lemma 4.2.5 and the definition of the bi-multiplicative map $B(\alpha, \beta)$ allow us to assume that $\epsilon(\pm\alpha, \pm\alpha) = (-1)^{(\alpha,\alpha)}$ and $\epsilon(\alpha, \beta) \in \{\pm 1\}$.

Definition 5.5.6. *There is an involution acting on the vertex algebra V_L called the Cartan involution given by:*

$$\omega(1 \otimes e^\alpha) = (-1)^{\frac{1}{2}(\alpha,\alpha)}(1 \otimes e^{-\alpha}), \quad \omega(h(n) \otimes 1) = -h(n) \otimes 1,$$

$$\alpha \in L, h \in H, n \in \mathbf{Z}.$$

In fact, it is easier to find the adjoint operator of the endomorphism u_n for arbitrary $n \in \mathbf{Z}$ than just for $n = 0$. First we give the adjoint of the map e^α.

Lemma 5.5.7. *For any α, the adjoint of the endomorphism e^α is $(-1)^{(\alpha,\alpha)}e^{-\alpha}$.*

Proof. For $\alpha, \beta, \gamma \in L$, $s, t \in S$,

$$(e^\alpha(s \otimes e^\beta), e^\alpha(t \otimes e^\gamma)) = \epsilon(\alpha, \beta)\epsilon(\alpha, \gamma)(s \otimes e^\beta, t \otimes e^\gamma)$$

$$= \begin{cases} 0 & \text{if } \beta \neq \gamma \\ (s \otimes e^\beta, t \otimes e^\beta) & \text{if } \beta = \gamma \end{cases}$$

since $e(\alpha, \beta)^2 = 1$ (see Remark 5.5.4). The result follows since $\epsilon(\alpha, -\alpha) = (-1)^{(\alpha,\alpha)}$ by assumption (see Remark 5.5.5). $\square$

Lemma 5.5.8. *For any $n, m \in \mathbf{Z}$, $v \in (V_L)^m$, the adjoint of the endomorphism v_n (with respect to the bilinear form defined in Theorem 4.3.3) is:*

$$(-1)^m \sum_{j=0}^\infty \frac{1}{j!}(L_1^j\omega(v))_{2m-j-n-2}.$$

Proof. We only need to prove this for the generators $h_1(m_1) \cdots h_r(m_r) \otimes e^\alpha$ of the vector space V_L, where $h_i \in H$, $m_i < 0$ and $\alpha \in L$. We will write ϕ^* for the adjoint of the operator ϕ on V_L.

First suppose that $a = 1 \otimes e^\alpha$, $\beta \in L$, $u \in (V_\Lambda)_\beta$, $v \in (V_\Lambda)$. As the adjoint of the endomorphism $\alpha(j)$ is $\alpha(-j)$ by definition of the bilinear form (see Theorem 4.3.3),

$$Y(1 \otimes e^\alpha u, v) = (e^\alpha e^{-\sum_{j<0} \frac{z^{-j}}{j}\alpha(j)} e^{-\sum_{j>0} \frac{z^{-j}}{j}\alpha(j)} z^{(\alpha,\beta)} u, v)$$

$$= z^{(\alpha,\beta)}(-1)^{(\alpha,\alpha)}(u, e^{-\alpha} e^{-\sum_{j>0} \frac{z^{-j}}{j}\alpha(-j)} e^{-\sum_{j<0} \frac{z^{-j}}{j}\alpha(-j)} v)$$

$$= z^{(\alpha,\beta)}(-1)^{(\alpha,\alpha)}(u, e^{-\alpha} e^{-\sum_{j<0} \frac{z^j}{j}(-\alpha)(j)} e^{-\sum_{j>0} \frac{z^j}{j}(-\alpha)(j)} v).$$
$$\tag{1}$$

Now, by definition of the bilinear form, $(Y(1 \otimes e^\alpha, z)u, v) = 0$ unless $v \in (V_L)_{\alpha+\beta}$. Suppose that $v \in (V_L)_{\alpha+\beta}$. Then,

$$(-1)^{(\alpha,\alpha)}(1 \otimes e^{-\alpha})_n v$$

is the coefficient of $z^{k+(\alpha,\beta)}$ on the right hand side of (1), where

$$-k - (\alpha, \alpha + \beta) = -n - 1.$$

Hence, the coefficient $((1 \otimes e^\alpha)_n)^* v$ of z^{-n-1} on the right hand side of (1) is equal to

$$(-1)^{(\alpha,\alpha)}(1 \otimes e^{-\alpha})_{(\alpha,\alpha)-n-2} v$$

since $k + (\alpha, \beta) = -n - 1$ implies that $-k - (\alpha, \alpha + \beta) = -(-n + (\alpha, \alpha) - 2) - 1$. This holds for all $\beta \in L$ and so for all $v \in V_L$. Therefore,

$$((1 \otimes e^{\alpha})_n)^* = (-1)^{(\alpha,\alpha)}(1 \otimes e^{-\alpha})_{(\alpha,\alpha)-n-2}$$
$$= (-1)^{\frac{1}{2}(\alpha,\alpha)}\omega(1 \otimes e^{\alpha})_{(\alpha,\alpha)-n-2}$$

since $L_0(1 \otimes e^{\alpha}) = \frac{1}{2}(\alpha, \alpha)(1 \otimes e^{\alpha})$ and by Definition 5.5.6, $\omega(1 \otimes e^{\alpha}) = (-1)^{\frac{1}{2}(\alpha,\alpha)}(1 \otimes e^{\alpha})$. Hence the result holds for $a = 1 \otimes e^{\alpha}$.

Assume next that the result holds for $b \in (V_L)_\alpha^l$. We compute the adjoints of the endomorphisms a_n, where $a = h(-1)b$, $h \in H$. By Theorem 4.2.16 and Corollary 4.1.32, for $u, v \in V_L$,

$(Y(a, z)u, v)$

$$= (u, \sum_{k<0} z^{-k-1} \sum_{n \in \mathbf{Z}} z^{-n-1}(-1)^l \sum_{j=0}^{\infty} \frac{1}{j!}(u, (L_1^j \omega(b))_{2l-j-n-2}h(-k)v)$$

$$+ (u, \sum_{k\geq 0} z^{-k-1} \sum_{n \in \mathbf{Z}} z^{-n-1}(-1)^l \sum_{j=0}^{\infty} \frac{1}{j!}(u, h(-k)(L_1^j \omega(b))_{2l-j-n-2}v).$$

$$(2)$$

So,

$(a_n)^*$

$$= \sum_{k>0}(-1)^l \sum_{j=0}^{\infty} \frac{1}{j!}(L_1^j \omega(b))_{2l-j-n-k-1}h(k)$$

$$+ \sum_{k\leq 0}(-1)^l \sum_{j=0}^{\infty} \frac{1}{j!}h(k)(L_1^j \omega(b))_{2l-j-n-k-1}.$$

Now, for any vector $u \in V_L$ and $m \in \mathbf{Z}$,

$$[h(0), u_m] = (h(0)u)_m.$$

Hence,

$$(a_n)^* = \sum_{k\geq 0}(-1)^l \sum_{j=0}^{\infty} \frac{1}{j!}(L_1^j \omega(b))_{2l-j-n-k-1}h(k)$$

$$+ \sum_{k<0}(-1)^l \sum_{j=0}^{\infty} \frac{1}{j!}h(k)(L_1^j \omega(b))_{2l-j-n-k-1}$$

$$+ (-1)^l \sum_{j=0}^{\infty} \frac{1}{j!}(h(0)L_1^j \omega(b))_{2l-j-n-1}$$

$$= (-1)^l \sum_{j=0}^{\infty} \frac{1}{j!}(h(-1)L_1^j b)_{2l-j-n} \quad \text{by Definition 4.1.27}$$

$$+ (-1)^l \sum_{j=0}^{\infty} \frac{1}{j!} (h, \alpha)(L_1^j \omega(b))_{2l-j-n-1}$$

$$= (-1)^l \sum_{j=0}^{\infty} \frac{1}{j!} (L_1^j h(-1)b)_{2l-j-n}$$

since by Corollary 4.1.26 (i),

$$[L_1, (h(-1) \otimes 1)_{-1}] = [\zeta_2, (h(-1) \otimes 1)_{-1}]$$

$$= \sum_{j=0}^{\infty} \binom{2}{j} (L_{j-1}(h(-1) \otimes 1))_{1-j}$$

$$= (L_{-1}(h(-1) \otimes 1))_1 + 2h(0)$$

$$= (h(-2) \otimes 1)_1 + 2h(0)$$

$$= -h(0) + 2h(0) \quad \text{by Lemma 4.1.31}$$

$$= h(0).$$

Moreover, the vector a is an eigenvector of the operator L_0 with eigenvalue $l + 1$ and by definition of the involution ω (see Definition 5.5.6), $\omega(h(-1)b) = -h(-1)\omega(b)$. Hence the result holds for $h(-1)b \, \forall h \in H$, if it holds for b.

Assume by induction that the result holds for $h(-m)b$. We finally show that it holds for $a = mh(-m-1)b$, where $m \in \mathbf{N}$. Since

$$a = d(h(-m)b) - h(-m)db,$$

and $d(h(-m)b)_n = -n(h(-m)b)_{n-1}$, the result holds for a by induction. □

Corollary 5.5.9. *The vector space P^1/N is a Lie algebra.*

Proof. Now $(dP^0, P^1) = (P^0, L_1 P^1) = 0$ by definition of the bilinear form (see Theorem 4.3.3) and Theorem 4.3.6. Hence $dP^0 \leq N$ and so Corollary 4.3.11 and Lemma 5.5.4 imply that the result will follow if N/dP^0 is an ideal of the Lie algebra P^1/dP^0. For any $v \in N$, $u, u' \in P^1$,

$$\begin{aligned}
([u, v], u') &= (u_0(v), u') \\
&= (v, (u_0)^*(u')) \\
&= -(v, -\omega(u)_0(u') \quad \text{by Lemma 5.5.8.}
\end{aligned} \tag{1}$$

Now, $\omega(u) \in P$ as the map ω clearly commutes with the operators L_j for $j > 0$. Moreover, $L_0\omega(u) = \omega(u)$ since $1 \otimes e^{-\alpha}$ and $1 \otimes e^{\alpha}$ are eigenvectors for the operator L_0 with the same eigenvalue. Hence, by Lemma 4.3.10 $((\omega(u))_0)v \in P^1$. And so (1) gives

$$([u, v], u') = 0.$$

In other words, $[u, v] \in N$, proving the result. □

The Lie algebra P^1/N will be denoted F_L. We show that this is a BKM algebra. The bilinear form on the vertex algebra V_L induces a well defined symmetric bilinear form on the Lie algebra F_L. We will keep the same notation for the bilinear form on $(.,.)$.

Lemma 5.5.10. *The Cartan involution induces an involution on the Lie algebra F_L and the bilinear form $(.,.)$ is contravariant on F_L with respect to this involution.*

Proof. The involution ω commutes with the action of the operators L_i, $i \geq 0$. Moreover by Lemma 5.5.7, $\omega(N) = N$: Indeed for any $u \in N_\alpha$, $v \in (P^1)_\alpha$, $(u, v) = 0$, i.e. for any $t \in S$ such that $t \otimes 1 \in (V_L)^{1 - \frac{1}{2}(\alpha, \alpha)}$ and $s \in S$ such that $s \otimes e^\alpha \in N$,

$$(s \otimes e^\alpha, t \otimes e^\alpha) = 0.$$

Hence, $(e^{-\alpha}u, e^{-\alpha}v) = 0$ gives

$$(s \otimes 1, t \otimes 1) = 0$$

Now $\omega(s \otimes e^\alpha) = \pm(s \otimes e^{-\alpha})$ and $t \otimes e^\alpha \in P^1$ if and only if $t \otimes e^{-\alpha} \in P^1$. It follows that for $t \otimes e^{-\alpha} \in P^1$,

$$
\begin{aligned}
(\omega(s \otimes e^\alpha), t \otimes e^{-\alpha}) &= (-1)^{(\alpha,\alpha)}(e^\alpha \omega(s \otimes e^\alpha), e^\alpha(t \otimes e^{-\alpha})) \\
&= \pm(s \otimes 1, t \otimes 1) \quad \text{by (1)} \\
&= 0.
\end{aligned}
$$

Therefore the involution ω induces an involution on the Lie algebra F_L. By Lemma 5.5.8, the bilinear form on F_L is contravariant with respect to the involution ω since for all $g, x, y \in P^1$,

$$
\begin{aligned}
([g, x], y) &= (g_0(x), y) \\
&= -(x, \omega(g)_0(y)) \quad \text{by Lemma 5.5.8} \\
&= -(x, [\omega(g), y]).
\end{aligned}
$$

$\square$

We keep the same notation for the induced involution on the Lie algebra F_L. Contravariance of the form $(.,.)$ leads us to consider the form

$$K(x, y) = -(\omega(x), y), \quad x, y \in F_L$$

(see section 2.2).

Lemma 5.5.11. *The form K is bilinear, symmetric, invariant and non-degenerate on the Lie algebra F_L.*

Proof. Symmetry follows from the fact that the action of the involution ω leaves the form $(.,.)$ invariant. This is a direct consequence of the definition of ω. Invariance is a result of the contravariance of $(.,.)$. Since the latter is non-degenerate on F_L, so is K. $\square$

Since $H^1 \cap N = 0$, the bilinear form on H being non-degenerate, we will use the same notation H^1 for the image of $\{h(-1) \otimes 1 : h \in H\}$ in F_L. We now use properties of Lorentzian spaces to prove the existence of a regular element and to simplify Condition 5 of Theorem 2.5.4.

Lemma 5.5.12. *Suppose that the lattice L is Lorentzian. The abelian Lie subalgebra H^1 contains a regular element having negative norm.*

Proof. Let $h \in H$ be a negative norm vector. Consider a basis $\alpha_1, \cdots, \alpha_n$ for L and its dual basis $\alpha'_1, \cdots, \alpha'_n$ in H. Set $h = \sum_{i=1}^n x_i \alpha'_i$. The vector $\alpha = \sum_{i=1}^n y_i \alpha_i \in H$ is orthogonal to h if and only if $\sum_{1 \le i, j \le n} y_i x_i = 0$. Since the bilinear form on H is Lorentzian, the negative norm vectors form two open cones. Hence, we can choose the vector h so that

$$\sum_{i=1}^n y_i x_i \neq 0$$

for all $y_i \in \mathbf{Z}$, $1 \le i \le n$. Hence, $h^\perp \cap L = 0$. As a result, the centralizer of $h(-1) \otimes 1 \in H^1$ in F_L is H^1 since $[h(-1) \otimes s \otimes e^\alpha] = (h, \alpha)(s \otimes e^\alpha)$ for all $s \in S$, $\alpha \in L$. $\square$

Lemma 5.5.13. *Suppose that the lattice L is Lorentzian. Set $h_L \in H^1$ to be a regular element such that $h_L^2 < 0$. Let $\alpha, \beta \in L$ be such that $\alpha^2 \le 0$, $\beta^2 \le 0$, $(\alpha, h_L) > 0$ and $(\beta, h_L) > 0$. Then $(\alpha, \beta) \le 0$. If $(\alpha, \beta) = 0$, then β is a positive multiple of α.*

Proof. Let $\alpha, \beta \in L$ be as stated. Since the vector space H is Lorentzian $\alpha = ah + \alpha^+$ and $\beta = bh + \beta^+$ for some vectors $\alpha^+, \beta^+ \in h^\perp$ and some non-trivial scalars $a, b \in \mathbf{R}$. Suppose that $(\alpha, \beta) > 0$. Then, $\alpha - s\beta$ is a non-positive norm vector for all $s \ge 0$. Since $(h, \alpha) > 0$ and $(h, \beta) > 0$, $a > 0$ and $b > 0$. The vector subspace $h^\perp$ being positive definite, this implies that $\alpha^+ = s\beta^+$ for $s = a/b$. This in turn gives $\alpha = s\beta$ and so $(\alpha, \beta) = s^2 \beta^2 < 0$, contradicting assumption and proving that $(\alpha, \beta) \le 0$.

If $(\alpha, \beta) = 0$, then the previous argument forces the vector α to be a positive multiple of the vector β. $\square$

Hence to prove that the Lie algebra F_L is a BKM algebra, it only remains to check the orthogonality part of Condition 5 of Theorem 2.5.4 for roots of norm 0 that are positive multiples of each other.

Theorem 5.5.14. *If L is a Lorentzian lattice, then the Lie algebra F_L is a BKM algebra. All the roots with positive norm have norm 2.*

Proof. Applying Theorem 2.5.4, since Lemmas 5.5.1, 5.5.2, 5.5.8, 5.5.11–5.5.13 hold, we only need to check that for $\alpha \in L$ such that $(\alpha, \alpha) = 0$, $[(F_L)_\alpha, (F_L)_{s\alpha}] = 0$ for all $s > 0$ such that $s\alpha \in L$. Let $x \in P_{s\alpha}^1$. Since $\alpha^2 = 0$, $L_0(x) = x$ implies that

$$x = h(-1) \otimes e^{s\alpha}$$

for some $h \in H$. Moreover, by definition of the operator L_1 (see Theorem 4.3.6),

$$0 = L_1(x)$$
$$= \frac{1}{2}\sum_i(\sum_{n<0}\alpha_i(n)\alpha'_i(1-n) + \sum_{n\geq0}\alpha'_i(1-n)\alpha_i(n))(h(-1)\otimes e^{s\alpha})$$
$$= \frac{1}{2}\sum_i(\alpha'_i(1)\alpha_i(0) + \alpha'_i(0)\alpha_i(1))(h(-1)\otimes e^{s\alpha})$$
$$= \frac{1}{2}\sum_i(s\alpha'_i(1)(\alpha_i,\alpha)(h(-1)\otimes e^{s\alpha}) + \alpha'_i(0)(\alpha_i,h)(1\otimes e^{s\alpha}))$$
$$= \frac{1}{2}(s\alpha'(1)(h(-1)\otimes e^{s\alpha}) + h(0)(1\otimes e^{s\alpha}))$$
$$= s(\alpha,h)(1\otimes e^{s\alpha}),$$

where the vectors $(\alpha_i)_i$ form a basis for the lattice L and the vectors $(\alpha'_i)_i$ form the dual basis in H. Hence,

$$(\alpha,h) = 0.$$

Let $h_1, h_2 \in \alpha^\perp$. Then,

$$[h_1(-1)\otimes e^\alpha, h_2(-1)\otimes e^{s\alpha}]$$
$$= (h_1(-1)\otimes e^\alpha)_0(h_2(-1)\otimes e^{s\alpha})$$
$$= (\sum_{j<0}h_1(j)(1\otimes e^\alpha)_{-j-1} + \sum_{j\geq0}(1\otimes e^\alpha)_{-j-1}h_1(j))(h_2(-1)\otimes e^{s\alpha}).$$

This last equality follows from Definition 4.1.27 of a normally ordered product. By definition of the vertex algebra structure on V_L (see Theorem 4.2.16),

$$Y(1\otimes e^\alpha, z)(h_2(-1)\otimes e^{s\alpha}) = \epsilon(\alpha,s\alpha)e^{-\sum_{j<0}\frac{z^{-j}}{j}\alpha(j)}(h_2(-1)\otimes e^{(s+1)\alpha})$$

since $(\alpha, h_2) = 0$. Hence,

$$(h_1(-1)\otimes e^\alpha)_0(h_2\otimes e^{s\alpha}) = \epsilon(\alpha,s\alpha)(h_1,h_2)(\alpha(-1)\otimes e^{(s+1)\alpha})$$
$$= \epsilon(\alpha,s\alpha)(h_1,h_2)d(1\otimes e^{(s+1)\alpha})/(s+1)$$

by definition of the derivation d given in Lemma 4.2.11.   Since $L_n((1\otimes e^{(s+1)\alpha}) = 0$ for all integers $n \geq 0$, it follows that

$$[h_1(-1)\otimes e^\alpha, h_2(-1)\otimes e^{s\alpha}] \in dP^0.$$

Therefore, by Lemma 5.5.8,

$$[h_1(-1)\otimes e^\alpha, h_2(-1)\otimes e^{s\alpha}] \equiv 0 \quad (\text{mod } N),$$

proving the Theorem.   □

In section 5.4, we saw that it is very likely that the root lattices of the LBKM algebras associated to a vector valued modular form in the manner stated in

Theorem 5.3.6 have rank at most 26. The above manner of showing that the Lie algebra F_L is a BKM algebra by applying Theorem 2.5.4 does not seem to lead to the calculation of the root space multiplicities. It is possible to do so when the lattice L is even, Lorentzian of rank 26 by applying the No-Ghost Theorem. In fact, this gives an alternative proof of the fact that F_L is a LBKM algebra when the rank of the lattice L is less than 26. The No-Ghost Theorem thus implies that 26 is a critical dimension.

Remember from Theorem 2.2.9 that roughly speaking any $\mathbf{Z}$-graded Lie algebra with an almost positive definite contravariant bilinear form is a BKM algebra. So let us see whether we can find a suitable grading of the subspace P^1 by the lattice L so that all negative norm vectors will belong to the 0-th piece. Since L is a free abelian group, an adequate $\mathbf{Z}$-grading of the Lie algebra P^1/dP^0 will follow. We show that, when the even Lorentzian lattice is of rank at most 26, there indeed is an L-grading of V_L with this property. The central result which allows this to be possible is the No-Ghost Theorem [GodT], [Fren], [Bor7], [Jur2] proved in the context of theoretical physics in 1971 by Goddard and Thorn [GodT].

Let us first define the grading. Since the bilinear form on V_L is Hermitian, for reasons of simplicity, we will consider the base field to be $\mathbf{R}$ instead of $\mathbf{C}$. Basically, we take H to be $L \otimes_{\mathbf{Q}} \mathbf{R}$ instead of $L \otimes_{\mathbf{Q}} \mathbf{C}$. All results of Chapter 4 remain valid over $\mathbf{R}$ with Hermitian forms becoming bilinear forms. The space V_L has a natural L-grading given by

$$deg(s \otimes e^{\alpha}) = \alpha \quad \text{for} \quad \alpha \in L, \quad s \in S.$$

Lemma 5.5.15. *The subspace of P^1 of degree 0 is as follows:*

$$P_0^1 = \{h(-1) \otimes 1 | \ h \in H\}.$$

It contains negative norm vectors if and only if so does the lattice L.

Proof. From Theorem 4.3.6,

$$L_0(h_1(-n_1)...h_r(-n_r) \otimes 1) = (n_1 + ...n_r)h_1(-n_1)...h_r(-n_r) \otimes 1$$

for all $h_1, ..., h_r \in H$, $n_1, ..., n_r \in \mathbf{N}$. Hence if an eigenvector for L_0 is in P_0^1, then it is of the form $h(-1) \otimes 1$, $h \in H$. Again applying Theorem 4.3.6, for $j > 0$ and $h \in H$,

$$L_j(h(-1) \otimes 1) = \frac{1}{2} \sum_i (\sum_{n<0} \alpha_i(n)\alpha_i'(j-n) + \sum_{n \geq 0} \alpha_i'(j-n)\alpha_i(n))(h(-1) \otimes 1)$$

$$= \frac{1}{2} \sum_i (\alpha_i, h)\alpha_i'(j-1)(1 \otimes 1)$$

$$= 0$$

Hence

$$P_0^1 = \{h(-1) \otimes 1 | \ h \in H\}.$$

Furthermore
$$(h(-1) \otimes 1, h(-1) \otimes 1) = (h, h),$$
proving the Lemma. $\square$

In particular, the above Lemma shows that the 0-degree piece contains negative norm vectors when the lattice L is Lorentzian. However, the converse, i.e that negative norm vectors (ghosts) are confined (if they exist) to the 0-degree piece is trivial when the lattice L is either positive definite but not at all obvious for Lorentzian lattices (see Remark 4.3.7 (i)). This is the purpose of the No-Ghost Theorem.

Theorem 5.5.16 (The No-Ghost Theorem). *For all $\alpha \in L - \{0\}$, $v \in P_\alpha^1$, $(v, v) \geq 0$ if an only if the lattice L is Lorentzian of rank at most 26 or positive definite.*

We prove this Theorem in several stages. It is easy to deal with the case of a lattice L that is not even Lorentzian.

Lemma 5.5.17.

(i) *Suppose that L is a lattice of signature (n, m) with $m > 1$, then there are elements $\alpha \in L - \{0\}$ such that the bilinear form is not semi-positive definite on P_α^1.*

(ii) *If the lattice L is positive definite, then for all elements $\alpha \in L - \{0\}$, the bilinear form is positive definite on P_α^1.*

(iii) *If $(\alpha, \alpha) \equiv 1 \pmod 2$, then $P_\alpha^1 = 0$.*

Proof. (i): Since $m > 1$, there is an element $\alpha \in L$ such that $(\alpha, \alpha) = 0$. Then, by Remark 4.3.7 (i),
$$V_\alpha^1 = \{h(-1) \otimes e^\alpha : h \in H\}.$$

Let $h \in H$. Applying the operator L_j, $j > 0$ on the vector $v = h(-1) \otimes e^\alpha$, Theorem 4.3.6 implies that $L_j v = 0$ for all $j > 1$ and $L_1 v = (h, \alpha)(1 \otimes e^\alpha)$. Hence,
$$P_\alpha^1 = \{h(-1) \otimes e^\alpha : h \in H, (h, \alpha) = 0\}.$$

Furthermore, $(v, v) = (h, h)$ by definition of the bilinear form (see Theorem 4.3.3). Since $m > 1$, there are elements $h \in H$ such that $(h, \alpha) = 0$ and $(h, h) < 0$, proving part (i).

(ii) follows from Remark 4.3.7(i) and the definition of the bilinear form given in Theorem 4.3.3 since they imply that for any $\alpha \in L - \{0\}$,
$$P_\alpha^1 = \begin{cases} \mathbf{R}(1 \otimes e^\alpha) & \text{if } (\alpha, \alpha) = 2 \\ 0 & \text{otherwise} \end{cases}$$
and $(1 \otimes e^\alpha, 1 \otimes e^\alpha) = 1$.

(iii): If $(\alpha, \alpha) \equiv 1 \pmod 2$ and $v \in (V_L)_\alpha$ is an eigenvalue of the operator L_0, then its corresponding eigenvalue is not an integer (see Theorem 4.3.6). This proves (iii). $\square$

We may therefore assume in the rest of the proof of the No-Ghost Theorem that the Lattice L is even Lorentzian.

Fix a root $\alpha \in L - \{0\}$. We show that there is a positive definite subspace in P_α^1 whose orthogonal complement in P_α^1 coincides with the kernel N_α of the bilinear form on P_α^1 if $\dim L = 26$ and which contains negative norm vectors if $\dim L > 26$.

Lemma 5.5.18. *There is a vector $\beta \in H$ such that $(\alpha, \beta) \neq 0$ and $(\beta, \beta) = 0$.*

Proof. Case 1: $(\alpha, \alpha) \neq 0$.

There is an element $\gamma \in H$ such that $(\alpha, \gamma) = 0$ and $(\gamma, \gamma) = -(\alpha, \alpha)$. Set $\beta = \alpha + \gamma$.

Case 2: $(\alpha, \alpha) = 0$.

In this case there are orthogonal roots γ, μ of positive and negative norm respectively such that $\alpha = \gamma + \mu$. Take $\beta = \gamma - \mu$. Then $(\beta, \beta) = (\alpha, \alpha) = 0$ and $(\beta, \alpha) = 2(\gamma, \gamma) \neq 0$. $\qquad\square$

We fix $\beta \in H$ to be as in Lemma 5.5.18 and T_α to be the following subspace of $(V_L)_\alpha$:

$$T_\alpha = \{v \in (V_L)_\alpha : L_n v = 0 = \beta(n)v, n > 0\} \quad \text{and} \quad T_\alpha^1 = T_\alpha \cap P_\alpha^1.$$

In other words,

$$T_\alpha^1 = \{v \in (V_L)_\alpha : L_n v = 0 = \beta(n)v, n > 0, L_0 v = v\}.$$

We will show that this is the subspace we are looking for.

Lemma 5.5.19. *For all $m, n \in \mathbf{Z}$, $[\beta(n), L_m] = n\beta(n + m)$ and $[\beta(m), \beta(n)] = 0$.*

Proof. By Corollary 4.1.26 (i) and Theorem 4.3.6,

$$[\beta(n), L_m] = \sum_{j=0}^{\infty} \binom{n}{j} (\beta(j)\zeta)_{m+1+n-j}. \tag{1}$$

For $j \geq 0$,

$$\beta(j)\zeta = \frac{1}{2}\beta(j)(\sum_i \alpha_i(-1)\alpha_i'(-1) \otimes 1)$$

$$= \delta_{j,1}\frac{1}{2}\sum_i (\beta, \alpha_i)(\alpha_i'(-1) \otimes 1) + \delta_{j,1}\frac{1}{2}\sum_i (\beta, \alpha_i')(\alpha_i(-1) \otimes 1)$$

$$= \delta_{j,1}\beta(-1) \otimes 1.$$

So substituting this in (1), we get

$$[\beta(n), L_m] = n(\beta(-1) \otimes 1)_{m+n}$$
$$= n\beta(m + n).$$

The second equality follows from Lemma 4.2.14 since $(\beta, \beta) = 0$. $\qquad\square$

Set E to be the associative algebra generated by the operators L_i and $\beta(i)$, $i \in \mathbf{Z}$.

Lemma 5.5.20. *The operator L_0 is diagonalizable on T_α.*

Since the operator L_0 is diagonalizable on $(V_L)_\alpha$ by Lemma 5.5.3, to prove that L_0 is diagonalizable on T_α, elementary linear algebra arguments show that we only need to check that $L_0 T_\alpha \subseteq T_\alpha$. By Theorem 4.3.6, $[L_n, L_0] = nL_n$ and by Lemma 5.5.19, $[\beta(n), L_0] = n\beta(n)$. Therefore, for all $u \in T_\alpha$, $n \in \mathbf{N}$,

$$L_n L_0 u = nL_n u = 0 = n\beta(n)u = \beta(n)L_0 u.$$

Hence, $L_0 u \in T_\alpha$. □
Set

$$T_\alpha^\perp = \{v \in (V_L)_\alpha : (v, T_\alpha) = 0\}.$$

Lemma 5.5.21. *The bilinear form is non-degenerate on T_α and*

$$(V_L)_\alpha = U(E)T_\alpha.$$

More precisely, the elements

$$v_{r,s,t} = L_{-1}^{s_1} \cdots L_{-n}^{s_n} \beta(-1)^{r_1} \cdots \beta(-m)^{r_m} t,$$

where $n, m > 0$, $r_i, s_i \geq 0$, and t runs over a basis for T_α, form a basis of the subspace $(V_L)_\alpha$.

Proof. By Lemma 5.5.19, the elements $v_{r,s}$ stated in the Lemma generate the subspace $U(E)T_\alpha$.

We first prove that the vectors $v_{r,s,t}$ are linearly independent. From elementary linear algebra arguments, since L_0 acts diagonally on $(V_L)_\alpha$ (see Lemma 5.5.3), we only need to show this for the vectors $v_{r,s,t}$ in the same eigenspace of the operator L_0. By Remark 4.3.7 (i), the eigenvalues of L_0 on $(V_L)_\alpha$ are $\frac{1}{2}(\alpha, \alpha) + n$, $n \in \mathbf{Z}_+$.

We prove, by induction on n, that the vectors $v_{r,s}$ satisfying

$$L_0 v_{r,s,t} = (n + \frac{1}{2}(\alpha, \alpha))v_{r,s}$$

are linearly independent. By definition of the vectors $v_{r,s,t}$, this holds for $n = 0$. Suppose that it holds for all integers $0 \leq m < n$. Let

$$\sum_{r,s,t} x_{r,s,t} v_{r,s,t} = 0, \tag{1}$$

where $x_{r,s,t} \in \mathbf{R}$. Suppose that there is a term in the sum with a non-zero index s_j and let j be minimal with this property. Then, for any r, s, t,

$$v_{r,s,t} = L_j^{s_j} v_{r,s',t},$$

where $s = (s_j, s_{j+1}, \cdots, s_n)$ and $s' = (s_{j+1}, \cdots, s_n)$. By Lemma 5.5.19,

$$\beta(j) L^{s_j}_{-j} = L^{s_j}_{-j} \beta(j) + s_j L^{s_j - 1}_{-j} \beta(0).$$

Hence,

$$
\begin{aligned}
0 &= \beta(j) \sum_{r,s} x_{r,s,t} v_{r,s,t} \\
&= \sum_{r,s,t} x_{r,s,t} L^{s_j}_{-j} \beta(j) v_{r,s',t} + \sum_{r,s,t} x_{r,s,t} s_j L^{s_j - 1}_{-j} \beta(0) v_{r,s',t} \\
&= \sum_{r,s,t} x_{r,s,t} L^{s_j}_{-j} \beta(j) v_{r,s',t} + (\alpha, \beta) \sum_{r,s,t} x_{r,s,t} s_j L^{s_j - 1}_{-j} v_{r,s',t}
\end{aligned}
$$

since all the basis vectors are of degree α. The first sum on the right hand side is as a sum of vectors $v_{r,s,t}$ in which the operator L_j appears with exponent at least s_j and the second sum on the right hand side is as a sum of vectors $v_{r,s,t}$ in which the operator L_j appears with exponent $s_j - 1$. The right hand side being an eigenvector of L_0 with eigenvalue $n - j + \frac{1}{2}(\alpha, \alpha)$, induction and the fact that $(\alpha, \beta) \neq 0$ (see Lemma 5.5.18), therefore imply that

$$\sum_{r,s,t} x_{r,s,t} s_j L^{s_j - 1}_{-j} v_{r,s',t} = 0.$$

Applying induction once again, it follows that $x_{r,s,t} = 0$.

Therefore $x_{r,s,t} \neq 0$ in (1) forces $s_j = 0$ for all j. A similar argument applying the operator L_j (instead of $\beta(j)$) to both sides of equality (1) implies that all the coefficients $x_{r,s,t} = 0$.

We have therefore shown that the vectors $v_{r,s,t}$ form a basis of the vector space $U(E) T_\alpha$. For $n \geq 0$, set

$$C^n = \sum_{j=0}^{n-1} (U(E) T^{j + \frac{1}{2}(\alpha,\alpha)}_\alpha)^{n + \frac{1}{2}(\alpha,\alpha)}.$$

Then,

$$T^{n + \frac{1}{2}(\alpha,\alpha)}_\alpha \cap C^n = 0. \tag{2}$$

Otherwise there is a non-trivial vector in $T^{n + \frac{1}{2}(\alpha,\alpha)}_\alpha$ which a linear combinations of basis vectors $v_{r,s,t}$ with $t \in T^{j + \frac{1}{2}(\alpha,\alpha)}_\alpha$ for $j < n$. The above linear independence shows that this is not possible. Set

$$(C^n)^\perp = \{ v \in (V_L)^{n + \frac{1}{2}(\alpha,\alpha)}_\alpha : (v, C^n) = 0 \}.$$

We next show by induction on n that for all $n \geq 0$,

$$T^{n + \frac{1}{2}(\alpha,\alpha)}_\alpha = (C^n)^\perp. \tag{3}$$

By definition of T_α,

$$T_\alpha^{n+\frac{1}{2}(\alpha,\alpha)} \subseteq (C^n)^\perp$$

for all $n \geq 0$. So we only need to prove the inverse inclusion.

For $n = 0$, $C^0 = 0$ and so $(C^0)^\perp = (V_L)_\alpha^{\frac{1}{2}(\alpha,\alpha)}$. By Remark 4.3.7 (i), this is equal to $T_\alpha^{\frac{1}{2}(\alpha,\alpha)}$ and so (3) holds.

Suppose that (3) holds for all $0 \leq j \leq n-1$. We show that $(C^n)^\perp \subseteq T_\alpha^{n+\frac{1}{2}(\alpha,\alpha)}$. Let $u \in (C^n)^\perp$.

$$k > n \Rightarrow L_k u = 0$$

for otherwise, by Theorem 4.3.6, it would be an eigenvector for L_0 with eigenvalue

$$\frac{1}{2}(\alpha,\alpha) - k < \frac{1}{2}(\alpha,\alpha),$$

contradicting Remark 4.3.7 (i). For $1 \leq k \leq n$, since $[L_0, L_{-k}] = k L_{-k}$ by Theorem 4.3.6,

$$L_{-k}(C^{n-k} + T_\alpha^{n-k+\frac{1}{2}(\alpha,\alpha)}) \leq C^n$$

by definition of C^n. Hence,

$$(u, L_{-k}(C^{n-k} + T_\alpha^{n-k+\frac{1}{2}(\alpha,\alpha)})) = 0,$$

and so by definition of the bilinear form (see Theorem 4.3.3),

$$0 = (L_k u, C^{n-k} + T_\alpha^{n-k+\frac{1}{2}(\alpha,\alpha)}).$$

Therefore,

$$L_k u \in (C^{n-k})^\perp \cap (T_\alpha^{n-k+\frac{1}{2}(\alpha,\alpha)})^\perp,$$

and so by induction,

$$L_k u \in T_\alpha^{n-k+\frac{1}{2}(\alpha,\alpha)} \cap C^{n-k}.$$

Then, using (2), we get

$$L_k u = 0.$$

We have shown this holds for all $k > 0$. A similar argument shows that $\beta(k)u = 0$ for all $k > 0$.

As a consequence, by definition of the subspace T_α,

$$u \in T_\alpha^{n+\frac{1}{2}(\alpha,\alpha)},$$

proving that (3) holds for all $n \geq 0$.

(3) together with (2) imply that the bilinear form is non-degenerate on $T_\alpha^{n+\frac{1}{2}(\alpha,\alpha)}$ for all $n \geq 0$ and so it is non-degenerate on T_α by Corollary 4.3.8.

From (3) and (2), we can also deduce that for all $n \geq 0$, the bilinear form is non-degenerate on C^n and that

$$(V_L)_\alpha^{n+\frac{1}{2}(\alpha,\alpha)} = T_\alpha^{n+\frac{1}{2}(\alpha,\alpha)} \oplus C^n.$$

As a result, for all $n \geq 0$,

$$(V_L)_\alpha^{n+\frac{1}{2}(\alpha,\alpha)} = (U(E)T)^{n+\frac{1}{2}(\alpha,\alpha)}.$$

□

Remark 5.5.22. Notice that in the proof of $T^{n+\frac{1}{2}(\alpha,\alpha)} \cap C^n = 0$ in Lemma 5.5.21, the operators $\beta(n)$ play a crucial role. On cannot just consider the operators L_n. It is not always true that the intersection of $P_\alpha^{n+\frac{1}{2}(\alpha,\alpha)}$ with the subspace C'^n generated by the eigenvectors $L_{-1}^{s_1} \cdots L_{-m}^{s_m} p$ for the operator L_0 having eigenvalue $n + \frac{1}{2}(\alpha,\alpha)$, where $p \in P_\alpha$, is trivial. This is well illustrated by the following example.

Example 5.5.23. Let α be a vector in the lattice L of norm 0 and let us consider the case $n = 1$. Then,

$$P_\alpha^1 = \langle h(-1) \otimes e^\alpha : h \in H, L_1(h(-1) \otimes e^\alpha) = 0 \rangle$$

since $L_j(h(-1) \otimes e^\alpha) = 0$ for all $j \geq 2$ follows from the action of the operators L_j given in Theorem 4.3.6. Hence, since $L_1(h(-1) \otimes e^\alpha) = (h,\alpha)(1 \otimes e^\alpha)$,

$$P_\alpha^1 = \langle h(-1) \otimes e^\alpha : h \in H, (h,\alpha) = 0 \rangle.$$

By definition,

$$C'^1 = \mathbf{R}(L_{-1}(1 \otimes e^\alpha)) = \mathbf{R}(\alpha(-1) \otimes e^\alpha)$$

and so

$$C'^1 \leq P_\alpha.$$

Moreover,

$$(\alpha(-1) \otimes e^\alpha, P_\alpha^1) = 0.$$

Thus, the bilinear form is degenerate on P_α^1.

This example also explains why the No-Ghost Theorem states this subspace is semi-positive definite and not positive definite when the lattice L is Lorentzian of dimension at most 26.

Corollary 5.5.24. *For any $n \in \mathbf{Z}_+$,*

$$\dim T_\alpha^{n+\frac{1}{2}(\alpha,\alpha)} = p_{\dim L - 2}(n),$$

where $p_m(n)$ is the number of partitions of n into parts of m colours.

Proof. We first calculate the dimension of the space $(V_L)_\alpha^{n+\frac{1}{2}(\alpha,\alpha)}$ in two different ways.

For $0 \leq k \leq n$, fix a basis B_k of the subspace $T_\alpha^{n+\frac{1}{2}(\alpha,\alpha)}$. By remark 4.3.7 (i), the set $\cup_{k=0}^n B_k$ is a basis of the space T_α. Hence, Lemma 5.5.21 shows that

$$\{v_{r,s,t} : t \in B_k, \sum i(r_i + s_i) = n - k; \forall 0 \leq k \leq n\}$$

is a basis of the subspace $(V_L)_\alpha^{n+\frac{1}{2}(\alpha,\alpha)}$. Let the operators L_i and $\beta(-i)$ correspond to two colours, $\mathcal{L}$ and $\mathcal{B}$ respectively. An element $L_{-1}^{s_1} \cdots \beta(-1)^{r_1} \cdots$ corresponds to the partition $(s_1, \cdots, r_1 \cdots)$ of n where the s_i's and the r_i's are written in $\mathcal{L}$ and $\mathcal{B}$'s respectively. Therefore,

$$\dim(V_L)_\alpha^{n+\frac{1}{2}(\alpha,\alpha)} = \sum_{k=0}^{n} p_2(n-k) \dim T_\alpha^{k+\frac{1}{2}(\alpha,\alpha)}. \tag{1}$$

On the other hand, take a basis h_i, $1 \le i \le \dim L$ of the space H. Then, the elements

$$h_1(-l_1)^{j_1} \cdots h_{\dim L}(-l_{\dim L})^{j_{\dim L}} \otimes e^\alpha,$$

where $\sum_{i=1}^{\dim L} j_i l_i = n$ form a basis of the vector space $(V_L)_\alpha^{n+\frac{1}{2}(\alpha,\alpha)}$. Therefore

$$\dim(V_L)_\alpha^{n+\frac{1}{2}(\alpha,\alpha)} = p_{\dim L}(n). \tag{2}$$

Equating the right hand sides of (1) and (2), we get

$$\sum_{k=0}^{n} p_2(n-k) \dim T_\alpha^{k+\frac{1}{2}(\alpha,\alpha)} = p_{\dim L}(n). \tag{3}$$

We now prove that the Lemma holds by induction on n.

For $n = 0$, $V_\alpha^{\frac{1}{2}(\alpha,\alpha)} = T_\alpha^{\frac{1}{2}(\alpha,\alpha)}$ has dimension 1 as they are generated by the vector $1 \otimes e^\alpha$. Suppose that the result holds for all integers $0 \le k \le n-1$. Then, (3) becomes

$$\dim T_\alpha^{n+\frac{1}{2}(\alpha,\alpha)} = p_{\dim L}(n) - \sum_{k=0}^{n-1} p_2(n-k) p_{\dim_L - 2}(k) \tag{4}$$

since $p_2(0) = 1$. However

$$p_m(n) = \sum_{k=0}^{n} p_{m-2}(k) p_2(n-k)$$

as any partition of n into parts of m colours can be considered as a partition of k into parts of $m - 2$ colours and of $n - k$ into parts of 2 colours. The result then follows from (4). $\qquad\square$

We next show that the bilinear form in not only non-degenerate on T_α but positive definite.

Lemma 5.5.25. *The bilinear form on T_α is positive definite.*

Proof. Since the lattice L is Lorentzian and $(\beta, \beta) = 0$, the subspace $\beta^\perp$ in H is semi-positive definite of dimension m, where $m + 1 = \dim L$ and has an orthogonal basis $\beta, \alpha_1, \cdots, \alpha_{m-1}$ such that $(\alpha_i, \alpha_i) = 1$ for all $1 \le i \le m - 1$. Together with $\alpha_m = \alpha$, they form a basis of H. Hence the vectors

$$v_{n,l} = \beta(-n_0)^{l_0} \alpha_1(-n_1)^{l_1} \cdots \alpha_m(-n_m)^{l_m} \otimes e^\alpha,$$

where $n_i > 0$, $l_i \geq 0$, form a basis of the vector subspace $(V_L)_\alpha$. Set $n = (n_0, \cdots, n_m)$ and $l = (l_0, \cdots, l_m)$ and consider an arbitrary vector $v \in T_\alpha$. Then, $v = \sum_{n,l} x_{n,l} v_{n,l} \in T_\alpha$, for some scalars $x_{n,l} \in \mathbf{R}$. For all $k > 0$,

$$0 = \beta(k)v = (\beta, \alpha) \sum_{n,l} x_{n,l} l v_{n_0, \cdots n_{m-1}, k, l_0, \cdots, l_{m-1}, l_m - 1}$$

and so $x_{n,l} = 0$ if $l_m > 0$. When $l_m = 0 = l'_m$, $(v_{n,l}, v_{n',l'}) \geq 0$ since $(\alpha_i, \alpha_j) = \delta_{ij}$ and $(\beta, \beta) = 0 = (\beta, \alpha_i)$ for all $i \leq m - 1$. Therefore, the bilinear form is positive semi-definite on T_α. Since it is also non-degenerate on T_α by Lemma 5.5.21, the result follows. $\square$

Set U_α to be the subspace of $(V_L)_\alpha$ generated by the elements

$$L_{-1}^{s_1} \cdots L_{-n}^{s_n} \beta(-m)^{r_m} \cdots \beta(-1)^{r_1} u,$$

where $n, m > 0$, $r_i, s_i \geq 0$, $u \in T_\alpha$ with $s_1 + \cdots + s_n \neq 0$ and B_α the subspace generated by the above elements with all $s_i = 0$, i.e.

$$B_\alpha = \langle \beta(-m)^{r_m} \cdots \beta(-1)^{r_1} u : \quad m > 0, \quad r_i \geq 0, \quad u \in T_\alpha \rangle.$$

We next prove sone basic facts about these subspaces.

Lemma 5.5.26.

(i) $(V_L)_\alpha = U_\alpha \oplus B_\alpha$.
(ii) $U_\alpha = L_{-1}(V_L)_\alpha \oplus L_{-2}(V_L)_\alpha$.
(iii) Let $v \in V_L$. If $L_{-2}v = 0$, then $v = 0$.

Proof. (i) follows immediately from Lemma 5.5.21.
(ii): Since $[L_{-n}, L_{-1}] = -(n - 1)L_{-n-1}$, any operator L_{-n} for $n > 0$ can be written as a polynomial in L_{-1} and L_{-2}. Hence, $U_\alpha \leq L_{-1}(V_L)_\alpha + L_{-2}(V_L)_\alpha$. The inverse inclusion follows from Lemma 5.5.24. In order to complete the proof of part (ii), it only remains to check that the sum is direct. Suppose that for some $u_1, u_2 \in (V_L)_\alpha$,

$$L_{-1}u_1 = L_{-2}u_2. \tag{1}$$

From Remark 4.3.7 it follows that for all $v \in (V_L)_\alpha$, there is a minimal integer $n \geq 0$ such that $L_j v = 0$ for all $j \geq n$. Let $n \geq 0$ be minimal such that $L_n u_1 = 0$. Apply the operator L_n to both sides of equality (1). Then,

$$(n + 1)L_{n-1}u_1 = (n + 2)L_{n-2}u_2. \tag{2}$$

Since the operator L_0 acts diagonally on the subspace $(V_L)_\alpha$ by Lemma 5.5.3, we may take u_1 and u_2 to be eigenvectors for L_0 with eigenvalues c_1 and c_2 respectively. So, applying the operators $(L_0)^j$ for all $j \in \mathbf{N}$, to both sides of (1) and (2), we get

$$(c_1 + 1)^j L_{-1}u_1 = (c_2 + 2)^j L_{-2}u_2 \tag{3}$$

and

$$(n + 1)(c_1 - n + 1)^j L_{n-1}u_1 = (n + 2)(c_2 - n + 2)^j L_{n-2}u_2 \tag{4}$$

for all $j \in \mathbf{N}$. Since $L_{n-1}u_1 \neq 0$, equality (2) forces $L_{n-2}u_2 \neq 0$. Thus, equalities (4) imply that $c_1 = n - 1$ and $c_2 = n - 2$. Therefore, a similar argument applied to equalities (3) shows that either $L_{-1}u_1 = 0 = L_{-2}u_2$ or $c_1 = -1$ and $c_2 = -2$ and $n = 0$. If the former holds, then the sum is direct. So suppose that the latter holds. In this case,

$$L_j u_1 = 0 = L_j u_2$$

for all $j \geq 0$. Applying the operator L_2 to both sides of (1) then gives

$$0 = L_0 u_2 = -2u_2,$$

and so $u_2 = 0$. So equality (1) becomes $L_1 u_1 = 0$, and so the sum is again direct.

(iii): Suppose that $L_{-2}v = 0$ for some $v \in V_L$. Hence,

$$0 = L_1 L_{-2}v = 3L_{-1}v + L_{-2}L_1 v$$

and so from part (i) we can deduce that $v = 0$. $\qquad\square$

Lemma 5.5.27. $P_\alpha \cap B_\alpha = T_\alpha$.

Proof. Since $L_j v = 0$ for all $j > 0$ and $v \in P_\alpha$, by definition of T_α, $T_\alpha \subseteq P_\alpha$. By definition of B_α (taking $r_i = 0$ for all i), $T_\alpha \subseteq B_\alpha$.

Conversely, by definition of P_α, for all $v \in P_\alpha \cap B_\alpha$, $L_n v = 0$ for all $n > 0$. Since $(\beta, \beta) = 0$, $\beta(n)\beta(-m)^{r_m} \cdots \beta(-1)^{r_1}u = 0$ for all $m > 0$, $r_i \geq 0$, $u \in T_\alpha$. Hence $\beta(n)v = 0$ for all $n > 0$ and $v \in B_\alpha$. Therefore, $P_\alpha \cap B_\alpha \subseteq T_\alpha$. $\qquad\square$

Lemma 5.5.28. $(U_\alpha, P_\alpha) = 0$.

Proof. For all $u \in T_\alpha$, $v \in P_\alpha$, $m, n > 0$, $r_i, s_i \geq 0$, $s_1 + ...s_n > 0$,

$$(L_{-n}^{s_n} \cdots L_{-1}^{s_1}\beta(-m)^{r_m} \cdots \beta(-1)^{r_1}u, v) = (u, \beta(1)^{r_1} \cdots \beta(m)^{r_m}L_1^{s_1} \cdots L_n^{s_n}v) = 0$$

as $v \in P_\alpha$. $\qquad\square$

Lemma 5.5.29. For all $n > 0$, $L_n(B_\alpha) \subseteq B_\alpha$.

Proof. Since $[L_m, \beta(n)] = -n\beta(n+m)$ by Lemma 5.5.19, the generators of the subspace B_α are mapped onto elements of B_α by the operators $L(m)$ for all $m > 0$. $\qquad\square$

However, there is no similar result for U_α. For example, if $t \in T_\alpha$ then, $L_{-1}t \in U_\alpha$ but $L_1 L_{-1}t = 2L_0 t \notin U_\alpha$. We investigate what happens when one applies the operators L_n to U_α^1 in the next result. This is where the numerical constraint 26 is needed.

Lemma 5.5.30.

$$P_\alpha^1 \leq B_\alpha^1 \oplus U_\alpha^1.$$

Let $v \in P_\alpha^1$ and $b \in B_\alpha$, $u \in U_\alpha$ be such that $v = b + u$. Then,

$$L_1 b = 0 = L_1 u$$

and there is some $t \in B_\alpha^{-1}$ such that $-L_2 b = (\frac{1}{2} \dim L - 13)t = L_2 u$.
Furthermore, there exits $v_1, v_2 \in (V_L)_\alpha$ such that

$$u = L_{-1} v_1 + L_{-2} v_2.$$

Then,

$$L_1 v_1 = -3 v_2 \quad and \quad v_2 \equiv t \pmod{U_\alpha}.$$

Proof. The inclusion follows from Lemma 5.5.26 (i) and Lemma 5.5.3. Let $v \in P_\alpha^1$ and $b \in B_\alpha^1$, $u \in U_\alpha^1$ be such that

$$v = b + u. \tag{1}$$

By Lemma 5.5.26 (ii),
$$u = L_{-1} v_1 + L_{-2} v_2$$
for some vectors $v_1, v_2 \in (V_L)_\alpha$. Hence,

$$L_1 u = L_1 L_{-1} v_1 + L_1 L_{-2} v_2 = L_{-1} L_1 v_1 + 2 L_0 v_1 + L_{-2} L_1 v_2 + 3 L_{-1} v_2. \tag{2}$$

By Lemma 5.5.3, $L_{-1} v \in (V_L)_\alpha^1$ and $L_{-2} v \in (V_L)_\alpha^1$. So

$$L_0 v_1 = 0, \quad L_0 v_2 = -v_2. \tag{3}$$

Therefore, equation (2) becomes

$$L_1 u = L_{-1} L_1 v_1 + L_{-2} L_1 v_2 + 3 L_{-1} v_2 \tag{4}$$

By Lemma 5.5.26 (ii), the vectors $L_{-1}(L_1 v_1 + 3 v_2)$ and $L_{-2} L_1 v_2$ are linearly independent. Hence, by definition of the subspace U_α, $L_1 u \in U_\alpha$. We know from Lemma 5.5.29 that $L_1 b \in B_\alpha$. Since $v \in P_\alpha$, $L_1 v = 0$. We can then deduce from the fact that $U_\alpha \cap B_\alpha = 0$ by definition of these subspaces that

$$L_1 b = 0 = L_1 u.$$

Since $v \in P_\alpha$, $L_2 v = 0$. And so,

$$0 = L_2 b + L_2 L_{-1} v_1 + L_2 L_{-2} v_2 \tag{2}$$

$$= L_2 b + L_{-1} L_2 v_1 + 3 L_1 v_1 + L_{-2} L_2 v_2 + 4 L_0 v_2 + \frac{1}{2}(\dim L) v_2 \tag{3}$$

since $\dim L$ is the central charge of the Virasoro algebra. Since $L_1 u = 0$ and by Lemma 5.5.26 (ii), the vectors $L_{-1}(L_1 v_1 + 3 v_2)$ and $L_{-2} L_1 v_2$ are linearly independent, equality (4) implies that

$$L_{-1}(L_1 v_1 + 3 v_2) = 0.$$

As a result, by Lemmas 4.3.6 and 4.3.9, $L_1 v_1 = -3v_2$. Hence substituting this into equality (5) and using (3), we get

$$0 = L_2 b + L_{-1} L_2 v_1 + L_{-2} L_2 v_2 + (\frac{1}{2}(\dim L) - 13)v_2. \tag{6}$$

By Lemma 5.5.26 (i), there exist $t \in B_\alpha$ and $u' \in U_\alpha$ such that

$$v_2 = t + u'.$$

By Lemmas 5.5.27 and 5.5.29,

$$L_2 b + (\frac{1}{2}(\dim L) - 13)t \in B_\alpha.$$

By Lemma 5.5.26 (ii),

$$L_{-1} L_2 v_1 + L_{-2} L_2 v_2 + (\frac{1}{2}(\dim L) - 13)u' \in U_\alpha.$$

Since $B_\alpha \cap U_\alpha = 0$, equation (5) thus implies that

$$-L_2 b = (\frac{1}{2} \dim L - 13)t.$$

Since $v \in P_\alpha$, from (1) we get $L_2 u = -L_2 b$. This finishes the proof. $\qquad\square$

Remark 5.5.31. It is important to note the importance of the fact that any operator L_n for $n < 0$ can be written as a polynomial in L_{-1} and L_{-2}. It leads, via Lemma 5.5.26 (ii) to the crucial property given in Lemma 5.5.30, namely that for any vector $b + u \in P_\alpha^1$, where $b \in B_\alpha$, $u \in U_\alpha$, $L_2 u = (\frac{1}{2} \dim L - 13)t$ for some $t \in B_\alpha$.

We can now deduce what happens at the critical dimension 26.

Corollary 5.5.32. *Suppose that* $\dim L = 26$. *Then,* $P_\alpha^1 = T_\alpha^1 \oplus N_\alpha$.

Proof. Let $v \in P_\alpha^1$. By Lemma 5.5.30, there exists vectors $b \in B_\alpha^1$ and $u \in U_\alpha^1$ such that

$$v = b + u,$$

$L_1 b = 0 = L_1 u$ and moreover $L_2 b = 0 = L_2 u$ since $\dim L = 26$. For all $n > 0$, the operators L_n are polynomials in L_1 and L_2 since $[L_n, L_1] = (n-1)L_{n+1}$ by Lemma 4.3.6. Therefore,

$$L_n v_1 = 0 = L_n v_2 \quad \forall n > 0.$$

In other words, by definition of P_α, $b \in B_\alpha^1 \cap P_\alpha^1$ and $u \in U_\alpha^1 \cap P_\alpha^1$. Hence, by Lemma 5.5.27, $b \in T_\alpha^1$ and by Lemma 5.5.28, $u \in P_\alpha^\perp$. Since $L_0 u = u$, by definition of the null space N_α, $u \in N_\alpha$. It follows that

$$P_\alpha^1 = T_\alpha^1 + N_\alpha.$$

The sum is direct because the bilinear form is non-degenerate on T_α by Lemma 5.5.21. $\square$

The No-Ghost Theorem now follows when $\dim L \leq 26$.

Corollary 5.5.33. *If* $\dim L \leq 26$, *then the bilinear form is semi positive definite on* P_α^1.

Proof. Suppose that $\dim L \leq 26$. Extend L to a Lorentzian lattice M of dimension 26 by adding a positive definite lattice in necessary. For any $\alpha \in L$, $\alpha \in M$. Let $(P_\alpha^1)_L$ and $(P_\alpha^1)_M$ be the corresponding subspaces. By Corollary 5.5.32, the bilinear form is semi-positive definite on the space $(P_\alpha^1)_M$ and thus it remains so on $(P_\alpha^1)_L$ since it is a subspace of $(P_\alpha^1)_M$. $\square$

It only remains to prove the converse.

Corollary 5.5.34. *Suppose that* $\dim L > 26$. *If* $(\alpha, \alpha) < 0$, *then there exists a vector with negative norm in* P_α^1.

Proof. Let $(\alpha, \alpha) < 0$ and $v \in P_\alpha^1$. By Lemma 5.5.30, there exist vectors $b \in B_\alpha^1$ and $u \in U_\alpha^1$ such that

$$v = b + u$$

and $L_1 b = 0 = L_1 u$. Lemma 5.5.26 (ii) tells us that there exist $v_1, v_2 \in (V_L)_\alpha$ such that $u = L_{-1} v_1 + L_{-2} v_2$ and by Lemma 5.5.26 (i), we can find $u_2 \in U_\alpha^{-1}$ and $t \in B_\alpha^{-1}$ such that $v_2 = t + u_2$.

Let us calculate the norm of v. Applying Lemma 5.5.28, we get

$$(v, v) = (v, b) = (b, b) + (v_1, L_1 b) + (v_2, L_2 b) \tag{1}$$

by definition of the bilinear form (see Theorem 4.3.3). From Lemma 5.5.30 we know that

$$L_1 b = 0 \quad \text{and} \quad L_2 b = -(\frac{1}{2} \dim L - 13)t.$$

Therefore, equation (1) becomes

$$(v, v) = (b, b) - (\frac{1}{2} \dim L - 13)(t, t + u_2) = (b, b) - (\frac{1}{2} \dim L - 13)(t, t) \tag{2}$$

by Lemma 5.5.28. We want to find b, v_1, v_2 such that $(v, v) < 0$.

Since $t \in B_\alpha$ and $(\beta, \beta) = 0$, $(t, t) = 0$ unless $t \in T_\alpha$. Also, $t \in B_\alpha^{-1}$. As a consequence of Corollary 5.5.24, $T_\alpha^{-1} \neq 0$ since $(\alpha, \alpha) < 0$. So choose a non-trivial vector $t \in T_\alpha^{-1}$ and set

$$v_2 t \in T_\alpha^{-1}.$$

By Lemma 5.5.30,

$$L_1 v_1 = -3t. \tag{3}$$

The fact that $L_1 t = 0$ suggests that for v_1 we should consider a multiple of $L_{-1} t$. Indeed (3) is satisfied by

$$v_1 = \frac{3}{2} L_{-1} t.$$

With these values, v_1 and v_2 satisfy (3),

$$L_1 u = \frac{3}{2} L_1 L_{-1}^2 t + L_1 L_{-2} t = 3 L_0 L_{-1} t + 3 L_{-1} L_0 t + 3 L_{-1} t = 0$$

and

$$L_2 u = \frac{3}{2} L_2 L_{-1}^2 t + L_2 L_{-2} t = \frac{9}{2} L_1 L_{-1} t + 4 L_0 t + \frac{1}{2} \dim L t = (\frac{1}{2} \dim L - 13) t \quad (4)$$

since the Virasoro algebra has central charge $\dim L$.

Again by Lemma 5.5.30, we need

$$L_2 b = -(\frac{1}{2} \dim L - 13) t. \tag{5}$$

As $b \in B_\alpha$ and $t \in T_\alpha$, using Lemma 4.3.17,

$$L_2 \beta(-2) t = 2 \beta(0) t = 2(\alpha, \beta) t$$

since $t \in (V_L)_\alpha$. Hence as $(\alpha, \beta) \neq 0$ by assumption, (5) holds for

$$b = -\frac{1}{2(\alpha, \beta)} (\frac{1}{2} \dim L - 13) \beta(-2) t \in B_\alpha.$$

However, Lemma 5.5.30 also tells us that $L_1 b = 0$ and the above value gives

$$L_1 b = -\frac{1}{(\alpha, \beta)} (\frac{1}{2} \dim L - 13) \beta(-1) t.$$

Now, $L_1 \beta(-1)^2 t = 2(\alpha, \beta) \beta(-1) t$ and $L_2 \beta(-1)^2 t = 0$ since $(\beta, \beta) = 0$ by assumption. This leads us to set

$$b = \frac{1}{2(\alpha, \beta)} (\frac{1}{2} \dim L - 13)(\frac{1}{(\alpha, \beta)} \beta(-1) \beta(-1) - \beta(-2)) t$$

and so finally we take

$$v = \frac{1}{2(\alpha, \beta)} (\frac{1}{2} \dim L - 13)(\frac{1}{(\alpha, \beta)} \beta(-1) \beta(-1) - \beta(-2)) t + \frac{3}{2} L_{-1} t + L_{-2} t,$$

where $t \in T_\alpha^{-1}$. Let us check that $v \in P_\alpha$. The above calculations show that $L_1 b = 0$ and so from (4), $L_1 v = 0$. Again as b satisfies (5), together with (4), this gives $L_2 v = 0$. As for any $n > 0$, L_n is a polynomial in L_1 and L_2, $v \in P_\alpha$. Clearly $(b, b) = 0$ since $(\beta, \beta) = 0$. Hence, (2) becomes

$$(v, v) = -(\frac{1}{2} \dim L - 13)(t, t)$$

and so by Lemma 5.5.25 $(v, v) < 0$ as wanted.   □

This finishes the proof of the No-Ghost Theorem.

Note that to take the base field to be $\mathbf{C}$ again, we can instead consider $F_L \otimes_{\mathbf{R}} \mathbf{C}$. However, all results of Chapters 1 and 2 hold over $\mathbf{R}$ (with Hermitian forms becoming bilinear forms) when $S = \emptyset$ and so it makes no difference.

Corollary 5.5.35. *Suppose that the lattice L is even and that it is either Lorentzian of rank at most 26 or positive definite. Then, the bilinear form $(.,.)$ is positive definite on the root spaces $(F_L)_\alpha$, $\alpha \in L$ and it is contravariant with respect to the involution ω. If the lattice L is positive definite, then the Lie algebra F_L is a finite dimensional simple Lie algebra of type A_n, D_n, E_6, E_7, or E_8.*

Proof. When the lattice is Lorentzian, the result follows from Lemma 5.5.9 and Theorem 5.5.16. When the lattice L is positive definite we do not know that the Lie algebra F_L is a BKM algebra, so the result will also follow from these results, Lemma 5.5.11, and Theorem 2.5.9, once we have defined an adequate $\mathbf{Z}$-grading of the Lie algebra $F_L = \sum_i (F_L)_i$. It must satisfy the following properties:

$$\omega((F_L)_i) = (F_L)_{-i} \tag{2}$$

for all $i \in \mathbf{Z}$ and

$$\dim(F_L)_i < \infty \quad \forall i \neq 0. \tag{3}$$

Since $\omega(P_\alpha^1) = P_{-\alpha}^1$, it is reasonable to construct a grading such that for all $\alpha \in L$, $P_\alpha^1/N \leq (F_L)_i$ for some i. For $\alpha \in L$, if $(\alpha, \alpha) > 0$, then $P_\alpha^1 \neq 0$ only if $(\alpha, \alpha) = 2$ in which case $P_\alpha^1 = \mathbf{R}(1 \otimes e^\alpha)$ has dimension 1. Moreover there are only finitely many vectors of norm 2 in L since any compact discrete set is finite. We can label them $\pm \beta_i, i = 1, \cdots, m$.

The lattice L being positive definite, set

$$\deg((F_L)_{\beta_i}) = i, \quad \deg((F_L)_0) = 0.$$

Then, Theorem 5.5.16 implies that F_L is a BKM algebra. As all the roots have positive norm, by Exercise 2.3.10, it is finite dimensional semisimple. $\qquad\square$

From Corollaries 5.5.24 and 5.5.32, the dimension of root spaces follows immediately when the Lorentzian lattice L has rank 26.

Corollary 5.5.36. *Suppose that L is an even Lorentzian lattice or rank 26. Then, F_L is a BKM algebra and the multiplicity of the root $\alpha \in L$ is $p_{24}(1 - \alpha^2/2)$.*

Using the denominator formula for the Fake Monster Lie algebra given in Example 2.6.40, uniqueness of LBKM algebras of type F_L at rank 26 can be deduced.

Corollary 5.5.37. *There is a unique LBKM algebra F_L of rank 26 constructible from the lattice vertex algebra V_L in the above manner, namely the Fake Monster Lie algebra.*

When the even Lorentzian root lattice L has rank strictly greater than 26, Theorem 5.5.14 says that the Lie algebra F_L is a BKM algebra. So there is another grading, involution, and bilinear form on F_L satisfying the conditions of Theorem 2.2.9. The ones induced from the vertex algebra V_L do not because of Corollary 5.5.21. The reason is that the involution ω is not adequate. 26 is a critical dimension in the sense that it is maximal for the natural grading, involution, and bilinear form coming from the lattice vertex algebra to be the ones characterizing F_L as a BKM algebra.

Section 5.4 leads one to believe that LBKM algebras of higher ranks may not be describable explicitly. In particular, it may not be possible to calculate their root multiplicities, etc. Or rather if it is possible, then the method will be different from the one developed in this chapter and is not yet known.

The No-Ghost Theorem does not necessarily give a full description of the set of roots of the BKM algebra F_L when L is a Lorentzian lattice of rank at most 26, i.e. it does not say which roots are simple. Sections 5.3 and 5.4 indicate that this may be possible for some of them. The aim of the classification is, in the case of Lie algebras (i.e. Lie superalgebras with trivial odd part), to find all BKM algebras with even Lorentzian lattice of rank at most 26 to which a vector valued modular form can be associated in the manner of Theorem 5.3.6 and to describe their root systems explicitly. They can be constructed explicitly using the techniques explained in this section.

The above construction from the bosonic lattice vertex superalgebra only gives BKM algebras when the lattice L is even Lorentzian. When the lattice L is odd, there is another version of the No-Ghost Theorem that can be used to construct BKM algebras from the tensor product of the bosonic and fermionic lattice vertex superalgebras. Therefore, it is also possible to construct LBKM algebras with odd root lattices. In fact, in rank 10 this construction gives the even part of the Fake Monster Lie superalgebra. To get the full Fake Monster Lie superalgebra, we need to consider a rational lattice containing the Lorentzian lattice of rank 10 and use Γ-graded lattice vertex algebras, where Γ is an abelian group of order greater than 2. We omit this more complex construction as this book only intends to give the reader the basis to understand the classification and construction of LBKM algebras. The interested reader should read [Sch1,2]. Exercise 5.5.2 leads to the construction of its even part.

Open Problem 7. Once the LBKM superalgebras having an associated automorphic form satisfying Theorem 5.3.6 have been fully classified and constructed, their modules will need to be studied. In particular, they need to be constructed or described in some concrete manner.

Some recent papers on the representation theory of BKM algebras include [Jur3,4]. Modules over affine algebras, whose characters are modular functions, have been studied in [KacP1,2, KacWak1,2,3].

Exercises 5.5

1. Calculate the root multiplicities of the BKM algebra F_L when the rank of L is equal to 26. Deduce that F_L is then the Fake Monster Lie algebra given in Example 2.6.40.

2. Let L be a Lorentzian lattice and consider $V_L = V_L^f \otimes V_L^b$. We keep the notation of Exercises 4.3.1 and 4.3.2. Suppose that L is a Lorentzian lattice of rank at most 10.

(i) Show that for all $\alpha \in L - \{0\}$, $v \in G_{\frac{1}{2}} P_\alpha^{\frac{1}{2}}$, $(v, v) \geq 0$.

(ii) Let N be the kernel of the bilinear form on $G_{\frac{1}{2}} P_\alpha^{\frac{1}{2}}$. Show that $dP^0 \leq N$ and N/dP^0 is an ideal of the Lie algebra $G_{\frac{1}{2}} P_\alpha^{\frac{1}{2}} / dP^0$.

(iii) Deduce that the Lie algebra $G = G_{\frac{1}{2}} P_\alpha^{\frac{1}{2}} / N$ is a BKM algebra.

(iv) When the lattice L has rank 10, calculate the root multiplicities of the BKM algebra G.

For a solution, see [GodT] and [Sch2].

Appendix A

Orientations and Isometry Groups

Let V be a real vector space with a non-degenerate symmetric bilinear form $(.,.)$ having signature (n, m) and $O(V)$ be its group of isometries, i.e. automorphisms preserving the bilinear form. Let V^+ be a maximal positive definite subspace. Fix an ordered basis $\mathcal{B} = (v_1, v_2, \cdots, v_n)$ for V^+. Set $\pi : V \to V^+$ to be the projection of V onto V^+ along its orthogonal complement.

Lemma A.1. *For any maximal positive definite subspace U, $\ker \pi \cap U = 0$.*

Proof. Suppose that $\pi(u) = 0$ for $u \in U$. Then u belongs to the complement of V^+ and so $(u, u) \le 0$. Since the vector space U is positive definite, this forces $u = 0$. □

Corollary A.2. *Let $u_1, u_2, \cdots, u_n$ be a basis for a maximal positive definite subspace U. Then, $\pi(u_1), \cdots, \pi(u_n)$ is a basis for V^+.*

Corollary A.3. *Let $u_1, u_2, \cdots, u_n$ be a basis for a maximal positive definite subspace U and $A = (a_{ij})$, where $\pi(u_i) = \sum_{j=1}^{n} a_{ji}v_j$ for $1 \le i \le n$. Then,*

$$\pi(u_1) \wedge \cdots \wedge \pi(u_n) = (\det A)v_1 \wedge \cdots \wedge v_n.$$

Definition A.4. *An ordered basis $\mathcal{B}_U$ for a maximal positive definite subspace U is said to have the same orientation as the ordered basis $\mathcal{B}$ if $\det A > 0$, where A is the transition matrix from the basis $\mathcal{B}$ to the basis $(\pi(u) : u \in \mathcal{B})$.*

The following result is an immediate consequence of elementary properties of determinants.

Lemma A.5. *If the bases $\mathcal{B}_1$ and $\mathcal{B}_2$ have the same orientation, and the bases $\mathcal{B}_2$ and $\mathcal{B}_3$ also have the same orientation, then so do the bases $\mathcal{B}_1$ and $\mathcal{B}_3$.*

Hence there are two possible orientations on V. A basis of a maximal positive definite subspace with the same orientation as the basis $\mathcal{B}$ will be said to be positively oriented.

For any $v \in V$ with non-zero norm, let r_u denote the reflection through the orthogonal hyperspace to v, i.e.

$$r_v(u) = u - \frac{2(u, v)}{(v, v)} v, \ u \in V.$$

Clearly $r_v \in O(V)$.

Theorem A.6 *(Cartan-Dieudonné). For any element $\phi \in O(V)$, there exist $x_1, \cdots, x_s$ in V with non-zero norm such that $\phi = r_{x_1} \cdots r_{x_s}$.*

Proof. We only prove the theorem for isometries ϕ having a fixed point with non-zero norm as we will only need the result in this case. Consider the subspace F of fixed points of the map ϕ. Let F_1 be the complement of the kernel of the bilinear form on F. Since there exists $v \in F$ with $(v, v) \neq 0$, $F_1 \neq 0$. By definition of the subspace F_1, the bilinear form is non-degenerate on it. By induction on the dimension of V, the result is therefore true for the restriction of the map ϕ to the complement $F_1^\perp$ of F_1 in V. So there exist $x_1, \cdots, x_s \in F_1^\perp$ such that $\phi_{F_1^\perp} = r_{x_1} \cdots r_{x_s}$, where the maps r_{x_i} are reflections of $F_1^\perp$. Extend the involution r_{x_i} to V by setting $r_{x_i}(x_i) = -x_i$ and $r_{x_i}(x) = 0$ for all $x \in V$ orthogonal to x_i. Then r_{x_i} becomes naturally a reflection of V. Since, $x_i \in F_1^\perp$, for all $x \in F_1$, $r_{x_i}(x) = x$. Therefore $\phi = r_{x_1} \cdots r_{x_s}$ on $V = F_1 \oplus F_1^\perp$. $\square$

Lemma A.7. *The group of isometries of V keeping the orientation invariant is the subgroup of products $r_{x_1} \cdots r_{x_s}$ where the number of vectors x_i having positive norm is even.*

Proof. Suppose that $x \in V$ is a vector having negative norm. Let $\mathcal{B}$ be an ordered basis of a positive definite maximal subspace orthogonal to x. Then, the reflection r_x fixes $\mathcal{B}$. Hence, Lemma A.5 implies that it leaves the orientation invariant. Suppose next that $x \in V$ is a vector having positive norm. Let $\mathcal{B}$ be an ordered orthogonal basis of a positive definite maximal subspace containing x. Then, the reflection r_x sends $\mathcal{B}$ to a basis whose elements are the same as those in $\mathcal{B}$ except that x becomes $-x$. Hence, Lemma A.5 implies that it changes the positive orientation into the negative one. The result is then a consequence of Theorem A.6.

Appendix B

Manifolds

B.1   Some Elementary Topology

Definition B.1.1. *A topological space is a pair* $\{M, \mathcal{T}\}$ *of a non-empty set* M *together with a collection* $\mathcal{T}$ *of subsets of* M *satisfying*

(i) $M, \emptyset \in \mathcal{T}$

(ii) *For any* $M_1, M_2 \in \mathcal{T}$, $M_1 \cap M_2 \in \mathcal{T}$

(iii) *For any collection of sets* $M_i \in \mathcal{T}$, $i \in I$, $\cup_{i \in I} M_i \in \mathcal{T}$.

A subset of M is said to be open if it is in $\mathcal{T}$.

Definition B.1.2. *A topological space* M *is Hausdorff if for any* $a, b \in M$, *there exist open subsets* U, V *of* M *such that* $a \in U$, $b \in V$ *and* $U \cap V = \emptyset$.

Definition B.1.3. *A basis of the topological space* M *is a collection of open subsets* $\mathcal{B}$ *such that any open subset of* M *is the union of open subsets in* $\mathcal{B}$. *The topological space* M *space is second countable if it has a countable basis.*

Definition B.1.4. *A map* $f : M_1 \to M_2$ *of topological spaces with topologies* $\mathcal{T}_i$ *is continuous if for any* $U \in \mathcal{T}_2$, $f^{-1}(U) \in \mathcal{T}_1$. *The map* f *is a homeomorphism if it is bijective and both* f *and* f^{-1} *are continuous.*

Definition B.1.5. *A function* $f : U \to \mathbf{R}^n$, *where* U *is an open subset of* $\mathbf{R}^n$ *is a diffeomorphism if*

(i) f *is smooth, i.e.* $f_i = x_i f$ *is infinitely differentiable, where* $x_i : \mathbf{R}^n \to \mathbf{R}$, $(x_1, \cdots, x_n) \mapsto x_i$ *are the coordinate functions on* $\mathbf{R}^n$.

(ii) $f : U \to f(U)$ *is a homeomorphism and its inverse is smooth.*

B.2   Manifolds

Definition B.2.1.

(1) A chart on a topological space M is a pair (U, ϕ), where U is an open subset of M and $\phi : U \to \mathbf{R}^n$ is a homeomorphism onto $\phi(U)$. The components $\phi = (x_1, \cdots, x_n)$, where $x_i : U \to \mathbf{R}$, are the coordinates.

(2) An atlas for M is a collection $\mathcal{A} = \{(U_\alpha, \phi_\alpha) : \alpha \in A\}$ of charts of M such that:

(i) $\cup_{\alpha \in A} U_\alpha = M$ and

(ii) for any $\alpha, \beta \in A$, the maps $\phi_\beta \phi_\alpha^{-1} : \phi_\alpha(U_\alpha \cap U_\beta) \to \phi_\beta(U_\alpha \cap U_\beta)$ are diffeomorphisms.

Definition B.2.2. *A differentiable (smooth) manifold is a second countable Hausdorff topological space together with an atlas.*

Definition B.2.3. *Let M and N be differentiable manifolds with atlases $\mathcal{A} = \{(U_\alpha, \phi_\alpha) : \alpha \in A\}$ and $\mathcal{B} = \{(V_\beta, \psi_\beta) : \beta \in B\}$ respectively. A map $f : M \to N$ is smooth (differentiable) if for any $\alpha \in A$, $\beta \in B$,*

$$\psi_\beta f \phi_\alpha^{-1} \mid_{\phi_\alpha(U_\alpha \cap f^{-1}(V_\beta))} : \phi_\alpha(U_\alpha \cap f^{-1}(V_\beta)) \to \psi_\beta(V_\beta)$$

is infinitely differentiable.

Definition B.2.4. *Let M be a smooth manifold with atlas $\mathcal{A}$. A smooth curve is a smooth map $\gamma :] - \epsilon, \epsilon[\; (\subset \mathbf{R}) \to M$, i.e. for any chart $(U, \phi) \in \mathcal{A}$ such that $\gamma(t) \in U$ for some $t \in \mathbf{R}$, the map $\phi\gamma : \gamma^{-1}(U_\alpha) \to \mathbf{R}^n$ is smooth.*

Definition B.2.5. *Let M be a smooth manifold with atlas $\mathcal{A}$. Two curves γ_1, γ_2 passing through the point $a \in M$, i.e. $\gamma_i(t_i) = a$, have the same tangent at a if for any chart $(U, \phi) \in \mathcal{A}$ such that $a \in U$, $(\phi\gamma_1)'(t_1) = (\phi\gamma_2)'(t_2)$.*

Lemma B.2.6. *Definition B.2.5 gives an equivalence relation.*

The equivalence class of γ will be written $\dot{\gamma}$.

Definition B.2.7. *Let M be a smooth manifold. The tangent space TM_a at $a \in M$ is the set of equivalence classes of curves passing through a given in Definition B.2.4.*

Lemma B.2.8. *Let M be a smooth manifold with atlas $\mathcal{A}$ and $(U, \phi) \in \mathcal{A}$ such that $a \in U$ and $\phi : U \to \mathbf{R}^n$. The tangent space at $a \in M$ is a real vector space of dimension n.*

Proof. Set

$$(d\phi)_a : TM_a \to \mathbf{R}^n$$
$$\dot{\gamma} \mapsto (\phi\gamma)'(t),$$

where $\gamma(t) = a$. The map $(d\phi)_\alpha$ is injective by definition of the equivalence relation. It is surjective since for any $v \in \mathbf{R}^n$, $\gamma_v(t) = \phi^{-1}(\phi(a) + tv)$ for $t \in \mathbf{R}$ defines a smooth curve mapped onto v by $(d\phi)_a$. The vector space structure is then transferred from $\mathbf{R}^n$ by $(d\phi)_a$. $\square$

For a M be a smooth manifold with atlas $\mathcal{A}$, set

$$C^\infty(M) = \{f : M \to \mathbf{R}, f\phi^{-1} \text{ is smooth } \forall \ (U, \phi) \in \mathcal{A}\}.$$

Definition B.2.9. *Let M be a smooth manifold. A derivation at $a \in M$ is a linear map $C^\infty(M) \to \mathbf{R}$ such that for all $f, g \in C^\infty(M), D(fg) = D(f)g(a) + f(a)D(g)$.*

Lemma B.2.10. *Let M be a smooth manifold. The derivations at $a \in M$ form a vector space with basis $\frac{\delta}{\delta x_i}$, $i = 1, \cdots, n$ (in local coordinates) isomorphic to the tangent space TM_a.*

Proof. Let γ be a curve such that $\gamma(t) = a \in M$. Set

$$\dot\gamma \mapsto D,$$

where $D(f) = (f\gamma)'(t)$ for all $f \in C^\infty(M)$. Note that $f\gamma = f\phi^{-1}\phi\gamma$, where (U, ϕ) is a chart such that $a \in U$, so that $f\gamma$ is differentiable. Set $\phi_i(a) = \phi(a)_i = x_i$, where x_i is the i-th local coordinate of a. The above defined map is injective. This follows from considering $f = \phi_i$ for each $1 \le i \le n$. It is clearly linear. The derivations form a vector space of dimension n since the elements $\frac{\delta}{\delta x_i}$, $1 \le i \le n$ form a basis. This can be seen by taking the Taylor expansions (see Definition C.2.6) of $f\phi_i^{-1}$ around $\phi_i(a)$. Therefore the map gives the desired isomorphism. $\square$

From now on for a smooth manifold M, we consider the tangent space TM_a at $a \in M$ as the vector space of derivations X_a at a.

Definition B.2.11. *Let M be a smooth manifold and U an open subset of M.*

(1) For each $a \in U$, choose a tangent vector $X_a \in TM_a$. The family $X = \{X_a : a \in U\}$ is a smooth vector field on U if for any smooth function $f : U \to \mathbf{R}$, the map

$$a \mapsto X_a(f)$$

is smooth.

(2) A smooth s-tensor field on U is a family $g = (g_a)_{a \in U}$ of multilinear maps

$$g_a : TM_a \times \cdots \times TM_a \to \mathbf{R}$$

(s copies of TM_a) such that for any smooth vector fields $X_1, \cdots, X_s$ on U, the map

$$U \to \mathbf{R}$$
$$a \mapsto g_a(X_1, \cdots, X_s)$$

is smooth.

(3) A smooth tensor field on U is a family $g = (g_a)_{a \in U}$ of linear maps

$$g_a : TM_a \to TM_a$$

such that for any smooth vector field $X = \{X_a : a \in U\}$,

$$a \mapsto g_a(X_a)$$

is a smooth vector field.

Hermitian symmetric spaces and Riemannian metrics

Definition B.2.12. *Let M be a smooth manifold. A smooth 2-tensor field t is a Riemannian metric if for all $a \in M$, the map t_a is symmetric and positive definite. A smooth manifold with a Riemannian metric is a Riemannian manifold.*

Let t be a Riemannian metric on the smooth manifold M and $a \in M$. By Lemma 15, The dual vector space TM_a of the tangent space TM_a has basis (in local coordinates) dx_i, $i = 1, \cdots, n$, where

$$dx_i\left(\frac{\delta}{\delta x_j}\right) = \delta_{ij}.$$

The dual vector space of $TM_a \times TM_a$ is $(TM_a)^* \otimes (TM_a)^*$. This clearly follows by comparing dimensions. Hence, as t_a is a bilinear map $TM_a \times TM_a \to \mathbf{R}$, it can be written as follows

$$t_a = \sum_{i,j=1}^{n} t_{i,j}(a)dx_i \otimes dx_j,$$

where $t_{i,j}(a) \in \mathbf{R}$. We will write $dx_i dx_j$ for $dx_i \otimes dx_j$.

Definition B.2.13. *Let M and N be smooth manifolds. The derivative at $a \in M$ of a smooth map $F : M \to N$ is the linear map*

$$(dF)_a : TM_a \to TN_{F(a)}$$
$$X_a \mapsto (f \mapsto X_a(fF))$$

for any $f \in C^\infty(N)$.

Let M and N be smooth manifolds and $F : M \to N$ a smooth map. For any $b \in N$, let $Y_b^* \in (TN_b)^*$ be the dual of the vector $Y_b \in TN_b$. Let $dF_a^* : (TN_b)^* \to (TM)_a^*$ be the dual map of $(dF)_a$. Then, for any $a \in M$, $X_a \in TM_a$,

$$((dF)_a Y_{F(a)})(X_a) = Y_{F(a)}(dF)_a(X_a).$$

Definition B.2.14. *Let M be a Riemannian manifold. An isometry $F : M \to M$ is a smooth map preserving the Riemannian metric g, i.e. for any $a \in M$, $X_a, Y_a \in TM_a$,*

$$g_{F(a)}((dF)_a(X_a), (dF)_a(Y_a)) = g_a(X_a, Y_a).$$

$Isom(M)$ denotes the isometry group of M.

Lemma B.2.15. *Let M be a Riemannian manifold with Riemannian metric g. An isometry $F : M \to M$ is a smooth map such that $(dF)_a^*(g_{F(a)}) = g_a$ for all $a \in M$.*

Definition B.2.16. *A homogeneous space is a Riemannian manifold M on which the action of its isometry group $Isom(M)$ is transitive.*

Definition B.2.17. *A symmetric space M is a connected homogeneous space such that there exists an involution $\phi \in Isom(M)$ (i.e. $\phi^2 = 1$) with at least one isolated fixed point (i.e there exists a point $a \in M$ and a neighbourhood U of a such that for all $u \in U$, $\phi(u) = u$ if and only if $u = a$).*

Definition B.2.18. *An almost complex structure on a smooth manifold M is a smooth tensor field $(J_a)_{a \in M}$, $J_a : TM_a \to TM_a$ such that $J_a^2 = -1$ for all $a \in M$.*

Definition B.2.19. *Let M be a second countable Hausdorff topological space, U_α, $\alpha \in A$, a family of open subspaces such that $M = \cup_{\alpha \in A} U_\alpha$, and $u_\alpha : U_\alpha \to \mathbf{C}^n$ homeomorphisms onto $u_\alpha(U_\alpha)$. Then, M is a complex manifold if the maps $u_\alpha u_\beta^{-1}$ are analytic (i.e differentiable) for all $\alpha, \beta \in A$.*

Lemma B.2.20. *A complex manifold M is a smooth manifold with an almost complex structure.*

Proof. Consider M as a smooth manifold by using the obvious map $\mathbf{C}^n \to \mathbf{R}^{2n}$. Let $a \in M$, U an open subset such that $\alpha \in U$ and $f : U \to \mathbf{R}$ a smooth map. Define

$$J_a : TM_a \to TM_a$$

$$\frac{\delta}{\delta x_i} \mapsto \frac{\delta}{\delta y_i}, \quad \frac{\delta}{\delta y_i} \mapsto -\frac{\delta}{\delta x_i}$$

$\square$

Definition B.2.21. *A Hermitian metric on a complex manifold M is a Riemannian metric t satisfying $t(JX, JY) = t(X, Y)$ for all vector fields X, Y. A Hermitian manifold is a complex manifold with a Hermitian metric.*

Definition B.2.22. *A Hermitian symmetric space is a Hermitian manifold which is a symmetric space.*

Lemma B.2.23. *If M is a real space with a non-degenerate, symmetric, bilinear form $(.,.)$, then $\mathbf{P}(M \otimes_{\mathbf{R}} \mathbf{C})$ is a Hermitian manifold and the Riemannian metric is given by the bilinear form.*

Proof. We consider an orthogonal basis e_i, $1 \le i \le n+1$ for the vector space $M \otimes \mathbf{C}$ such that $e_i^2 = \pm 1$. For $z \in M \otimes_{\mathbf{R}} \mathbf{C}$, we write $z = (z_1, \cdots, z_n)$ with respect to this basis. Set $U_i = \{z \in \mathbf{P}(M \otimes \mathbf{C}) : z_i \ne 0\}$. Then,

$$\mathbf{P}(M \otimes_{\mathbf{R}} \mathbf{C}) = \bigcup_{i=1}^{n+1} U_i.$$

For any $z \in V$, there is a unique representative such that $z_i = 1$. Then, there is a natural well defined analytic map $\phi_i : U_i \to \mathbf{C}^n$. Furthermore,

$$\phi_i \phi_j^{-1} : \phi_j(U_i \cap U_j) \to \phi_i(U_i \cap U_j)$$
$$(z : z_i = 1, z_j = 1) \mapsto (z : z_i = 1, z_j = 1)$$

is analytic. Hence, $\mathbf{P}(M \otimes_{\mathbf{R}} \mathbf{C})$ is a complex manifold. Considering it as a real manifold and setting $z_j = x_j + iy_j$, the Riemannian metric on U_i is given by

$$(\frac{\delta}{\delta x_j}, \frac{\delta}{\delta x_k}) = (\frac{\delta}{\delta y_j}, \frac{\delta}{\delta y_k}) = \delta_{jk} e_i^2; \ (\frac{\delta}{\delta x_j}, \frac{\delta}{\delta y_k}) = 0 \ \forall 1 \le j, k \le n+1, \ j, k \ne i.$$

It follows from Lemma B.2.20 that $\mathbf{P}(M \otimes_{\mathbf{R}} \mathbf{C})$ is a Hermitian manifold. $\square$

Definition B.2.24. *The hyperbolic space of dimension n is the following submanifold of the Lorentzian space $\mathbf{R}^{n,1}$ of rank $(n, 1)$:*

$$\mathbf{H}^n = \{u \in \mathbf{R}^{n,1} : (u, u) = -1, x_{n+1} > 0\},$$

where $(.,.)$ is the bilinear form on $\mathbf{R}^{n,1}$ and for $x = (x_1, \cdots, x_{n+1}) \in \mathbf{R}^{n,1}$, $(x, x) = x_1^2 + \cdots + x_n^2 - x_{n+1}^2$.

Note the similarity with the definition of the n-sphere in Euclidean space.

Lemma B.2.25. *When $n = 2$, $\mathbf{H}^2$ is the upper half plane.*

Lemma B.2.26. *The upper half plane $\mathcal{H}$ is a Hermitian symmetric space with the Poincaré metric given by $\frac{dx^2 + dy^2}{y^2}$ and isometry group $SL_2(\mathbf{R})$.*

Proof. $\mathcal{H}$ is clearly a complex manifold and $\frac{dx^2 + dy^2}{y^2}$ gives a Hermitian metric. Let $F : \mathcal{H} \to \mathcal{H}$ be a smooth map and $z = x + iy \in \mathcal{H}$. Then,

$$(dF)_z^*(dF(z)) = (dF(z))(dF)_z$$

and for any $f \in C^\infty(M)$,

$$(dF)_z(\frac{\delta f}{\delta z}) = \frac{\delta}{\delta z} f(F(z)) = \frac{\delta f(F(z))}{\delta F(z)} \frac{\delta F(z)}{\delta z}.$$

Therefore,

$$(dF)_z(\frac{\delta}{\delta z}) = \frac{\delta F(z)}{\delta z} \frac{\delta}{\delta F(z)},$$

and so

$$(dF)_z^*(dF(z))(\frac{\delta}{\delta z}) = \frac{\delta F(z)}{\delta z}.$$

As a result,

$$(dF)_z^*(dF(z)) = \frac{\delta F(z)}{\delta z} dz.$$

Let $A = \begin{pmatrix} a & b \\ c & d \end{pmatrix} \in SL_2(\mathbf{R})$ and $F(z) = Az$. This is clearly a smooth map and

$$(dF)_z^*(dF(z)) = \frac{dz}{(cz+d)^2}.$$

Hence

$$(dF)_{(x,y)}^*((dF(x))^2 + (dF(y))^2) = (dF)_z^*(dF(z))(dF)_z^*(dF(\bar{z}))$$

$$= \frac{dz d\bar{z}}{|cz+d|^2}.$$

Moreover, writing $y = y(z)$, we get

$$yF(z) = \frac{1}{2i}(F(z) - F(\bar{z}))$$

$$= \frac{1}{2i}\left(\frac{az+b}{cz+d} - \frac{a\bar{z}+b}{c\bar{z}+d}\right)$$

$$= \frac{y}{|cz+d|^2}.$$

It follows that

$$(dF)_{(x,y)}^*\left(\frac{(dF(x))^2 + (dF(y))^2}{F(y)^2}\right) = \frac{dx^2 + dy^2}{y^2}.$$

Therefore $SL_2(\mathbf{R})$ acts as a group of isometries of the Riemannian manifold $\mathcal{H}$. As its action is transitive, $\mathcal{H}$ is a homogeneous space. The map $z \mapsto \frac{-1}{z}$ is an involution with a unique fixed point in $\mathcal{H}$, namely i. So $\mathcal{H}$ is a Hermitian symmetric space. $\qquad\square$

B.3   Fibre Bundles and Covering Spaces

Definition B.3.1. *A covering map on a topological space M is a continuous surjective map f from a topological space N onto M with the property that for any $x \in M$, there exists an open neighbourhood U of x such that $f^{-1}(U)$ is the union of disjoint open sets, each homeomorphic to U. The space N is a covering space of M. If N is connected then it said to be a universal cover. For $x \in M$, $f^{-1}(x)$ is the fibre over x. If for any $x \in M$, $f^{-1}(x)$ contains two elements, the topological space N is a double cover of M. The triple (N, M, f) is a fibre bundle. If the topological spaces N, M are smooth manifolds and f is a smooth map, then (N, M, f) is a smooth fibre bundle.*

Definition/Lemma B.3.2. *Let M be a topological space.*

(i) *Two covering spaces $f_i : N_i \to M$, $i = 1, 2$ of M are equivalent if there is a homeomorphism $F : N_1 \to N_2$ such that $f_1 = f_2 F$. This gives an equivalence relation.*

(ii) *If M has a universal cover then it is unique.*

Definition B.3.3. *Let M be a topological space.*

  (i) *A loop with base point $x_0 \in M$ is a continuous function $f : [0,1] \to M$ such that $f(0) = x_0 = f(1)$.*
  (ii) *Two loops f, g with base point x_0 are homotopy equivalent if there exists a continuous map $F : [0,1] \times [0,1] \to M$ such that $F(t,0) = f(t)$, $F(0,t) = g(t), F(0,t) = x_0 = F(1,t)$ for all $t \in [0,1]$.*

Lemma B.3.4. *Let M be a topological space.*

  (i) *Homotopy equivalence defines an equivalence relation on the set of all paths based at $x_0 \in M$. The equivalence classes are called homotopy classes.*
  (ii) *If f, g are two loops based at x_0,*

$$fg : [0,1] \to M$$
$$t \mapsto f(2t), t \in [0, 1/2],$$
$$t \mapsto f(2t-1), t \in [1/2, 1]$$

is a loop based at x_0.

Definition/Lemma B.3.5. *Let M be a topological space. The set of homotopy classes of loops based at $x_0 \in M$ together with the multiplication induced from Lemma B.3.4 (ii) forms a group, called the fundamental group at x_0, written $\pi_1(M, x_0)$.*

Lemma B.3.6. *Let M be a connected topological space. Then, $\pi_1(M, x) \cong \pi_1(M, y)$ for all $x, y \in M$.*

Theorem B.3.7. *Let M be a connected topological space. There is a bijective correspondence between equivalence classes of covering spaces and subgroups of the group $\pi_1(M, x)$ for any $x \in M$.*

Sketch of Proof. Let N be a covering space of M and $F : N \to M$ a covering map. Choose a point $y_0 \in N$ such that $f(y_0) = x_0$. Define

$$F_* : \pi_1(N, y_0) \to \pi_1(M, x_0)$$
$$[f] \mapsto [Ff].$$

Then, F_* is a monomorphism and equivalence classes of covering spaces induce the same subgroup of $\pi_1(M, x_0)$. Hence, this defines a map from the set of equivalence classes of covering spaces to the set of subgroups of $\pi_1(M, x_0)$.

  Let f be a loop in M based at x_0. Then there is a unique path $\hat{f}$ in N based at y_0 such that $F\hat{f} = f$. Therefore one can naturally construct an inverse to the above map. This shows that it is bijective.   □

Theorem B.3.8. *The Fundamental group of $SL(2, \mathbf{R})$ is infinite cyclic.*

Sketch of Proof. Let **1** denote the identity matrix. For $n \in \mathbf{Z}$,

$$f_n : [0, 1] \to SL_2(\mathbf{R})$$
$$t \mapsto \begin{pmatrix} \cos(2\pi i n t) & \sin(2\pi i n t) \\ -\sin(2\pi i n t) & \cos(2\pi i n t) \end{pmatrix} \cdot$$

This gives an isomorphism from $\mathbf{Z}$ to $\pi_1(SL_2(\mathbf{R}), \mathbf{1})$. □

Definition B.3.9. *A principal bundle is a fibre bundle* (P, B, π) *together with a right action* $P \times G \to P$ *by a Lie group* G *in such a way that for any* $b \in B$, $\pi^{-1}(b)g = \pi^{-1}(b)$ *for all* $g \in G$ *and* G *acts transitively and freely (i.e. without fixed points) on all the fibres.*

Appendix C

Some Complex Analysis

C.1   Measures and Lebesgue Integrals

Definition C.1.1. *Let X be a set. A σ-algebra $\mathcal{M}$ is a collection of subsets of X with the following properties:*

(i) $X \in \mathcal{M}$;

(ii) For any $A \in \mathcal{M}$, $X \backslash A \in \mathcal{M}$;

(iii) For $i = 1, 2, \cdots$, if $A_i \in \mathcal{M}$, then $\bigcup_{i=1}^{\infty} A_i \in \mathcal{M}$.

Lemma C.1.2. *Let X be a set and $\mathcal{T}$ be a subset of $P(X)$ (the set of all subsets of X). There is a smallest σ-algebra $\mathcal{M}$ such that $\mathcal{T} \subseteq \mathcal{M}$.*

Proof. $\bigcap_{\substack{\mathcal{M} \\ \mathcal{T} \subseteq \mathcal{M}}} \mathcal{M}$ is the desired σ-algebra. $\qquad\qquad\qquad\qquad\qquad\qquad\qquad\quad\square$

Definition C.1.3. *Let X be a topological space. The elements of the smallest σ-algebra $\mathcal{B}$ containing the open sets are the Borel sets.*

Definition C.1.4. *Let X be a set and $\mathcal{M}$ a σ-algebra A function $f : X \to [0, \infty]$ is measurable if for any open set $U \subseteq [0, \infty]$, $f^{-1}(U) \in \mathcal{M}$.*

In the topological space $[0, \infty]$ the open sets are the usual ones together with the sets $[0, a)$ and $(a, \infty]$, $a \in \mathbf{R}_+$.

Definition C.1.5. *Let X be a set. A function $s : X \to \mathbf{R}$ is simple if it takes only finitely many values.*

For any $E \in P(X)$, set $\chi_E : X \to \mathbf{R}$ to be the characteristic function of E, i.e.

$$\chi_E(x) = \begin{cases} 1 & \text{if } x \in E; \\ 0 & \text{otherwise.} \end{cases}$$

Lemma C.1.6. *Let X be a set and $\mathcal{M}$ a σ-algebra. Let s be a simple function $X \to \mathbf{R}$, $a_1, \cdots, a_n$ the values of s and $X_i = \{x \in X : s(x) = a_i\}$. Then, s is measurable if and only if $X_i \in \mathcal{M}$ for all $1 \le i \le n$.*

Proof. The result follows from the fact that $s = \sum_{i=1}^{n} a_i \chi_{X_i}$. $\square$

Definition C.1.7. *Let X be a set and $\mathcal{M}$ a σ-algebra. A function $\mu : \mathcal{M} \to [0, \infty]$ is a positive measure if for any mutually disjoint countable collection of set X_i, $i = 1, \cdots$, $\mu(\bigcup_{i=1}^{\infty} X_i) = \sum_{i=1}^{\infty} \mu(X_i)$.*

Definition C.1.8. *A measure space is a triple $(X, \mathcal{M}, \mu)$, where X is a set, $\mathcal{M}$ a σ-algebra, and μ a positive measure of X.*

If $f, g : X \to [-\infty, \infty]$ and $f(x) \leq g(x)$ for all $x \in X$, we will write $f \leq g$.

Definition C.1.9. *Let $(X, \mathcal{M}, \mu)$ be a measure space and $s = \sum_{i=1}^{n} a_i \chi_{X_i}$ a simple measurable function of X, where $a_i \in \mathbf{R}$ for $i = 1, \cdots, n$. Then the Lebesgue integral of s over $E \in \mathcal{M}$ is:*

$$\int_E s d\mu = \sum_{i=1}^{n} a_i \mu(X_i \cap E).$$

More generally, if $f : X \to [0, \infty]$ is a measurable function, then the Lebesgue integral of f over $E \in \mathcal{M}$ is:

$$\int_E f d\mu = \sup\{\int_E s d\mu : s \quad \text{simple}, \quad 0 \leq s \leq f\}.$$

Theorem C.1.10. *Let X be a locally compact, Hausdorff topological space. Let $C_c(X)$ be the set of all functions $f : X \to \mathbf{R}$ with compact support, i.e. $\{x \in X : f(x) \neq 0\}$ is compact, and Λ be linear map $C(X) \to \mathbf{C}$ such that for any $f \geq 0$, $\Lambda(f) \geq 0$. Then, there is a σ-algebra $\mathcal{M}$ which contains all the Borel sets and a positive measure μ on X satisfying $\Lambda(f) = \int_X f d\mu$ for all $f \in C_c(X)$.*

Proof. If V is an open subset of X, then define

$$\mu(V) = \sup\{\Lambda f : f \in C_x(X), f(x) : X \to [0, 1], f(x) \neq 0 \Rightarrow x \in V\}.$$

For $E \in P(X)$, define

$$\mu(E) = \inf\{\mu(V) : E \subseteq V, V \quad \text{open}\}.$$

Set $\mathcal{M}_1$ be the subset of $P(X)$ consisting of subsets E of X for which $\mu(E) = \sup\{\mu(V) : V \subseteq E, V \quad \text{compact}\}$ and $\mu(E) < \infty$. Then, define

$$\mathcal{M} = \{E \in P(X) : E \cap F \in \mathcal{M}_1, \forall F \in \mathcal{M}_1\}.$$

$\square$

Since $\mathbf{R}^n$ is locally compact, i.e. all points have a compact neighbourhood, so are all manifolds. Therefore the following result is now immediate since manifolds are second countable and Hausdorff.

Corollary C.1.11. *Let M be a Riemannian manifold with metric g. Set $G = \det(g_{ij})$. Let (U_α, ϕ_α), $\alpha \in \mathcal{A}$ be an atlas. Let $\phi_\alpha = (x_1, \cdots, x_n)$ be the local coordinates. Then, for any function $f \in C_c(M)$,*

$$\int_M f d\mu := \sum_{\alpha \in \mathcal{A}} \int_{\phi(U_\alpha)} f \circ \phi_\alpha^{-1} \sqrt{G \circ \phi_\alpha^{-1}} dx_1 \cdots dx_n,$$

defines a σ-algebra containing the Borel sets and a positive measure on M which is independent of the atlas chosen.

Let us apply Corollary C.1.11 to the upper half plane $\mathcal{H}$.

Corollary C.1.12. *$d\mu = dxdy/y^2$ gives a positive measure on $\mathcal{H}$.*

Proof. By Lemma B.2.23, $g_{11} = 1/y^2 = g_{22}$ and $g_{12} = 0 = g_{21}$. Hence the result. $\qquad\square$

C.2 Complex Functions

Definition C.2.1. *A function f on $\mathbf{C}$ is analytic or holomorphic at a point $\tau \in \mathbf{C}$ if there is a neighbourhood of z such that the derivative of f exists at all points of it. More generally, a function on $\mathbf{C}^n$ is analytic or holomorphic at a point $(z_1, \cdots, z_n) \in \mathbf{C}^n$ if there exists a neighbourhood of z such that all the partial derivatives of f exist at all points of it. A singular point or singularity of a function f on $\mathbf{C}^n$ is a point at which f is not analytic. If there is an integer $n > 0$ such that $(z - z_0) \lim\limits_{z \to z_0}^{n} f(z)$ is a non-zero real, then $z = z_0$ is a pole of order n. A function is meromorphic in a region $\mathcal{R}$ of the complex plane if it is analytic everywhere on $\mathbf{R}$ except at a finite number of poles.*

The notion of singularities for functions on Hermitian symmetric domains (see Definition B.2.22) can easily be derived from that of functions on $\mathbf{C}^n$.

Definition C.2.2. *For $n \in \mathbf{Z}$, let f_n be real or complex functions. Let*

$$s_n(s) = \sum_{m=N}^{n} f_m(s)$$

be the partial sums. The series $\sum_{n \geq N} f_n(s)$ converges uniformly towards the function f if $\forall \epsilon > 0$, there is an integer M (depending only on ϵ) such that for all $n \geq M$, $|s_n(s) - f(s)| < \epsilon$.

The difference with ordinary convergence is that the integer M does not depend on s.

Lemma C.2.3.(Cauchy's test). *The series $\sum_{n \geq N} f_n(s)$ converges uniformly if and only if $\forall \epsilon > 0$, there is an integer M (depending only on ϵ) such that for all $n, m \geq M$, $|s_n(s) - s_m(s)| < \epsilon$.*

Lemma C.2.4. (Weierstrass M-test). *If for all n, there exists $M_n \in \mathbf{R}$ such that in a region $\mathcal{S}$, $\left| f_n(s) \right| \leq M_n$, where M_n is independent of s and the series $\sum_n M_n$ converges, then the series $\sum_n f_n(s)$ converges uniformly in $\mathcal{S}$.*

Lemma C.2.5. *Suppose that for all n, the functions f_n are analytic on a region $\mathcal{S}$ of $\mathbf{C}$ and the series $\sum_n f_n(x)$ converges uniformly on all compact subsets of $\mathcal{S}$. Then the function $f(s) = \sum_n f_n(s)$ is analytic on $\mathcal{S}$ and $f'(s) = \sum_n f'_n(s)$ on all compact subsets.*

Theorem/Definition C.2.6. *If f is complex valued function on the complex plane, analytic in $D(s_0, R) = \{ s \in \mathbf{C} : \left| s - s_0 \right| < R \}$, then*

$$f(s) = \sum_{n=0}^{\infty} \frac{f^n(s_0)}{n!} (s - s_0)^n$$

for all $\tau \in D(s_0, R)$. This is the Taylor series of f at s_0. It converges uniformly in $D(s_0, R)$. If for all $n \in \mathbf{Z}$, $f^n(s_0) \in \mathbf{R}$, then it is said to be real analytic at s_0. Otherwise it is complex analytic.

Corollary C.2.7. *If f is a meromorphic complex valued function on the complex plane, with a pole of order m at s_0, then*

$$f(s) = \sum_{n=-m}^{\infty} a_n (s - s_0)^n,$$

where $a_n \in \mathbf{C}$ for all $n \geq -m$. This series is uniquely defined and is the Laurent series of f at s_0. There exists $R_1 \leq R_2$ such that it converges uniformly in $R_1 < \left| s - s_0 \right| < R_2$. The coefficient a_{-1} is called the residue of f at $s = s_0$.

C.3   Integration

Lemma C.3.1. *Let f be a complex valued function on $\mathbf{R}$.*

(i) *If the function f is continuous on the interval $[a, b]$, then the integral $\int_a^b f(x)dx$ exists.*

(ii) *The integral $\int_a^{\infty} f(x)dx$ converges if there exists $M, N \in \mathbf{R}$ with $N > 1$ such that $0 < \left| f(x) \right| \leq M/x^N$ for all $x \geq a$.*

Proof.   We only prove (ii).   Suppose that M, N are as given and $0 < \left| f(x) \right| \leq M/x^N$ for all $x \geq a$. Set $g(b) = \int_a^b f(x)dx$. For $b \geq a$, we then get

$$|g(b)| \leq M \int_a^b dx/x^N$$

$$= [\frac{M}{(-N+1)x^{N-1}}]_a^b$$

$$= \frac{M}{(N-1)}(\frac{1}{a^{N-1}} - \frac{1}{b^{N-1}})$$

$$\leq \frac{M}{(N-1)a^{N-1}}.$$

Hence in the interval $[a, \infty)$, $|g|$ is a bounded increasing function and so $\lim_{b \to \infty} g(b)$ exists. $\qquad\qquad\qquad\qquad\qquad\qquad\qquad\qquad\qquad\quad\square$

Lemma C.3.2. *If $\sum_{n=0}^{\infty} f_n(x)$ is a uniformly converging series on the closed interval $[a, b]$ and each function f_n is continuous on this interval, then*

$$\int_a^b (\sum_{n=0}^{\infty} f_n(x))dx = \sum_{n=0}^{\infty}(\int_a^b f_n(x)dx).$$

C.4 Some Special Functions

(i). The Gamma function

For any $x \in \mathbf{R}$, $x > 0$, any $s \in \mathbf{C}$, $x^s = e^{s \log(x)}$ by definition. By Lemma C.3.1.(ii), the following result holds.

Definition/Lemma C.4.1. *The Gamma function is the function defined on the region $\mathcal{R}(s) > 0$ by*

$$\Gamma(s) = \int_0^{\infty} e^{-x}x^{s-1}dx.$$

It is analytic on this region.

Lemma C.4.2. *For any $s \in \mathbf{C}$ with $\mathcal{R}(s) > 0$, $\Gamma(s+1) = s\Gamma(s)$.*

Proof.

$$\Gamma(s+1) = \int_0^{\infty} e^{-x}x^s dx$$

$$= -e^{-x}x^s|_0^{\infty} + s \int_0^{\infty} e^{-x}x^{s-1}dx$$

$$= s\Gamma(s).$$

$\qquad\qquad\qquad\qquad\qquad\qquad\qquad\qquad\qquad\qquad\qquad\qquad\qquad\qquad\quad\square$

This shows that the Gamma function generalizes the factorials.

Corollary C.4.3. $\Gamma(n) = n!$ *for all $n \in \mathbf{N}$.*

Proof. $\Gamma(1) = \int_0^\infty e^{-x} dx = 1$ and so the result follows from Lemma C.4.2. $\square$

By using Lemma C.4.2, the Gamma function can also be analytically continued.

Corollary C.4.4. *The function Γ can be extended to a meromorphic function on $\mathbf{C}$ satisfying*

$$\Gamma(s) = \frac{\Gamma(s+n+1)}{s(s+1)\cdots(s+n)}$$

for any $s \in \mathbf{C}$. It is analytic everywhere except at $s = 0, -1, -2, \cdots$, where it has simple poles.

Proof. From Lemma C.4.2,

$$\Gamma(s) = \Gamma(s+1)/s = \Gamma(s+2)/s(s+1) = \cdots = \Gamma(s+n+1)/s(s+1)\cdots(s+n).$$

The rest of the proof then follows. $\square$

Lemma C.4.5.

(i) If $\mathcal{R}(s) > 0$, $\Gamma(s) = 2\int_0^\infty e^{-x^2} x^{-1/2} dx$
(ii) $\Gamma(\frac{1}{2}) = \sqrt{\pi}$

Proof. Setting $x = y^2$, we get

$$\Gamma(s) = \int_0^\infty e^{-x} x^{s-1} dx = 2\int_0^\infty e^{-y^2} y^{2s-1} dy,$$

proving (i).

In particular,

$$\Gamma(\frac{1}{2}) = 2\int_0^\infty e^{-y^2} dy.$$

Hence, using polar coordinates $x = r\cos\theta$, $y = r\sin\theta$, we get $dxdy = rdrd\theta$.

$$(\Gamma(\frac{1}{2}))^2 = 4\int_0^\infty \int_0^\infty e^{-x^2-y^2} dxdy = 4\int_{\theta=0}^{\pi/2} \int_{r=0}^\infty e^{-r^2} rdrd\theta = 2\int_{\theta=0}^{\pi/2} d\theta = \pi.$$

$\square$

Corollary C.4.6. (Replication formula). **If $\mathcal{R}(s) > 0$,**

$$2^{2s-1}\Gamma(s)\Gamma(s+\frac{1}{2}) = \sqrt{\pi}\Gamma(2s).$$

Proof. Applying Lemma C.4.5 (i),

$$\Gamma(s)\Gamma(s+\frac{1}{2}) = 4\int_0^\infty e^{-x^2-y^2} x^{2s-1} y^{2s} dxdy.$$

Substituting polar coordinates,

$$\Gamma(s)\Gamma(s+\frac{1}{2}) = 4\int_{\theta=0}^{\pi/2}\int_{r=0}^{\infty} e^{-r^2} r^{4s} \cos(\theta)^{2s-1}\sin(\theta)^{2s} dr d\theta$$

$$= 4\int_{r=0}^{\infty} e^{-r^2} r^{4s} \int_{\theta=0}^{\pi/2} \cos(\theta)^{2s-1}\sin(\theta)^{2s} d\theta.$$

Integrating by parts several times gives

$$\int_{\theta=0}^{\pi/2} \cos(\theta)^{2s-1}\sin(\theta)^{2s} d\theta = 2^{-2s-1}\int_{\theta=0}^{\pi/2} \cos(\theta)^{4s-1} d\theta.$$

Hence by Lemma C.4.5, the formula follows □

Differentiating both sides of the replication formula we get the following equality.

Corollary C.4.7. *If* $\mathcal{R}(s) > 0$,

$$\log(2)2^{2s}\Gamma(s)\Gamma(s+\frac{1}{2}) + 2^{2s-1}G'(s)\Gamma(s+\frac{1}{2}) + 2^{2s-1}\Gamma(s)\Gamma'(s+\frac{1}{2}) = 2\sqrt{\pi}\Gamma'(2s).$$

Lemma C.4.8. *For any* $x \in \mathbf{R}$, $e^x = \lim_{n\to\infty}(1+\frac{x}{n})^n$.

Proof. We know that for $x \in \mathbf{R}$,

$$\frac{1}{x} = \log'(x) = \lim_{h\to 0}\frac{\log(x+h)-\log(x)}{h}.$$

In particular,

$$1 = \lim_{h\to 0}\frac{\log(1+h)}{h}.$$

Hence, for any $x \in \mathbf{R}$,

$$x = \lim_{h\to 0} x\frac{\log(1+h)}{h} = \lim_{l\to\infty} l\log(1+x/l),$$

where $l = x/h$. Hence, for n integers,

$$x = \lim_{n\to\infty}\log(1+x/n)^n.$$

Taking exponentials of both sides of this equality gives the desired result. □

Set $\gamma = \lim_{N\to\infty}(\sum_1^N \frac{1}{n} - \log(N))$ to be the Euler-Mascheroni constant.

Corollary C.4.9. *(Weierstrass formula).* *For* $s \in \mathbf{C}$,

$$\frac{1}{\Gamma(s)} = se^{\gamma s}\prod_{n=1}^{\infty}(1+\frac{s}{n})e^{\frac{-s}{n}}.$$

Proof. Since for $s \in \mathbf{C}$,

$$\Gamma(s) = \int_0^\infty e^{-x} x^{s-1} dx$$

$$= \int_0^\infty \lim_{n \to \infty} (1 - \frac{x}{n})^n x^{s-1} dx$$

$$= \lim_{n \to \infty} \int_0^n (1 - \frac{x}{n})^n x^{s-1} dx$$

$$= \lim_{n \to \infty} n^s \int_0^1 (1 - y)^n y^{s-1} dy$$

$$= \lim_{n \to \infty} n^s (\frac{1}{s} y^s (1 - y)^n |_0^1 + \frac{n}{s} \int_0^1 y^s (1 - y)^{n-1} dy).$$

Hence applying induction on n to the calculation of the integral part in the last expression, we get

$$\Gamma(s) = \lim_{n \to \infty} \frac{n! n^s}{s(s+1) \cdots (s+n)}$$

$$= \frac{1}{s} \prod_{n=1}^\infty (1 + \frac{s}{n})^{-1} n^s .$$

It follows that

$$\frac{1}{\Gamma(s)} = s \lim_{n \to \infty} (1 + s)(1 + \frac{s}{2}) \cdots (1 + \frac{s}{n}) e^{-\log(n)s}$$

$$= s \lim_{n \to \infty} (1 + s) e^{-s} (1 + \frac{s}{2}) e^{\frac{-s}{2}} \cdots (1 + \frac{s}{n}) e^{\frac{-s}{n}} e^{(1 + \frac{1}{2} + \cdots \frac{1}{n} - \log(n))s}.$$

This leads to the result. □

Corollary C.4.10.

$$\Gamma'(1) = \gamma.$$

Proof. Taking the logarithm of both sides of the Weierstrass formula, we get

$$- \log(\Gamma(s)) = \log(s) + \gamma s + \sum_{n=1}^\infty (\log(1 + \frac{s}{n}) - \frac{s}{n}).$$

Differentiating both sides of this equality. we find

$$\frac{\Gamma'(s)}{\Gamma(s)} = -\frac{1}{s} - \gamma + \sum_{n=1}^\infty (\frac{1}{n} \frac{1}{1 + \frac{s}{n}} + \frac{1}{n})$$

$$= -\frac{1}{s} - \gamma + \sum_{n=1}^\infty (\frac{1}{n} - \frac{1}{n+s}).$$

Setting $s = 1$ and using Corollary C.4.3 we can deduce that

$$\Gamma'(1) = -1 - \gamma + \sum_{n=1}^{\infty} (\frac{1}{n} - \frac{1}{n+1}).$$

Since $\sum_{n=1}^{\infty} (\frac{1}{n} - \frac{1}{n+1}) = 1$, this proves the result.   □

(ii). The Riemann Zeta function

Definition/Lemma C.4.11. *The Riemann zeta function is the function defined on the region $\{s \in \mathbf{C} : \mathcal{R}(s) > 1\}$ by*

$$\zeta(s) = \sum_{n=1}^{\infty} \frac{1}{n^s}.$$

It is analytic in this region and

$$\zeta'(s) = -\sum_{n=1}^{\infty} \frac{\log n}{n^s}.$$

Proof. A compact subset of $\{s \in \mathbf{C} : \mathcal{R}(s) > 1\}$ is contained in subset of the type $\{s \in \mathbf{C} : \mathcal{R}(s) > 1 + \epsilon\}$, where ϵ is a positive number. Each term $\frac{1}{n^s}$ is analytic. If $\mathcal{R}(s) \geq 1 + \epsilon$ then,

$$\left| \frac{1}{n^s} \right| = \frac{1}{n^{(s)}} \leq \frac{1}{n^{1+\epsilon}}.$$

The series $\sum_{n=1}^{\infty} \frac{1}{n^{1+\epsilon}}$ converges by the integral test. Hence the series $\sum_{n=1}^{\infty} \frac{1}{n^s}$ converges uniformly by Lemma C.2.4. The result now follows from Lemma C.2.5.   □

Lemma C.4.12.
$$\zeta(s) = (1 - 2^{1-s})^{-1} \sum_{n=1}^{\infty} \frac{(-1)^n}{n^s},$$

and the series on the right hand side converges conditionally if $\mathcal{R}(s) > 0$ and $\zeta'(s)$ is equal to the series obtained by differentiating each term of the above series.

Proof. Since

$$(1 - 2^{1-s})\zeta(s) = \sum_{n=1}^{\infty} \frac{1}{n^s} - \sum_{n=1}^{\infty} \frac{2}{(2n)^s},$$

the above formula follows. The alternating series test shows that the series $\sum_{n=1}^{\infty} \frac{(-1)^n}{n^s}$ is conditionally convergent when $\zeta'(s)$ for then the sequence $1/n^s$ is decreasing. The statement about $\zeta'(s)$ can be derived in a similar manner.

□

Lemma C.4.13. *For* $|s| < 1$ *and* $z = 1$,

$$\log(1 + s) = \sum_{n=1}^{\infty} (-1)^{n+1} s^n / n$$

and the series converges in the above region.

Proof. The result follows by applying Theorem 2.6 around the point $s = 0$. Since

$$\lim_{n \to \infty} \left| \frac{(-1)^{n+2} s^{n+1}}{n+1} \frac{n}{(-1)^{n+1} s^n} \right| = \frac{n}{n+1} |s| = |s|,$$

by the ratio test, the series converges absolutely when $|s| < 1$. When $s = 1$, the series converges conditionally by the alternating series test. $\square$

Corollary C.4.14. *The zeta function can be analytically continued to a meromorphic function on the region given by* $\mathcal{R}(s) > 0$. *It is analytic at all points of this region except at* $s = 1$, *where it has a simple pole with residue 1. Moreover the constant coefficient of its Laurent series at* $s = 1$ *is equal to* $-\Gamma'(1)$.

Proof.   From Lemma C.2.5, if s is a singular point of $\zeta(s)$ then, $e^{\log(2)((1-\mathcal{R}(s))-i\mathcal{I}(s))} = e^{\log(2)(1-s)} = 2^{1-s} = 1$. Hence $s = 1 + 2\pi i m / \log(2)$. Similarly as in Lemma C.4.12, it can be shown that

$$\zeta(s) = (1 - 3^{1-s})^{-1} \sum_{n=1}^{\infty} \frac{\epsilon(n)}{n^s},$$

where

$$\epsilon(n) = \begin{cases} 1 & \text{if } n \equiv \pm 1 \pmod 3 \\ -2 & \text{otherwise.} \end{cases}$$

This series also converges conditionally when $\mathcal{R}(s) > 0$ and so if s is a singular point of $\zeta(s)$ then, $s = 1 + 2\pi i m / \log(3)$ where $m \in \mathbf{Z}$. Hence, together with the above, if s is a singularity of $\zeta(s)$ then $s = 1$. Moreover, Lemma C.4.12 shows that $(s - 1)\zeta(s)$ is analytic at $s = 1$ and that

$$\lim_{s \to 1} (s-1)\zeta(s) = \lim_{s \to 1} \frac{(s-1)}{-\log(2)(1-s)(1 + (\log(2)(1-s))/2! + \cdots)} \sum_{n \geq 1} \frac{(-1)^n}{n} = 1$$

by Lemma 4.12. Hence, the ζ has a unique pole in $\mathcal{R}(s) > 0$. It is at $s = 1$ and the pole is simple. Furthermore, the residue of $\zeta(s)$ at $s = 1$ is 1.

   It remains to calculate the constant of the Laurent expansion at $s = 1$. This is equal to

$$\lim_{s \to 1} (\zeta(s) - \frac{1}{s-1}). \tag{1}$$

For $\mathcal{R}(s) > 1$,

$$\zeta(s) = \sum_{n=1}^{\infty} \frac{1}{n^s}$$

$$= \sum_{n=1}^{\infty} s \int_n^{\infty} \frac{dt}{t^{s+1}}$$

$$= \sum_{n=1}^{\infty} \sum_{k=n}^{\infty} \int_k^{k+1} \frac{dt}{t^{s+1}}$$

$$= s \sum_{k=1}^{\infty} \sum_{n=1}^{k} \int_k^{k+1} \frac{dt}{t^{s+1}}$$

$$= s \sum_{k=1}^{\infty} k \int_k^{k+1} \frac{dt}{t^{s+1}} \qquad (2)$$

$$= s \sum_{k=1}^{\infty} \int_k^{k+1} \frac{k\,dt}{t^{s+1}}$$

$$= s \int_1^{\infty} \frac{[t]}{t^{s+1}} dt$$

$$= s \int_1^{\infty} \frac{1}{t^{s+1}} dt - s \int_1^{\infty} \frac{\{t\}}{t^{s+1}} dt$$

$$= \frac{s}{s-1} - s \int_{k=1}^{\infty} \frac{\{t\}}{t^{s+1}} dt,$$

where $[t]$ is the integer part of $t \in \mathbf{R}$ and $\{t\} = t - [t]$. In the above we have changed the order of summation, which is possible since the series is absolutely convergent for $\mathcal{R}(s) > 1$. Substituting (2) into the expression (1), we see that the constant term of the Laurent series is equal to

$$1 - \int_{k=1}^{\infty} \frac{\{t\}}{t^2} dt = 1 - \lim_{N \to \infty} \sum_{n=1}^{N} \int_n^{n+1} \frac{t-n}{t^2} dt$$

$$= 1 - \lim_{N \to \infty} \left(\int_{n=1}^{N} \frac{dt}{t} - \sum_{n=1}^{N} \frac{1}{n+1} \right)$$

$$= \lim_{N \to \infty} \left(\sum_{n=1}^{N} \frac{1}{n} - \log(N) \right)$$

$$= \gamma.$$

The result then follows from Corollary C.4.13. $\qquad \square$

Lemma C.4.15. $\zeta(2) = \pi^2/6$.

Proof. Consider the function $f(x) = x(x-1)$. It is continuous on the interval

$[0, 1]$ and $f(0) = f(1)$. Hence, by Corollary D.3,

$$f(x) = \sum_{n=-\infty}^{\infty} \hat{f}(n) e^{2\pi i n x},$$

where the Fourier coefficients are given by

$$
\begin{aligned}
\hat{f}(n) &= \int_0^1 x(x-1) e^{-2\pi i n x} \, dx \\
&= \frac{1}{-2\pi i n} e^{-2\pi i n x} x(x-1)\big|_0^1 + \frac{1}{2\pi i n} \int_0^1 (2x-1) e^{-2\pi i n x} \, dx \\
&= \frac{1}{4\pi^2 n^2} (2x-1) e^{-2\pi i n x}\big|_0^1 - \frac{1}{2\pi^2 n^2} \int_0^1 e^{-2\pi i n x} \, dx \\
&= \frac{1}{2\pi^2 n^2}
\end{aligned}
$$

for $n \neq 0$ and

$$\hat{f}(0) = \int_0^1 x(x-1) \, dx = \frac{-1}{6}.$$

For $x = 0$, we get

$$0 = \frac{-1}{6} + \sum_{\substack{n \in \mathbf{Z} \\ n \neq 0}} \frac{1}{2\pi^2 n^2}.$$

Equivalently,

$$\sum_{n=1}^{\infty} \frac{1}{\pi^2 n^2} = \frac{1}{6},$$

proving the result.   □

Appendix D

Fourier Series and Transforms

Definition D.1. *The Fourier series of a (real or complex valued) function f on $\mathbf{R}$ is*

$$\sum_{n=-\infty}^{\infty} \hat{f}(n)e^{-2\pi i n x},$$

where the Fourier coefficients are given by

$$\hat{f}(n) = \int_0^1 f(x)e^{2\pi i n x}dx.$$

Our aim is to represent f by its Fourier series. Supposing that this series converges absolutely and is equal to f, then by Lemma C.3.2,

$$\int_0^1 f(x)e^{2\pi i n x}dx = \sum_{m=-\infty}^{\infty} \int_0^1 \hat{f}(n)e^{-2\pi i(m-n)x} = \hat{f}(n).$$

Lemma D.2. *Let f be a periodic continuous function on $\mathbf{R}$ having period 1 and with continuous derivative, then the Fourier series of f converges uniformly on $\mathbf{R}$ and is equal to f.*

Any continuous function on $[0, 1]$ satisfying $f(0) = f(1)$ can be periodically continued to a continuous function on $\mathbf{R}$ to a function having period 1. So the next result is immediate.

Corollary D.3. *Let f be a continuous function on the interval $[0, 1]$ such that $f(0) = f(1)$. Then, it Fourier series is equal to f on this interval.*

Corollary D.4. $\sum_{\substack{n \in \mathbf{Z} \\ n \neq 0}} \frac{e^{2\pi i n x}}{n^2} = 2\pi(x^2 - |x| + 1/6)$ *for* $-1 \leq x \leq 1$.

Proof. Consider the function $f(x) = x(x-1)$. It is continuous on the interval $[0,1]$ and $f(0) = f(1)$. Hence, by Corollary D.3,

$$f(x) = \sum_{n=-\infty}^{\infty} \hat{f}(n)e^{2\pi inx},$$

where the Fourier coefficients are given by

$$\hat{f}(n) = \int_0^1 x(x-1)e^{-2\pi inx}dx$$

$$= \frac{1}{-2\pi in}e^{-2\pi inx}x(x-1)\big|_0^1 + \frac{1}{2\pi in}\int_0^1 (2x-1)e^{-2\pi inx}dx$$

$$= \frac{1}{4\pi^2 n^2}(2x-1)e^{-2\pi inx}\big|_0^1 - \frac{1}{2\pi^2 n^2}\int_0^1 e^{-2\pi inx}dx$$

$$= \frac{1}{2\pi^2 n^2}$$

for $n \neq 0$ and

$$\hat{f}(0) = \int_0^1 x(x-1)dx = \frac{-1}{6}.$$

Hence, the result holds for $0 \le x \le 1$. For the case $-1 \le x \le 0$, consider

$$f(x) = \sum_{n=-\infty}^{\infty} \overline{f}(n)e^{-2\pi inx},$$

where

$$\overline{f}(n) = \int_0^1 x(x-1)e^{2\pi inx}dx$$

$$= -\frac{1}{(2\pi in)^2}(2x-1)e^{-2\pi inx}\big|_0^1$$

$$= \frac{1}{2\pi^2 n^2}.$$

Hence the result follows as above.   □

Definition D.5. *Given the function f on $\mathbf{R}$, the Fourier transform of f is the function*

$$\hat{f}(y) = \int_{-\infty}^{\infty} f(x)e^{2\pi ixy}dx.$$

Theorem D.6. *Suppose that the integral $\int_{-\infty}^{\infty}|f(x)|dx$ exists. Then the Fourier inversion formula holds for f:*

$$f(x) = \int_{-\infty}^{\infty} \hat{f}(y)e^{-2\pi ixy}dy.$$

Corollary D.7. *Let g be a function on $\mathbf{R}$ for which the integral $\int_{-\infty}^{\infty} |g(x)| dx$ exists. Then, for any $x \in \mathbf{R}$,*

$$\sum_{n=-\infty}^{\infty} g(x+n) = \sum_{n=-\infty}^{\infty} \hat{g}(n) e^{2\pi i n x}.$$

Proof. Set $f(x) = \sum_{n=-\infty}^{\infty} g(x+n)$. Then, $f(x+1) = f(x)$. Hence f has a Fourier series expansion. All we need to show is that the Fourier coefficients $\hat{f}(n)$ of f are given by the Fourier transform of g. Indeed,

$$\begin{aligned}
\hat{f}(n) &= \int_0^1 f(x) e^{-2\pi i n x} dx \\
&= \int_0^1 \sum_{k=-\infty}^{\infty} g(x+k) e^{-2\pi i n x} dx \\
&= \sum_{k=-\infty}^{\infty} \int_k^{k+1} g(x) e^{-2\pi i n x} dx \\
&= \int_{-\infty}^{\infty} g(x) e^{-2\pi i n x} dx \\
&= \hat{g}(n).
\end{aligned}$$

$\square$

The following useful consequence is then immediate.

Corollary D.8. (Poisson summation formula). *Let g be a function on $\mathbf{R}$ for which the integral $\int_{-\infty}^{\infty} |g(x)| dx$ exists. Then,*

$$\sum_{n=-\infty}^{\infty} g(n) = \sum_{n=-\infty}^{\infty} \hat{g}(n).$$

We next generalize the concept of Fourier transforms to higher dimensions.

Lemma/Definition D.9. *Let L be an integral lattice with bilinear form $(.,.)$ and $f : L \otimes_{\mathbf{Z}} \mathbf{R} \to \mathbf{R}$ a continuous function with continuous derivatives, having period L, i.e. $f(x+y) = f(x)$ for all $x \in L \otimes_{\mathbf{Z}} \mathbf{R}$, $y \in L$. Then, the Fourier series*

$$\frac{1}{|L^*/L|} \sum_{\lambda \in L^*} \hat{f}(\lambda) e^{-2\pi i (x,\lambda)},$$

where

$$\hat{f}(\lambda) = \int_{(L \otimes_{\mathbf{Z}} \mathbf{R})/L} f(x) e^{2\pi i (x,\lambda)} dx,$$

$\lambda \in L^*$, *converges uniformly to f.*

In the previous Lemma, the region of integration is given as follows: Suppose $\dim L = n$. Take $\mathcal{B}$ to be a basis for the lattice L, then with respect to this basis, f becomes a function $\mathbf{R}^n \to \mathbf{R}$ and the region of integration is $[0, 1)^n$. As the integral is independent of the basis chosen, it can be written as above.

Definition D.10. *Let $f : \mathbf{R}^n \to \mathbf{R}$ a function for which $\int_{\mathbf{R}^n} f(y)dy$ exists. Then, the Fourier transform of f is the function*

$$\hat{f}(y) = \int_{\mathbf{R}^n} f(x)e^{2\pi i x_i y_i}dx,$$

where $x = (x_1, \cdots, x_n)$.

The generalized version of the Poisson summation formula is then as follows.

Theorem D.11. *Let L be an integral positive definite lattice with bilinear form $(.,.)$. Let g be a function on $L \otimes_{\mathbf{Z}} \mathbf{R}$ satisfying the conditions of Definition D.10. Then, for $x \in L \otimes_{\mathbf{Z}} \mathbf{R}$,*

$$\sum_{\lambda \in L} g(x + \lambda) = \frac{1}{|L^*/L|} \sum_{\lambda \in L^*} \hat{g}(\lambda)e^{2\pi i n(x,\lambda)}.$$

Lemma D.12. *Let V be a real vector space of dimension n.*

(i) If $a \in \mathbf{R}$ and $a > 0$, then the Fourier transform of $f(ax)$ is $a^{-n}f(x/a)$.
(ii) If V is positive definite, then the Fourier transform of $e^{-\pi x}$ is $e^{-\pi x}$.

Proof. (i): Set $g(x) = f(ax)$. By definition,

$$\hat{g}(y) = \int_V f(ax)e^{2\pi i xy}dx$$

$$= \frac{1}{a^n} \int_V f(x)e^{2\pi i xy/a}dx$$

$$= \frac{1}{a^n}\hat{g}(y/a)$$

for $a > 0$.

(ii): We prove the result for $n = 1$. The general case is an immediate consequence.

$$\hat{f}(y) = \int_{-\infty}^{\infty} e^{-\pi x^2}e^{2\pi i xy}dx.$$

Hence

$$\frac{d\hat{f}}{dy}(y) = \int_{-\infty}^{\infty} 2\pi i x e^{-\pi x^2}e^{2\pi i xy}dx$$

$$= -i(e^{-\pi x^2}e^{2\pi i xy})_{-\infty}^{\infty} - 2\pi y \int_{-\infty}^{\infty} e^{\pi i x^2 \tau}e^{2\pi i xy}dx$$

$$= -2\pi y \hat{f}(y).$$

Set $g(x) = e^{-\pi x^2}$. Now,

$$\frac{dg}{dx}(x) = -2\pi x g(x).$$

It follows that

$$\hat{f}'(x)g(x) - g(x)'(x)\hat{f}(x) = 0.$$

And so,

$$\frac{\hat{f}'}{g}(x) = 0.$$

As a result,

$$\hat{f}(x) = Cg(x)$$

for some constant C. We next calculate C. Setting $y = 0$, we get

$$C = \int_{-\infty}^{\infty} e^{-\pi x^2} dx = 1.$$

$\square$

Lemma D.13. *Let V be a real vector space of dimension n with a positive (resp. negative) definite bilinear form. For $\tau \in \mathcal{H}$, the Fourier transform of the complex valued function $f(x) = e^{\pi i x^2 \tau}$ (resp. $e^{\pi i x^2 \bar{\tau}}$) on V is the function $(i/\tau)^{n/2} e^{-\pi i x^2/\tau}$ (resp. $(-i/\bar{\tau})^{n/2} e^{\pi i x^2/\bar{\tau}}$).*

Proof. The convergence criterion needed for the existence of the Fourier transform holds. This is why in the positive (resp. negative) definite case, we need $Im(\tau) > 0$ (resp. $Im(\tau) < 0$). Suppose first that the bilinear form is positive definite. We prove the result for $n = 1$. The general case is an immediate consequence.

$$\hat{f}(y) = \int_{-\infty}^{\infty} e^{\pi i x^2 \tau} e^{2\pi i x y} dx.$$

Hence

$$\frac{d\hat{f}}{dy}(y) = \int_{-\infty}^{\infty} 2\pi i x e^{\pi i x^2 \tau} e^{2\pi i x y} dx$$

$$= \frac{1}{\tau}(e^{\pi i x^2 \tau} e^{2\pi i x y})_{-\infty}^{\infty} - \frac{2\pi i y}{\tau} \int_{-\infty}^{\infty} e^{\pi i x^2 \tau} e^{2\pi i x y} dx$$

$$= -\frac{2\pi i y}{\tau} \hat{f}(y).$$

Set $g(x) = e^{\frac{-\pi i x^2}{\tau}}$. Now,

$$\frac{dg}{dx}(x) = -\frac{2\pi i x}{\tau} g(x).$$

It follows that

$$\hat{f}'(x)g(x) - g(x)'(x)\hat{f}(x) = 0.$$

And so,

$$\frac{\hat{f}'}{g}(x) = 0.$$

As a result,

$$\hat{f}(x) = Cg(x)$$

for some constant C. We next calculate C. Setting $y = 0$, we get

$$C = \int_{-\infty}^{\infty} e^{\pi i x^2 \tau} dx = (i/\tau)^{\frac{1}{2}}.$$

For the negative definite case, set $x_- = ix$. Then,

$$\exp(\pi i x^2 \bar{\tau}) = \exp(\pi i x_-^2 (-\bar{\tau})).$$

Hence the result follows from the positive definite case. □

Lemma D.14. *Let V be a real vector space with non-degenerate bilinear form* $(.,.)$.

   (i) For $a, x \in V$, the Fourier transform of $f(x)e^{2\pi i(x,a)}$ is $\hat{f}(x+a)$.
   (ii) For $x, a \in V$, the Fourier transform of $f(x-a)$ is $e^{2\pi i(x,a)}f(x)$.

Proof. Without loss of generality, we may assume that $x, a \in \mathbf{R}$, i.e. that $\dim V = 1$.

$$\int_{-\infty}^{\infty} f(x)e^{2\pi i x a}e^{2\pi i x y}dx = \int_{-\infty}^{\infty} f(x)e^{2\pi i x(y+a)}$$
$$= \hat{f}(y+a)$$

and

$$\int_{-\infty}^{\infty} f(x-a)e^{2\pi i x y}dx = \int_{-\infty}^{\infty} f(x)e^{2\pi i(x+a)y}dx$$
$$= e^{2\pi i a y}\int_{-\infty}^{\infty} f(x)e^{2\pi i x y}dx$$
$$= e^{2\pi i a y}\hat{f}(y).$$

   □

Corollary D.15. *The Fourier transform of $e^{2\pi i(ax^2+bx+c)}$ for $x \in \mathbf{R}$, $a, b, c \in \mathbf{C}$, $Im(a) > 0$, is $(2a/i)^{\frac{-1}{2}}e^{2\pi i(-x^2/4a-bx/2a+c-b^2/4a)}$.*

Proof. Applying Lemma D.14 (i) for $\tau = 2a$ and using Lemma D.13, it follows that the Fourier transform of $e^{2\pi i(ax^2+bx+c)}$ is $(2a/i)^{\frac{-1}{2}}e^{2\pi i c}e^{-(x+b)^2/4a}$.

Notation

A indecomposable symmetric Cartan matrix with entries $a_{ij} = (h_i, h_j)$.

$\alpha_i \in Q$, $i \in I$, simple roots of G.

$B_+ = H \oplus N_+$ a (generalized) positive Borel sub-superalgebra

$B_- = H \oplus N_-$ a (generalized) negative Borel sub-superalgebra

$B_n(x)$ A Bernoulli piecewise polynomial $-n! \sum_{j \neq 0} \epsilon(jx)/(2\pi i j)^n$

$\mathbf{C}$ The complex numbers.

$c(n)$ coefficient of the q-expansion of the normalized modular invariant J, i.e. $c(-1) = 1$ and $J(q) = q^{-1} + \sum_{n \geq 1} c(n) q^n$

$\text{ch}(V))$ character of G-module $V \in \mathcal{O}$

$d(x)$ degree of homogeneous element $x \in G$.

$\mathcal{C}$ The positive open cone in the Lorentzian lattice L.

δ_n^m 1 if $m = n$, 0 otherwise.

Δ set of roots.

Δ_0 set of even roots

Δ_1 set of odd roots

Δ^+ set of positive roots

$\Delta(\tau)$ The delta function, $\Delta(\tau) = q \prod_{n>0}(1 - q^n)^{24}$

$\mathbf{e} = \exp(2\pi i)$

$\mathbf{e}(x)$ $\exp(2\pi i x)$.

e_γ An element of a basis of $\mathbf{C}[M^*/M]$.

E_k An Eisenstein series of weight k, equal to $1 - (2k/B_k) \sum_{n>0} \sigma_{k-1}(n) q^n$ if $k \geq 2$.

E_2 The non-holomorphic modular form $E_2(\tau) - 3/\pi \Im(\tau)$ of weight 2.

$\eta(\tau) = q^{1/24} \prod_{n>0}(1 - q^n)$

$\mathrm{F}[X] = \{\sum_{i=0}^n a_i X^i, a_i \in F\}$

$\mathrm{F}[X] = \{\sum_{i=0}^\infty a_i X^i, a_i \in F\}$

$F(X) = \{\sum_{i=-N}^\infty a_i X^i, a_i \in F, N \geq 0.\}$

$\mathcal{F}$ Fourier group

f_Q natural map from Q to H taking α_i to h_i

g_Q natural map from Q to H^* taking α_i to the linear functional β_i in H^* mapping h to (h, h_i)

f_γ A component of the vector valued modular form F.

F either a vector valued modular form with components f_γ or the fake monster lie algebra

F_L = P^1/N, the BKM algebra derived from the lattice vertex algebra V_L

G = $G(A, H, S)$ BKM superalgebra with generalized symmetric Cartan matrix A, generalized Cartan subalgebra H and indexing set S for the odd generators

G' = $[G, G]$

$\tilde{G}$ = $\tilde{G}(A, H, S)$ is the Lie superalgebra generated by the Lie subalgebra H and elements e_i, f_i, $i \in I$ satisfying relations $(1) - (3)$ of Definition 2.1.7.

G_α ($\alpha \in Q$, H or H^*) root space. For $\alpha \in \Delta^+$, $e_{\alpha,i}$, $1 \leq i \leq \dim G_\alpha$ basis for the root space G_α and $f_{\alpha,i}$, $1 \leq i \leq \dim G_\alpha$ dual basis for $G_{-\alpha}$

$\mathcal{G}^+$ If $\mathcal{G}$ is a subgroup of a real orthogonal group then G^+ means the elements of G fixing the chosen orientation

Γ finite abelian group

$\Gamma(z)$ Euler's gamma function

$G(M)$ Grassmannian of maximal positive definite subspaces of real vector space $M \otimes_{\mathbf{Z}} \mathbf{R}$

$H_{\mathbf{R}}$ real vector space with a non-degenerate symmetric real valued bilinear form $(.,.)$ and elements $h_i, i \in I$ such that

(i) $(h_i, h_j) \leq 0$ if $i \neq j$,

(ii) If $(h_i, h_i) > 0$, then $\frac{2(h_i, h_j)}{(h_i, h_i)} \in \mathbf{Z}$ for all $j \in I$

(iii) If $(h_i, h_i) > 0$ and $i \in S$, then $\frac{(h_i, h_j)}{(h_i, h_i)} \in \mathbf{Z}$ for all $j \in I$

H = $H_{\mathbf{R}} \otimes_{\mathbf{Z}} \mathbf{C}$

$\tilde{H}$ affinization of $H = L \otimes_{\mathbf{Z}} \mathbf{C}$

$\mathcal{H}$ The upper half plane

$h(n)$ = $h \otimes t^n \in \tilde{H}$, $h \in H$, $n \in \mathbf{Z}$, or homomorphism generated by this element on V_L.

$\mathrm{ht}\,(\lambda)$ = $\sum_{i \in I} x_i$ for $\lambda = \sum_{i \in I} x_i \alpha_i \in H$, $x_i \in \mathbf{C}$, i.e. height of λ.

I finite set $\{1, ..., n\}$ or countably infinite one usually identified with $\mathbf{Z}_+$

$Inn(G)$ group of inner automorphism of G

$\mathcal{I}(\tau)$ The imaginary part of the complex number τ.

$II_{m,n}$ unimodular even lattice of rank (m, n)

J = $\{i \in I : a_{ii} > 0\}$

$\mathcal{J}$ Jacobi group

j The elliptic modular function $j(\tau) = q^{-1} + 744 + 196884q + \cdots$, or an integer.

K negative definite lattice of rank $b^- - 1$

∇^2 hyperbolic Laplacian

L lattice, in Chapter 5 assumed to be Lorentzian and such that $L = K \otimes_{\mathbf{Z}} II_{1,1}$

L^* dual of lattice L

M = $L \otimes_{\mathbf{Z}} II_{1,1}$ lattice of signature $(2, b^-)$.

λ_V orthogonal projection of a vector λ onto a subspace V.

$L(\Lambda)$ irreducible G-module of highest weight Λ

$M(\Lambda)$ Verma G-module of highest weight Λ

$m_0(\alpha) = \dim G_\alpha \cap G_{\bar{0}}$

$m_1(\alpha) = \dim G_\alpha \cap G_{\bar{1}} = \mathrm{mult}(\alpha) - m_0(\alpha)$

G_p (even) Lie subalgebra generated by the elements e_i, f_i, $i \in I \backslash S$ and $[e_i, e_i]$, $[f_i, f_i]$, $i \in S$

$\mathbf{N} = \{n \in \mathbf{Z} : n > 0\}$

$N = \{x \in P^1 : (x, P^1) = 0\}$, the kernel of the bilinear form restricted to the subspace P^1 of V_L

$\mathcal{O}$ category of G-modules defined in Definition 2.6.15

Ω generalized Casimir operator

ω Chevalley automorphism

ω_0 compact automorphism

$P^+ = \{\Lambda \in H : \forall i \in I, (\alpha, \alpha_i) \geq 0, \frac{2(\Lambda, \alpha_i)}{(\alpha_i, \alpha_i)}$ (resp. $\frac{(\Lambda, \alpha_i)}{(\alpha_i, \alpha_i)}) \in \mathbf{Z}_+$ if $a_{ii} > 0$ and $i \in I \backslash S$ (resp. $i \in S$)$\}$

P set of elements $Z_M = X_M + iY_M \in M \otimes_{\mathbf{Z}} \mathbf{C}$ with norm 0 and for which X_M, Y_M form an oriented basis for a maximal positive definite subspace of $M \otimes_{\mathbf{Z}} \mathbf{R}$.

$P^n = \{x \in V_L : L_0 x = nx\, L_j x = 0 \,\forall j \geq 1\}$

$P(V) = \{\lambda \in H : V_\lambda \neq 0\}$ for $V \in \mathcal{O}$

$p(a)$ parity of homogeneous element $a \in V_L$

$p(\alpha)$ parity $1 \otimes e^\alpha \in V_L$

$\Phi_M(v^+, F)$ theta transform of vector valued modular form F, $v^+ \in G(M)$

Ψ_M automorphic form on $G(M)$

Ψ_z A restriction of Ψ_M to the hermitian symmetric space $L \otimes_{\mathbf{Z}} \mathbf{R} + i\mathcal{C}$.

Π base of the set of roots Δ of the BKM superalgebra G.

r_α reflection corresponding to the root α

r_i reflection corresponding to the simple root α_i or corresponding inner automorphism

q $e^{2\pi i \tau}$

$\mathbf{Q}$ The rational numbers.

Q formal root lattice. For $\alpha, \beta \in Q, H$ or H^* or $\beta = 0$, write $\beta < \alpha$ if $\alpha - \beta$ is a sum of simple roots

$$R = \frac{\prod_{\alpha \in \Delta_0^+}(1 - e(-\alpha))^{m_0(\alpha)}}{\prod_{\alpha \in \Delta_1^+}(1 + e(-\alpha))^{m_1(\alpha)}}$$

$$R' = \frac{\prod_{\alpha \in \Delta_0^+}(1 - e(-\alpha))^{m_0(\alpha)}}{\prod_{\alpha \in \Delta_1^+}(1 - e(-\alpha))^{m_1(\alpha)}}$$

ρ Weyl vector

$\rho(M, W, F)$ A Weyl vector.

ρ_M A representation of $Mp_2(\mathbf{Z})$ on the group ring $\mathbf{C}[M^*/M]$.

$\mathcal{R}$ The real part of a complex number.

$\mathbf{R}$ The real numbers.

SL A special linear group.

$\operatorname{supp}(\alpha)$ support of the root α.

$(.,.)$ bilinear form on G, Q, H or H^*.

$\operatorname{sch}(M(\Lambda))$ super-character of G-module $V \in \mathcal{O}$

S Either subset of I indexing the odd simple roots or
$$= \left(\begin{pmatrix} 0 & -1 \\ 1 & 0 \end{pmatrix}, \sqrt{\tau} \right) \in Mp_2(\mathbf{R}) \text{ or } \begin{pmatrix} 0 & -1 \\ 1 & 0 \end{pmatrix} \in SL_2(\mathbf{R})$$

$$T = \left(\begin{pmatrix} 1 & 1 \\ 0 & 1 \end{pmatrix}, 1 \right) \in Mp_2(\mathbf{R}) \text{ or } \begin{pmatrix} 1 & 1 \\ 0 & 1 \end{pmatrix} \in SL_2(\mathbf{R})$$

τ A complex number $x + iy$ with positive imaginary part y.

Θ_L generalized or Siegel Theta function on lattice L

V_L Γ-graded (bosonic) lattice vertex algebra

$V(\Lambda)$ highest weight G-module of highest weight Λ

V_L^f fermionic lattice vertex superalgebra

V_λ , weight space of weight λ of G-module V ($\lambda \in Q$ or H or H^*)

$v_\lambda \in V_\lambda$ weight vector of weight λ of G-module $V \in \mathcal{O}$

$v^\perp$ Orthogonal complement of a vector (or sublattice) in a lattice or vector space V.

v^+ element in $G(M)$, v^- it complement in $M \otimes_{\mathbf{Z}} \mathbf{R}$

Vir Virasoro algebra

W Weyl group or Weyl chamber

W_E even Weyl group.

X $\in L \otimes_{\mathbf{Z}} \mathbf{R}$, the real part of Z

X_M the real part of Z_M

Y $\in C$, the imaginary part of Z

Y_M The imaginary part of Z_M.

$Y(a, v)$ generalized vertex operator on V_L for $a \in V_L$

z A primitive norm 0 vector of M.

z' A vector of M' such that $(z, z') = 1$.

Z The element $Z = (\begin{pmatrix} -1 & 0 \\ 0 & -1 \end{pmatrix}, i)$ generating the center of order 4 of $Mp_2(\mathbf{Z})$, or the element $X + iY \in M \otimes_{\mathbf{Z}} \mathbf{C}$.

Z_M $= (Z, 1, -Z^2/2 - z'^2/z) = X_M + iY_M$

$\mathbf{Z}$ The integers

$\mathbf{Z}_+$ $= \{n \in \mathbf{Z} : n \geq 0\}$

ζ The Riemann zeta function

Bibliography

[AbS]   M. Abramowitz, I.A. Stegun (Eds.), "Handbook of Mathematical Functions with Formulas, Graphs, and Mathematical Tables", New York, Dover, 1972

[Ar]   J. Arthur, "On some problems suggested by the trace formula", Lie group representations, II (College Park, Md., 1982/1983), Springer, Berlin, 1984, 1–49

[BaeGN]   O. Baewald, R.W. Gebert, H. Nikolai, On the imaginary simple roots of the Borcherds algebra $G(II_{9,1})$, *preprint Hep-th/9705144*

[Bar]   A.G. Barnard, The singular theta correspondence, Lorentzian lattices and Borcherds-Kac-Moody algebras, Berkeley PhD thesis, 2003

[BenKM1] G.M. Benkart, S.-J. Kang, K.C. Misra, Indefinite Kac-Moody algebras of special linear type, *Pacific Jour. Math.*

[BenKM2] G.M. Benkart, S.-J. Kang, K.C. Misra, Indefinite Kac-Moody algebras of classical type, *Adv. Math.* **105** (1994), 76–101

[BenM]   G.M. Benkart, R.V. Moody, Derivations, central extensions and affine Lie algebras, *Algebras, Groups, and Geometries* **3** (1986), 456–492

[BerM]   S. Berman, R.V. Moody, Lie algebra multiplicities, *Proc. Amer. Math. Soc.* **76** (1979), 223–228

[Borc1]   R.E. Borcherds, The Leech lattice, *Proc. R. Soc. Lond. A* **398** (1985), 365–376

[Borc2]   R.E. Borcherds, Vertex algebras, Kac-Moody algebras, and the Monster, *Proc. Natl. Acad. Sci. USA* **83** (May 1986), 3068–3071

[Borc3]   R.E. Borcherds, Automorphism groups of Lorentzian lattices, *Journal of Algebra* **111** (1) (1987), 133–153

[Borc4]   R.E. Borcherds, Generalized Kac-Moody Algebras, *Journal of Algebra* **115** (2) (1988), 501–512

[Borc5]   R.E. Borcherds, The Monster Lie algebra, *Advances in Math.* **83** (1) (Sept. 1990), 30–47

[Borc6]   R.E. Borcherds, Central extensions of generalized Kac-Moody algebras, *Journal of Algebra* **140** (1991), 330–335

[Borc7]   R.E. Borcherds, Monstrous Moonshine and monstrous Lie superalgebras, *Invent. Math.* **109** (1992), 405–444

273

[Borc8]   R.E. Borcherds, A characterization of Generalized Kac-Moody algebras, *Journal of Algebra* **174** (3) (1995), 1073–1079

[Borc9]   R.E. Borcherds, Automorphic forms on $O_{s+2,2}(\mathbf{R})$ and infinite products, *Invent. Math.* **120** (1995), 161–213

[Borc10] R.E. Borcherds, Automorphic forms on $O_{s+2,2}(\mathbf{R})$ and generalized Kac-Moody algebras, *Proceedings of the International Congress of Mathematicians*, **Vol. 1,2** (Zurich 1994), 744–752, Birkhauser, Basel, 1995

[Borc11] R.E. Borcherds, Automorphic forms with singularities on Grassmannians, *Invent. Math.* **132** (1998), 491–562

[Borc11] R.E. Borcherds, Automorphic forms and Lie algebras, *Current Developments in Mathematics* (1996), International Press, 1998

[Borc12] Reflection group of Lorentzian lattices, *Duke Math J.* **104** 2000 (no 2), 319–366

[Br]   J.H. Bruinier, "Borcherds products on $O(2, l)$ and Chern Classes of Heegner divisors", Springer-Verlag, Berlin, (Lecture notes in mathematics 1780) 2000

[BrB]   J.H. Bruinier, M. Bundschuh, On Borcherds products associated with lattices of prime discriminant, *Raman. J.* to appear

[BuLS]   M. Burger, J.S. Li, P. Sarnak, On Ramanujan duals and automorphic spectrum *Bull. Amer. Math. Soc. (N.S.)* **26** 1992 (no 2), 253–257

[BuS]   M. Burger, P. Sarnak, Ramanujan duals II *Invent. Math.* **106** 1991 (no 1) 1–11

[Bur]   W. Burnside, "Group theory", Cambridge University Press, 1921

[Con1]   J.H. Conway, A characterisation of Leech's lattice, *Invent. Math.* **7** (1969), 137–142.

[Con2]   J.H. Conway, The automorphism group of the 26-dimensional even Lorentzian lattice, *J. Algebra* **80** (1983), 159–163.

[Con3]   J.H. Conway, The automorphism group of the 26-dimensional even unimodular Lorentzian lattice, *J. Algebra* **80** (1983), 159–163

[ConN]   J.H. Conway, S.P. Norton, Monstrous Moonshine, *Bull. London Math. Soc.* **11** (1979), 308–339

[ConPS]   J.H. Conway, R.A. Parker, N.J.A. Sloane, The covering radius of the Leech lattice, *Proc. R. Soc. Lond. A* **380** (1982), 261–290

[ConS1]   J.H. Conway, N.J.A. Sloane, Twenty-three constructions for the Leech lattice, *Proc. R. Soc. Lond. A* **381** (1982), 275–283

[ConS2]   J.H. Conway, N.J.A. Sloane, Lorentzian forms for the Leech lattice, *Bull. Am. Math. Soc.* **6** (1982), 215–217

[ConS3]   J. Conway, Sloane, "Sphere packings and lattices and groups", Grundlehren der Mathematischen Wissenschaten 290, Springer Verlag, New York-Berlin, 1988

[Don]   C. Dong, Vertex algebras associated to even lattices, *J. Algebra* **160** (1993), 245–265

[DonL]   C. Dong, J. Lepowsky, "Generalized vertex algebras and relative vertex operators", Birkhauser, 1993

[DonLM1] C. Dong, H. Li, G. Mason, Regularity of rational vertex operator
          algebras, *Advances in Math.* **132** (1997), 305–321
[DonLM2] C. Dong, H. Li, G. Mason, Twisted representations of vertex oper-
          ator algebras, *Math. Ann.* **310** (1998), 571–600
[Dys]   F. Dyson, Missed opportunities, *Bull. Amer. Math. Soc.* **78** no 5
          (1972), 635–652
[EicZ]   M. Eichler, D. Zagier, "The theory of Jacobi forms", Progress in
          math. 55, Birkhäuser, Boston, 1985
[FeinF]   A. Feingold, I. Frenkel, A hyperbolic Kac-Moody algebra and the
          theory of Siegel modular forms of genus 2, *Math. Ann.* **263** (1983)
          no. 1, 87–144.
[FeinFR] A. Feingold, I.B. Frenkel, J.F.X. Ries, Spinor constructions of ver-
          tex operator algebras, Triality, and $E_8^{(1)}$, *Contemporary mathemat-
          ics* **Vol 121** (1991), A.M.S.
[FeinL]   A. Feingold, J. Lepowsky, The Weyl-Kac character formula and
          power series identities, *Advances in Math.* **29** (1978), 271–309
[FeinN]   A. Feingold, H. Nicolai, Subalgebras of hyperbolic Lie algebras,
          Kac-Moody Lie Algebras and Related Topics, eds. N. Sthanumoor-
          thy, K.C. Misra, *Contemporary Math, Amer. Math. Soc.* **343** (2004)
[FeitT]ref-1 W. Feit, J.G. Thompson, Solvability of groups of odd
          order, *Pacific J. Math.* **13** (1963), 775–1029
[FisLT]   B. Fischer, D. Livingstone, M.P. Thorne, The characters of the
          "Monster" simple group, Birmingham (1978)
[Frei]   E. Freitag, "Siegelsche Modulfunktionen", Die Grundlehren der
          mathematischen Wissenschaften 254, Springer–Verlag, Berlin Hei-
          delberg New York 1983
[Fren]   I.B. Frenkel, "Representations of Kac-Moody algebras and Dual
          resonance models", *Lectures in Applied mathematics* **21** (1985),
          325–353
[FrenB]   E. Frenkel, D. Ben-Zvi, "Vertex Algebras and Algebraic Curves",
          2nd Edition, Mathematical Survey and Monographs 88, A.M.S.,
          2004
[FrenK]   I.B. Frenkel, V.G. Kac, Basic representations of affine Lie algebras
          and dual resonance models, *Invent. Math.* **62** (1980), 23–66
[FrenHL] I.B. Frenkel, Y.-Z. Huang, J. Lepowsky, "On Axiomatic approaches
          to Vertex Operator Algebras and Modules", *Memoirs Amer. Math.
          Soc.* **104** (1993)
[FrenLM1] I.B. Frenkel, J. Lepowsky, J. Meurman, A natural representation
          of the Fischer-Griess Monster with the modular function J as char-
          acter, *Proc. Natl. Acad. Sci. USA* **81** (1984), 3256–3260
[FrenLM2] I.B. Frenkel, J. Lepowsky, J. Meurman, "Vertex operator algebras
          and the Monster", Academic Press, 1988
[FrenZ]   I.B. Frenkel, Y. Zhu, Vertex operator algebras associated to repre-
          sentations of affine and Virasoro algebras, *Duke Math. J.* **66** (1992),
          123–168

[FucRS]   J. Fuchs, U. Ray, C. Schweigert, Some automorphisms of general-
           ized Kac-Moody algebras, *J. of Algebra*, **191** (1997), 518–540

[GabK]   O. Gabber, V. G. Kac, On defining relations of certain infinite-
           dimensional Lie algebras, *Bull. Amer. Math. Soc.* **5** (1981), 185–189

[Gan]   T. Gannon, "Moonshine beyond the Monster: The bridge connect-
           ing algebra, modular forms, and physics", Cambridge University
           Press

[Gar1]   H. Garland, The Arithmetic Theory of Loop Algebras, *J. Algebra*
           53 (1978)

[Gar2]   H. Garland, The arithmetic theory of loop groups, *J. Algebra*

[GarL]   H. Garland, J. Lepowsky, Lie algebra homology and the Macdonald-
           Kac Formulas, *Inv. Math.* **34** (1976) 37–76

[GebN]   R.W. Gebert, H. Nikolai, On the imaginary simple roots of the
           Borcherds algebra $G(II_{9,1})$, *preprint Hep-th/9703084*

[Geb]   R.W. Gebert, Introduction to vertex algebras, Borcherds Algebras,
           and the monster, *International Jour. Modern Physics A* (1993)

[GebT]   R.W. Gebert, J. Teschner, On the fundamental representation of
           Borcherds algebras with one imaginary simple root, *Lett. Math.
           Phys.*

[GodT]   P. Goddard, C.B. Thorn, Compatibility of the dual pomeron with
           unitary and the absence of ghosts in the dual resonance model,
           Physics Letters **40B** (26 June 1972) no.2, 235–237

[Gor]   D. Gorenstein, "Finite simple groups", Plenum Press, New York,
           1982

[Gri]   R.L. Griess, Jr., The friendly Giant, *Invent. Math.* **69** (1982), 1–102

[Grit]   V.G. Gritsenko, Jacobi functions of n variables, *J. Soviet Math.* **53**
           (1991), 243–252.

[GriN1]   V.A. Gritsenko, V.V. Nikulin, Igusa modular forms and the "sim-
           plest" Lorentzian Kac-Moody algebras, *Sbornik Math* **187** no. 11
           (1996), 1601–1641

[GritN2]   V.A. Gritsenko, V.V. Nikulin, Siegel automorphic form corrections
           of some Lorentzian Kac-Moody Lie algebras, *Amer. J. Math.* **119**
           (1997) 181–223

[GritN3]   V.A. Gritsenko, V.V. Nikulin, Automorphic forms and Lorentzian
           Kac-Moody algebras, Part I, *Intern. J. Math.* **9** no 2 (1998), 153–
           199

[GriN4]   V.A. Gritsenko, V.V. Nikulin, Automorphic forms and Lorentzian
           Kac-Moody algebras, Part I, *Intern. J. Math.* **9** no 2 (1998), 153–
           199

[GritN5]   V.A. Gritsenko, V.V. Nikulin, Automorphic forms and Lorentzian
           Kac-Moody algebras, Part II, *Intern. J. Math.* **9** no 2 (1998), 201–
           275

[Hard]   G.H. Hardy, Ramanujan's function $\tau(n)$, *Twelve Lectures on Sub-
           jects Suggested by his Life and Works*, third ed., Chelsea, New-York
           (1999), 63 and 161–185

[HarvM]   J. Harvey, G. Moore, Algebras, BPS States, and Strings, *Nuclear
           Phys. B* **463** (1996), no 2-3, 315–368

[HoeS] G. Hoehn, N. Scheithauer, A natural construction of Borcherds' fake baby monster Lie algebra, *Amer. J. Math.* **125** No 3 (2003), 1139–1172

[Hua1] Y.-Z. Huang, Geometric interpretation of vertex operator algebras, *Proc. Natl. Acad. Sci. USA* **88** (1991), 9964–9968

[Hua2] Y.-Z. Huang, Vertex operator algebras and conformal field theory, *International Jour. Modern Phys. A* **7** (1992), 2109–2151

[HuaL1] Y.-Z. Huang, J. Lepowsky, A Theory of tensor products for module categories for a vertex operator algebra, I, *Selecta Mathematica* **1** (1995), 699–756
Y.-Z. Huang, "Two-Dimensional Conformal Field Theory and Vertex Operator Algebras", *Progress in Math.* **Vol. 148**, Birkhauser, Boston 1997

[HuaL2] Y.-Z. Huang, J. Lepowsky, A Theory of tensor products for module categories for a vertex operator algebra, II, *Selecta Mathematica* **1** (1995), 657–786

[HuaL3] Y.-Z. Huang, J. Lepowsky, Tensor products of modules for a vertex operator algebra and vertex tensor categories, *Lie Theory and Geometry, in honor of Bertram Kostant,* ed. R. Brylinski, J.-L. Brylinski, V. Guillemin, V. Kac, Birkhauser, Boston, 1994, 349–383

[Hum] "Linear Algebraic Groups", Springer-Verlag, New York, Heidelber, Berlin, 1975

[Jac] N. Jacobson, "Lie algebras", New-York, Dover, 1979

[Jeu] J. Van der Jeugt, Character formulae for the Lie superalgebra $C(n)$, *Comm. Algebra* **19** no 1 (1991), 199–222

[JeuHKT] J. Van der Jeugt, J. W. B. Hughes, R. C. King, J. Thierry-Mieg, A character formula for singly atypical modules of the Lie superalgebra $sl(m/n)$, *Comm. Algebra* **18** no 10 (1990), 3453–3480

[Jur1] E. Jurisch, An exposition of generalized Kac-Moody algebras, *Contemporary Math.* **194** (1996)

[Jur2] E. Jurisch, Generalized Kac-Moody Lie algebras, free Lie algebras and the structure of the Monster Lie algebra *J. Pure Appl. Algebra* **122** (1997), 233–266

[Jur3] E. Jurisch, A resolution for standard modules of Borcherds Lie algebras, *J. Pure Appl. Algebra* **192** (2004), no 1–3, 149–158

[Jur4] E. Jurisch, An Equivalence between Categories of Modules for Generalized Kac-Moody Lie algebras, *J. Lie Theory* **14** (2004), 141–150

[JurLW] E. Jurisch, J. Lepowsky, R. Wilson, Realizations of the Monster Lie algebra, *Selecta Mathematica*, New Series **1** (1995), 129–161

[Kac1] V.G. Kac, Simple graded Lie algebras of finite growth, (English translation) *Funct. Anal. Appl.* **1** (1967), 328–329

[Kac2] V.G. Kac, Graded Lie algebras and symmetric spaces, (English translation) *Funct. Anal. Appl.* **2** (1968), 183–184

[Kac3] V.G. Kac, Simple Irreducible graded Lie algebras of finite growth, (English translation) *Math. USSR-Izvestija* **2** (1968), 1271–1311

[Kac4]   V.G. Kac, Automorphisms of finite order of semisimple Lie algebras
            (English translation) *Funct. Anal. Appl.* **3** (1969), 252–254

[Kac5]   V.G. Kac, Infinite-dimensional Lie algebras and Dedekind's η-
            Function, (English translation) *Funct. Anal. Appl.* **8** (1974), 68–70

[Kac6]   V.G. Kac, Lie Superalgebras, *Advances in Mathematics* **26** (1977),
            8–96

[Kac7]   V.G. Kac, Infinite-Dimensional Algebras, Dedekind's η-Function,
            Classical Möbius Function and the Very Strange Formula, *Advances
            in Mathematics* **30** (1978), 85–136

[Kac8]   V.G. Kac, "Representations of classical Lie superalgebras", Lecture
            Notes in Math. 676 (1978), 597–626

[Kac9]   V.G. Kac, On simplicity of certain infinite-dimensional Lie algebras,
            Bull. Amer. Math. Soc. **2** (1980), 311–314

[Kac10]   V.G. Kac, A remark on the Conway-Norton conjecture about the
            "Monster" simple group, *Proc. Natl. Acad. Sci. USA* **77** (1980),
            5048–5049

[Kac11]   V.G. Kac, An elucidation of "Infinite dimensional algebras... and
            the very strange formula". $E_8^{(1)}$ and the cube root of the modular
            invariant j, *Advances in Math.* **35** (1980), 264–273

[Kac12]   V.G. Kac, Laplace operators of infinite-dimensional Lie algebras
            and theta functions, *Proc. Nat'l. Acad. Sci. USA* **81** (1984), 645–
            647

[Kac13]   V.G. Kac, Constructing groups associated to infinite-dimensional
            Lie algebras, *Proceedings of the conference on Infinite-dimensional
            groups*, Berkeley 1984, MSRI publ. **4**, 1985, 167–216

[Kac14]   V.G. Kac, "Infinite dimensional Lie algebras", third ed., Cambridge
            University Press, 1990

[Kac15]   V.G. Kac, "Vertex algebras for beginners", University Lecture Se-
            ries, 10. American Mathematical Society, Providence, RI, 1997

[KacB1]   V.G. Kac, B. Bakalov, Twisted modules over lattice vertex algebras,
            in Lie theory and its applications to physics V, eds. H.-D. Doebner
            and V.K. Dobrev, World Sci. 2004, pp 3-26. math.QA/0402315

[KacB2]   V.G. Kac, B. Bakalov, Generalized vertex algebras, in Lie theory
            and its applications to physics VI, eds.V.K. Dobrev et al, Heron
            Press, Sofia, 2006. math.QA/0602072

[KacKan]   V.G. Kac, S.-J. Kang, Trace formula for graded Lie algebras and
            Monstrous Moonshine, *Canad. Math. Soc. Conf. Proc.* **16** (1995),
            141–14

[KacKazLW] V.G. Kac, D.A. Kazhdan, J. Lepowsky, R.L. Wilson, Realization
            of the basic representation of the Euclidean Lie algebras, *Advances
            in Math.* **42** (1981), 83–112

[KacMW]   V.G. Kac, R. Moody, W. Wakimoto, On $E_1 0$, *Proceedings of
            the 1987 conference on differential-geometrical methods in physics*,
            Kluwer, 1988, 102–128

[KacP1]   V.G. Kac, D.H. Peterson, Affine Lie algebras and Hecke modular
            forms, *Bull. Amer. Math. Soc.* **3** (1980), 1057–1061

[KacP2]   V.G. Kac, D.H. Peterson, Infinite dimensional Lie algebras, theta functions and modular forms, *Advances in Math.* **53** (1984), 125–264

[KacP3]   V.G. Kac, D.H. Peterson, Unitary structure in representations of infinite-dimensional groups and a convexity theorem, *Invent.Math.* **76** (1984), 1–14

[KacP4]   V.G. Kac, D.H. Peterson, Regular functions on certain infinite-dimensional groups in "Arithmetic and Geometry" (M. Artin and J. Tate, Eds). *Progress in Math.*, **Vol. 36** 141–166, Bikhauser, Boston, 1983

[KacP5]   V.G. Kac, D.H. Peterson, Defining relations of certain infinite-dimensional groups, in "Proceedings, Cartan Conference, Lyon, 1984", *Astérisque* (1985), 165–208

[KacP6]   V.G. Kac, D.H. Peterson, Infinite flag varieties and conjugacy theorems, *Proc. Natl. Acad. Sci., USA* **80** (1983), 1778–1782

[KacP7]   V.G. Kac, D.H. Peterson, 112 constructions of the basic representation of the loop group of E_8, Proceedings of the conference "Anomalies, geometry, topology" Argonne, 1985. World Sci., 1985, 276–298

[KacP8]   V.G. Kac, D.H. Peterson, On geometric invariant theory for infinite-dimensional groups, in "Proceedings, Algebraic Groups" 107–142, Lecture notes in Math., **Vol. 1271**, Springer-Verlag, New York/Berlin, 1987

[KacR]   V.G. Kac, A. Raina, Bombay lectures on highest weight representations of infinite dimensional Lie algebras, *World scientific* (1987)

[KacT]   V.G. Kac, I. Todorov, Affine orbifolds and RCFT extensions of $W_{1+\infty}$, *Comm. Math.Phys.* **190** (1997), 57-111

[KacWak1]   V.G. Kac, W. Wakimoto, Modular and conformal invariance constraints in representation theory of affine algebras, *Advances in Math.* **70** (1988), 156–234

[KacWak2]   V.G. Kac, W. Wakimoto, Modular invariant representations of infinite dimensional Lie algebras and superalgebras, *Proc. Nat'l. Acad. Sci. USA* **85** (1988), 4956–4960

[KacWak3]   V.G. Kac, W. Wakimoto, Classifications of modular invariant representations of affine algebras, in Infinite dimensional Lie algebras and groups, Adv. Ser. Math. Phys. **7**, World Scientific, 1989, 138–177

[KacWak4]   V.G. Kac, W. Wakimoto, Integrable highest weight modules over affine lie superalgebras and number theory, *Progress in Math.* **123** (1994), 415–456

[KacWak5]   V.G. Kac, W. Wakimoto, Integrable highest weight modules over affine superalgebras and Appell's function, *Comm. Math. Phys.* **215** (2001), 631–682

[KacWan1]   V.G. Kac, W. Wang, On automorphisms of Kac-Moody Lie algebras and groups, *Advances in Math.* **92** (1992), 129–195

[KacWan2]   V.G. Kac, W. Wang, Vertex Operator Superalgebras and their Representations, *Contemporary mathematics* **Vol. 175** (1994) 161–191

[Kang1]   S.-J. Kang, Root multiplicities of the hyperbolic Lie algebra
            $HA(1)1$, *J. Algebra* **160** (1993), 492–523

[Kang2]   S.-J. Kang, Root multiplicities of Kac-Moody algebras, *Duke Math.
            J.* **74** (1994), 645–666

[Kang3]   S.-J. Kang, Root multiplicities of the hyperbolic Lie algebra
            $HA(1)n$, *J. Algebra* **170** (1994), 277–299

[KnoM]   M.I. Knopp, G. Mason, Vector valued modular forms and their
            coefficients, *preprint*

[Kon]   M. Kontsevich, Product formulas for modular forms on $O(2,n)$,
            Séminaire Nicolas Bourbaki **821** November 1996. alg-geom/9709006

[Kos]   B. Kostant, Lie Algebra Cohomology and the Generalized Borel-
            Weyl Theorem, *Ann. of Math.* **74** (1961), 329

[Kum1]   S. Kumar, Demazure character formula in arbitrary Kac-Moody
            setting, *Invent. Math.* **89** (1987), 395–423

[Kum2]   S. Kumar, "Kac-Moody Groups, Their Flag Varieties and Rep-
            resentation Theory", *Progress in Math.* **Vol. 204**, Birkhauser,
            Boston, 2002

[LeiS]   D. Leites, V. Serganova, Defining relations for simple Lie superal-
            gebras I. Lie superalgebras with Cartan matrix, *Proc. of the conf.
            Topological methods in physics 1991*, World Scientific, (1992), 194–
            201

[Lep1]   J. Lepowsky, Calculus of twisted vertex operators, *Proc. Natl. Acad.
            Sci. USA.* **82** (1985), 8295–8299

[Lep2]   J. Lepowsky, The Work of Richard E. Borcherds, in "The Math-
            ematical Work of the 1988 Fields Medallists", *Notices A.M.S.* **46**
            (1999), 17–19

[LepL]   J. Lepowsky, H. Li "Introduction to Vertex Operator Algebras and
            Their Representations", Birkhauser, 2003

[LepM]   J. Lepowsky, R.V. Moody, Hyperbolic Lie algebras and quasi reg-
            ular cusps on Hibert modular surfaces, *Math. Ann.* **245** (1979),
            63–88

[LepW]   J. Lepowsky, R.L. Wilson, Construction of the affine Lie algebra
            $A_1^{(1)}$, *Commun. Math. Phys.* **62** (1978), 43–53

[Leu]   J.W. van de Leur, A classification of contragredient Lie superalge-
            bras of finite growth, *Communications in Algebra* **17** (1989), 1815–
            1841

[Li1]   H.S. Li, Symmetric linear bilinear forms on vertex operator alge-
            bras, *J. Pure Appl. Algebra* **109** (1994) 279–297

[Li2]   H.S. Li, Local systems of vertex operators, Vertex superalgebras
            and modules *J. Pure Appl. Algebra* **109** (1996) 143–195

[Liu]   L.-S. Liu, Kostant's formula in Kac-Moody algebras, *J. Algebra*
            149 (1992), 155–178

[Mac]   I.G. Macdonald, Affine root systems and Dedekind's η-function,
            Invent. Math. **15** (1972), 91–143

[Man]   S. Mandelstam, Dual resonance models, *Phys. Rep.* **13** (1974), 259–
            353

[Mi]   T. Miyake, "Modular forms", Springer-Verlag, Berlin-New York, 1989. ISBN: 3-540-50268-8

[Mo1]   R.V. Moody, Lie algebras associated with generalized Cartan matrices, *Bull. Amer. Math. Soc.* **73** (1967), 217–221

[Mo2]   R.V. Moody, A new class of Lie algebras, *J. Algebra* **10** (1968), 211–230

[Mo3]   R.V. Moody, Euclidean Lie algebras, *Can. J. Math.* **21** (1969), 1432–1454

[Mo4]   R.V. Moody, Simple quotients of Euclidean Lie algebras, *Can. J. Math.* **22** (1970), 839–846

[Mo5]   R.V. Moody, Macdonald identities and Euclidean Lie algebras, *Proc. Amer. Math. Soc.* **48** (1975), 43–52

[Mo6]   R.V. Moody, Root systems of hyperbolic type, *Advances in Mathematics* **33** (1979), 144–160

[Mo7]   R.V. Moody, generalized root systems and characters, *Contemporary Mathematics* **45** (1985), 245–269

[MoPa]   R.V. Moody, J. Patera, Fast recursion formula for weight multiplicities, *Bull. Amer. Math. Soc.* **7** (1982), 237–242

[MoPi]   R.V. Moody, A. Pianzola, "Lie algebras with triangular decompositions", Wiley-Interscience, 1995

[MoY]   R.V. Moody, T. Yokonuma, Root systems and Cartan matrices, *Can. J. Math.* **34** (1982), 63–79

[Na]   S. Naito, The Strong Bernstein-Gelfand-Gelfand resolution for generalized Kac-Moody algebras, I, *Publ. of the Research Institute for Mathematical Sciences* **29** (1993), 709–730

[Nie]   P. Niemann, Some generalized Kac-Moody algebras with known root multiplicities, *Memoirs of the AMS* **Vol 157**, No 746, Amer. Math. Soc., Providence, RI 2002

[Nik1]   V.V. Nikulin, Integer symmetric bilinear forms and some of their geometric applications. (Russian) *Izv. Akad. Nauk SSSR Ser. Mat.* **43** (1979), no. 1, 111–177, 238. English translation in Mathematics of the U.S.S.R., *Izvestia*, **Vol 14** No. 1 1980, 103-167.

[Nik2]   V.V. Nikulin, A lecture on Kac-Moody Lie algebras of the arithmetic type, *Preprint Queen's University, Canada* **1994-16** (1994), *alg-geom* **9412003**

[Nik3]   V.V. Nikulin, A theory of Lorentzian Kac-Moody algebras, *alg-geom* **9810001**

[Rad]   H. Rademacher, The Fourier coefficients of the Modular invariant $j(\tau)$, *Amer. J. Math.* **60** (1938), 501–512

[Ray1]   U. Ray, A character formula for generalized Kac-Moody superalgebras, *J. Algebra* **177** (1995), 154–163

[Ray2]   U. Ray, Some subalgebras of the Monster Lie algebra and the modular polynomial, *J. Algebra* **191** (1997), 109–126

[Ray3]   U. Ray, A Characterization Theorem for a certain class of graded Lie superalgebras, *J. Algebra* **229** (2000), 405–434

[Ray4]   U. Ray, Generalized Kac-Moody algebras and related topics, *Bulletin A.M.S.*, Vol. 38, No 1, 2001, 1–42

[Ray5]   U. Ray, Finite dimensional modules of Borcherds-Kac-Moody Lie superalgebras, *CRM preprint* **586** (June 2004)

[Ray6]   U. Ray, The super-character formula for a class of integrable highest weight module of Borcherds-Kac-Moody Lie superalgebras, *CRM preprint* **587** (June 2004)

[Sch1]   N.R. Scheithauer, Vertex algebras, Lie algebras, and Superstrings, *J. Alg.* **200** (1998), 363–403

[Sch2]   N.R. Scheithauer, The fake monster superalgebra, *Adv. in Math.* **Vol 151** no 2 (2000), 226–269

[Sch3]   N.R. Scheithauer, Twisting the fake monster superalgebra, *Preprint* QA/0001160 v2 26 Jul 2000

[Sch4]   N.R. Scheithauer, Automorphic forms, Fake Monster algebras, and hyperbolic reflection groups, *Adv. Math.* **164** (2001) (no 2), 301–324

[Serg1]   V.V. Serganova, Automorphisms of simple Lie superalgebras, *Izv. Akad. Nauk. SSSR Ser. Mat.* **48** (1984), no. 3, 585–598

[Serg2]   V. Serganova, Kazhdan-Lusztig polynomials and character formula for the Lie superalgebra $gl(m \mid n)$, *Selecta Math.* (N.S.) **2** no 4 (1996), 607–651

[Serr1]   J.-P. Serre, "Algèbres de Lie semisimples complexes", W. A. Benjamin, New York, 1966

[Serr2]   J.-P. Serre, "A course in arithmétic", Springer-Verlag, New York, 1973, Translated from the French, Graduate texts in Mathematics, no 7

[Shi]   G. Shimura, "Introduction to the arithmetic theory of automorphic functions", Princeton University Press, Princeton, New Jersey, 1971

[Th]   J.G. Thompson, Some numerology between the Fischer-Griess monster and the elliptic modular function, *Bull. London. Math. Soc.* **11** (1979), 352–353.

[Ti1]   J. Tits, Résumé de cours, *Annuaire du Collège de France* (1980-81), 75–87

[Ti2]   J. Tits, Résumé de cours, *Annuaire du Collège de France* (1981-82), 91–106

[Ti3]   J. Tits, Uniqueness and presentation of Kac-Moody groups over fields, *J. Algebra* **105** (1987), 542–573

[Ti4]   J. Tits, Groupes associés aux algèbres de Kac-Moody, *Séminaire Bourbaki 700*, Astérisque **vol. 177-178**, société mathématique de France, 1989, 249–286

[Ts]   H. Tsukada, Vertex operator superalgebras, *Comm. Alg.* **18** (1990) 2249–2274

[VinK]   E.B. Vinberg, I.M. Kaplinskaja, On the groups $O_{18,1}(\mathbf{Z})$ and $O_{19,1}(\mathbf{Z})$, *Soviet Math.* **19** No 1 (1978) 194–197.

[Wan]   Z.-X. Wan, Introduction to generalized Kac-Moody algebras, World Scientific, Singapore, 1991

[WhiW]   E.T. Whittaker, G.N. Watson, "A Course in Modern Analysis", 4th ed., Cambridge University Press, 1990

[Z]   Y. Zhu, Modular invariance of characters of vertex operator algebras, *Journal AMS* **9** (1996), 237–302.

Index

Algebra and Applications

1. C.P. Millies and S.K. Sehgal: *An Introduction to Group Rings*. 2002
   ISBN Hb 1-4020-0238-6; Pb 1-4020-0239-4
2. P.A. Krylov, A.V. Mikhalev and A.A. Tuganbaev: *Endomorphism Rings of Abelian Groups*. 2003   ISBN 1-4020-1438-4
3. J.F. Carlson, L. Townsley, L. Valero-Elizondo and M. Zhang: *Cohomology Rings of Finite Groups.* Calculations of Cohomology Rings of Groups of Order Dividing 64. 2003   ISBN 1-4020-1525-9
4. K. Kiyek and J.L. Vicente: *Resolution of Curve and Surface Singularities*. In Characteristic Zero. 2004   ISBN 1-4020-2028-7
5. U. Ray: *Automorphic Forms and Lie Superalgebras*. 2006
   ISBN 1-4020-5009-7